Variational Methods for Engineers with Matlab®

Series Editor
Piotr Breitkopf

Variational Methods for Engineers with Matlab®

Eduardo Souza de Cursi

iSTE

WILEY

First published 2015 in Great Britain and the United States by ISTE Ltd and John Wiley & Sons, Inc.

ISTE Ltd
27-37 St George's Road
London SW19 4EU
UK

www.iste.co.uk

John Wiley & Sons, Inc.
111 River Street
Hoboken, NJ 07030
USA

www.wiley.com

Library of Congress Control Number: 2015948071

British Library Cataloguing-in-Publication Data
A CIP record for this book is available from the British Library
ISBN 978-1-84821- 914-4

Contents

Introduction

Variational methods may be seen as a new language leading to new formulations and new methods of solution of equations, where the central concept is the physical notion of work. In this language, we do not say that a quantity is null, but we say that its work is null for all the possible values of a connected variable. For instance, we do not say that a displacement is null, but that the associated work is null for any force. However, we do not say that a force is null, but that its work is null for any displacement. In the simplest situation, we do not say that $x = 0$, but we say that "x is a real number such that its product for any other real number is equal to zero", e.g. $\in \mathbb{R}$ and $xy = 0, \forall y \in \mathbb{R}$. Although this modification and the equivalence between both the formulations seem trivial, it implies a deep conceptual change, as it will be seen in the following. Moreover, despite the view expressed by Richard Feynman – from the standpoint of Physics – in [FEY 85], these two formulations may not be equivalent – from the standpoint of Mathematics – in certain situations: it may become necessary to adopt a complex theoretical framework in order to obtain such an equivalence – that will never be complete.

The beginning of the history of variational methods is often brought to the works of the Greek philosopher Aristotle regarding the *lever problem*. Nowadays, lever analysis is an exercise for students, which automatically applies the physical principles of momentum and force equilibrium in order to explain how a lever works. So, it may be difficult to conceive that such knowledge was built up patiently over more than a millenium and has intrigued mankind for a long time. For more than ten centuries, the reason why a small force was able to move a large load has remained obscure and has been the subject of passionate discussions. For instance, the text *Problemata Mechanica* (*Problems of Mechanics*, attributed to Aristotle, but whose real author is uncertain) starts by saying that [ARI 38]:

> *Among the problems included in this class are included those concerned with the lever. For, it is strange that a great weight can be moved by a small force, and that, too, when a greater weight is involved. For the very same weight, which a man cannot move without a lever, he quickly moves by applying the weight of the lever.*

In this text, the author studies the lever by considering fictitious possible motions – e.g. *virtual* motions – of the system. An explanation is proposed by considering these possible motions and selecting a particular motion among all: the analysis of fictitious movements furnishes the real one. This approach contains the seeds of the modern variational methods: define the virtual motions and find the real motion by analyzing the virtual motions. It is interesting to notice that restrictions are taken into account: it is observed that the extremities of the lever move on a circle – there are *natural* and *unnatural* motions: the natural motion of a load placed on the extremity of the lever corresponds to the tangent direction, while, in fact, the load moves on a circle.

A second ancient author is Heron Alexandrius, who formulated a principle of economy (or of minimum violence of Nature on itself). This principle leads to the principle that light rays follow the shortest path between two points. In the Middle Ages, Jordanus Nemorarius studied the motion of bodies on inclined planes and associated it with the statics of levers having arms of unequal lengths and obtained a result of equality of virtual works [RAD 98, SIM 12].

These ancient texts do not consider the notion of *work*, unknown in ancient times and which needed many centuries to emerge and to be formalized. Tools such as *linear functionals*, *limits* and *differential calculus* were not available at those times, which limited the development and the formalization of variational methods.

The situation changed radically with the invention of differential and integral calculus. Gottfried Wilhelm Leibniz, Jacques Bernoulli, Jean Bernoulli, Leonhard Euler, Pierre Varignon, Pierre Louis de Maupertuis, Charles-Augustin Coulomb, Jean le Rond D'Alembert and others seized these new tools and used them to develop a new theory of mechanics, which culminated with the publication of Joseph-Louis Lagrange's *Analytical Mechanics* and the formalization of the Principle of Virtual Works.

Lagrange's formalism was extended by William Rowan Hamilton, who introduced a unified point of view connecting optics and dynamics. Hamilton's approach led to new variational formulations in Physics, usually referred to as *Hamiltonian Mechanics*, which is a highly reducing expression – in fact, the works of Hamilton led to formulations involving many fields, such as, for instance, classical mechanics, quantum mechanics, thermodynamics, electromagnetism, etc. It furnished a unified point of view, which may be considered as not completely exploited at this date. Moreover, significant works by Charles Jacobi, Joseph Liouville, Henri Poincaré, Alexandre Lyapounov, Lev Pontryagin, Andrei Kolmogorov and many other researchers extended Hamilton's theory, namely to applications in automatics and control.

The development of information technology has popularized numerical methods based on variational formulations such as, for instance, finite elements, finite volumes, fundamental solutions, smooth particle hydrodynamics and so on. A number of contributions led to these developments, namely the works of Boris Galerkin, Paul Dirac, Maurice Fréchet, Norbert Wiener, Sergei Sobolev, Solomon Bochner, Laurent Schwarz among a large set of researchers.

The revolution of variational methods is still under development, since new connections between variational formulations and probabilities have been highlighted by recent developments. The exploration of these relationships opens up new perspectives and gives us a glimpse of exciting prospects for the future.

1

Integrals

Variational methods are closely connected to integrals. So, before starting our variational adventures, let us recall some elements about integrals and their numerical evaluation.

The history of infinite sums is such as integrals may be brought to Zeno of Elea's paradox about the grain of millet. As described by Simplicius [FAI 98]:

> *Tell me, Protagoras, said he, does one grain of millet make a noise when it falls, or does the ten-thousandth part of a grain? On receiving the answer that it does not, he went on: Does a measure of millet grains make a noise when it falls, or not? He answered, it does make a noise. Well, said Zeno, does not the statement about the measure of millet apply to the one grain and the ten-thousandth part of a grain? He assented, and Zeno continued, Are not the statements as to the noise the same in regard to each? For as are the things that make a noise, so are the noises. Since this is the case, if the measure of millet makes a noise, the one grain and the ten-thousandth part of a grain make a noise.*

Here, Zeno considers the problem of the sum of small negligible components, what is the principle of the integration of infinitesimal contributions. In these ancient times, the concept of the limit was not known, so the notion of infinite sums and the evaluation of areas remained unsolved. In fact, many philosophers considered the question of limits, such as Antiphon the sophist, who argued that continuously doubling the number of sides of a polygon inscribed in a circle gives as result a polygon whose sides coincide with the circumference and having the same area as the circle [PEN 02]. Eudoxus of Cnidus proposed a generalization which led to the exhaustion *method*, which was extensively used for the evaluation of areas for about 2 thousand years. Archimedes applied this method to the evaluation of surfaces and volumes by considering sums of "lines" [BAU 09, BLO 11]. Archimedes' method was improved by Thabit ibn Qurra and inspired Bonaventura Cavalieri and Gilles de Roberval, namely the principle of *indivisibles*. Many researchers enriched the works of Cavalieri. For instance, Evangelista Torricelli introduced indivisibles having a thickness and Blaise Pascal considered triangular and pyramidal sums of indivisibles leading to the evaluation of double and triple integrals.

The differential and integral calculus introduced by Gottfried Leibniz and Isaac Newton opened up a new era. Agustin-Louis Cauchy introduced the concept of the definite integral and chose the notation proposed by Jean-Baptiste Joseph Fourier, which is used today. A complete formalization was proposed by Georg Riemann and led to the Riemann sums and integrals.

At this moment, the horizon seemed clear, but clouds appeared with the works of Karl Weierstrass and Richard Dedekind, which drew attention to fundamental inconsistencies in the theory. Georg Cantor, Camille Jordan and the proposition of a new set and measure theory furnished a new impulse. Emile Borel, René Baire and Henri Lebesgue formed the trinity that constructed a complete formalization of the integrals in the new framework and that gave the impression of taking away the difficulties raised by their predecessors.

But the adventure continues, since one of the most important foundations of the actual theory is the axiom of choice, which generates a paradox which has not yet been solved: the Banach-Tarski paradox, which establishes that a volume cannot be defined for a sphere of radius one. Research regarding this paradox will probably lead to new developments and evolutions in our understanding of these old concepts of areas and volumes, issues that have concerned humanity for millennia and whose complete solution eludes our sagacity every time we think we are near to their complete understanding.

1.1. Riemann integrals

The first formal theory concerning integrals was proposed by Riemann and was provoked by his interest in *Fourier series*. In 1807, a paper by Siméon-Denis Poisson [POI 08] mentioned that Joseph Fourier had proposed the representation of some functions by trigonometric series. The ideas of Fourier immediately aroused controversy. Much of the criticism was related to formal aspects. In his book *Théorie analytique de la chaleur* [FOU 22], Fourier made the following comment about trigonometric series:

> *One could doubt that there existed such a function, but this issue will be clarified later.*

In fact, as often in the history of science, the explanation furnished by Fourier was not complete and has raised fundamental questions, namely what are the functions that may be represented by a trigonometric series and the evaluation of integrals for the determination of the coefficients? Integrals were introduced by Isaac Newton and Gottfried Leibniz in the 17th Century, but their theory was the subject of discussion years before acceptation by the scientific community. About one century after, integrals were being taught in university courses, such as, for instance, those that Augustin-Louis Cauchy presented [CAU 23, CAU 29], but the formal theory waited for the works of Georg Friedrich Bernard Riemann about the questions raised by Fourier, namely his habilitation thesis [RIE 67]. In this thesis, Riemann introduced the fundamental elements for the definition of an integral. Basically, the evaluation of

$$I = \int_\Omega f(x)dx$$

requires a subdivision of Ω into a finite number of non-recovering subsets, i.e. a partition $\wp = \{\Omega_i : 1 \leq i \leq N\}$ of Ω:

$$\cup_{i=1}^{N} \Omega_i = \Omega \ , \quad \Omega_i \cap \Omega_j = \emptyset, if\ i \neq j \,.$$

For instance, when $x \in \mathbb{R}$ and $\Omega = (a, b) \subset \mathbb{R}$, we may consider a family of N subintervals $\Omega_i = (x_i, x_{i+1})$ $(i = 1, \ldots, N)$ such that $a = x_1 < x_2 < \cdots < x_N < x_{N+1} = b$. For each subinterval Ω_i, the function f has a maximum $\overline{f_i}$ and a minimum $\underline{f_i}$. Thus, we may consider the Riemann sums:

$$\overline{R}(f, \wp) = \sum_{i=1}^{N} \overline{f_i}(x_{i+1} - x_i) \ , \underline{R}(f, \wp) = \sum_{i=1}^{N} \underline{f_i}(x_{i+1} - x_i) \ ,$$

which verify $\underline{R}(f, \wp) \leq \ \overline{R}(f, \wp)$ (see Figure 1.1).

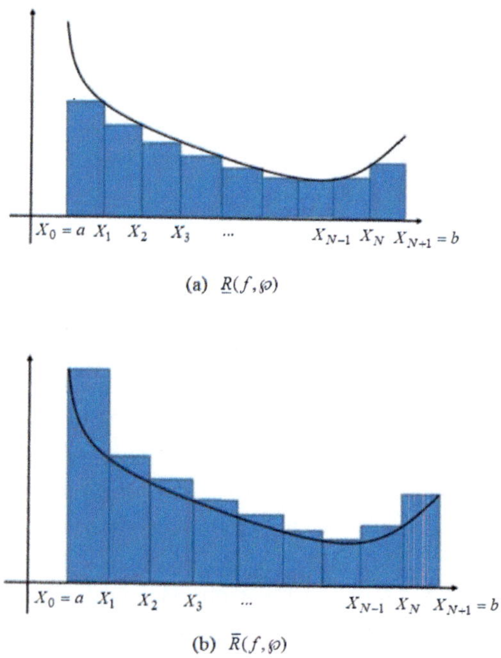

Figure 1.1. *Riemann sums: partitions of the horizontal axis*

Adding supplementary points to the partition increases $\underline{R}$ and decreases $\overline{R}$ (Figure 1.1), so that, $\underline{R}$ has as cluster point an upper bound $\underline{I}$ and $\overline{R}$ has as cluster point a lower bound $\overline{I}$, $\underline{R}(f,\wp) \leq \underline{I} = \overline{I} \leq \overline{R}(f,\wp)$. When both these values coincide, we say that their common value is $I = \underline{I} = \overline{I}$.

A practical estimation of the value of I is:

$$I \approx \frac{1}{2}\left(\underline{R}(f,\wp) + \overline{R}(f,\wp)\right),$$

i.e.

$$I \approx \sum_{i=1}^{N} \left(\frac{\overline{f_i}+\underline{f_i}}{2}\right)(x_{i+1} - x_i)$$

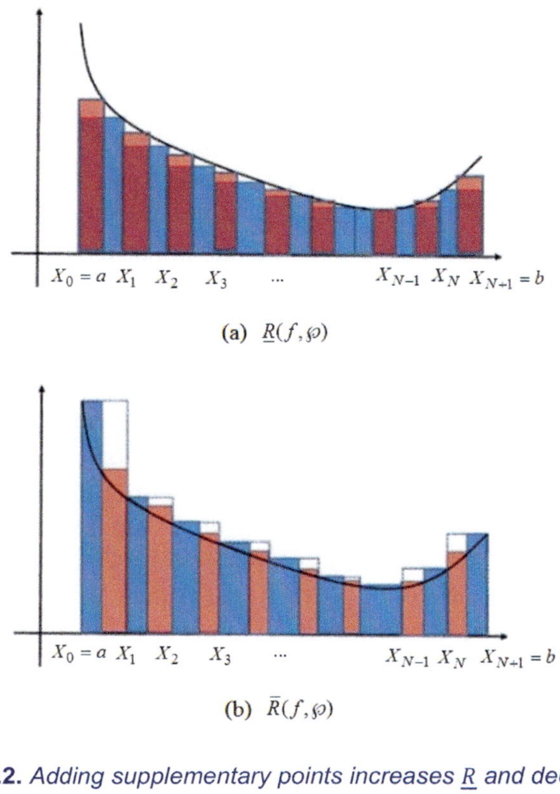

Figure 1.2. *Adding supplementary points increases $\underline{R}$ and decreases $\overline{R}$*

In practice, the values of $\overline{f_i}$ and $\underline{f_i}$ are not determined exactly, but are approximated by using the values $f_i = f(x_i)$ and $f_{i+1} = f(x_{i+1})$:

$$I \approx \sum_{i=1}^{N} \left(\frac{f_i + f_{i+1}}{2}\right)(x_{i+1} - x_i) \qquad [1.1]$$

This formula is known as the *trapezoidal rule*. A simple estimative is furnished by the *mean value approximation*:

$$I \approx \frac{b-a}{N} \sum_{i=1}^{N+1} f_i \ .$$

For $\Omega = (a_1, b_1) \times (a_2, b_2)$ or $\Omega = (a_1, b_1) \times (a_2, b_2) \times (a_3, b_3)$, we may consider families of intervals associated with each component and the Cartesian product of these families. Analogous Riemann sums $\underline{R}(f, \wp) \leq \overline{R}(f, \wp)$ may be defined in this case.

1.2. Lebesgue integrals

Riemann's theory was quickly challenged by the works of Richard Dedekind and Karl Weierstrass. Their work on the foundations of function theory led, on the one hand, to the development of set theory by Georg Cantor and, on the other hand, to measure theory initiated by Giuseppe Peano and Camille Jordan, then formalized by Emile Borel and René Baire. Henri Lebesgue adopted the point of view of measure theory in order to redefine integrals. The theory initiated by Lebesgue is not yet the end of integration theory, since measure theory produces sets that are not measurable (for instance, Vitali sets, presented in the pamphlet [VIT 05]) and, as a consequence, functions that are not measurable. These elements result straight from the axiom of choice (see, for instance, [SOU 10]) and lead to fundamental difficulties, from those presented Felix Hausdorff [HAU 14], and Stefan Banach and Alfred Tarski [BAN 24]. These difficulties have not been solved to date and promise interesting developments in the future.

The numerical evaluation of the Lebesgue integral

$$I = \int_\Omega f$$

requires more information than the evaluation of a Riemann integral; we need a partition $\wp = \{\Omega_i : 1 \le i \le N\}$ of Ω and a partition $\mathcal{F} = \{\mathcal{F}_i : 1 \le i \le M\}$ of the image $f(\Omega)$.

For instance, let us consider the situation where both Ω and $f(\Omega)$ are intervals. We define $\mathcal{F}$ by taking points $y_{min} = y_1 < y_2 < \cdots < y_N < y_{M+1} = y_{max}$, where $y_{min} = min\{f(x) : x \in \Omega\}$ and $y_{max} = max\{f(x) : x \in \Omega\}$. In practice, y_{min} and y_{max} may be approximations of these values satisfying where $y_{min} \le min\{f(x) : x \in \Omega\}$ and $y_{max} \ge max\{f(x) : x \in \Omega\}$.

Let us denote by ℓ the Lebesgue measure, given by:

$$\ell\big((\alpha,\beta)\big) = \beta - \alpha, \quad \ell\big((\alpha_1,\beta_1) \times (\alpha_2,\beta_2)\big) = (\beta_1 - \alpha_1)(\beta_2 - \alpha_2), \ldots$$

and

$$\ell_i = \ell\big(f^{-1}([y_i, y_{i+1}))\big) = \ell(\{x \in \Omega : y_i \le f(x) < y_{i+1}\})$$

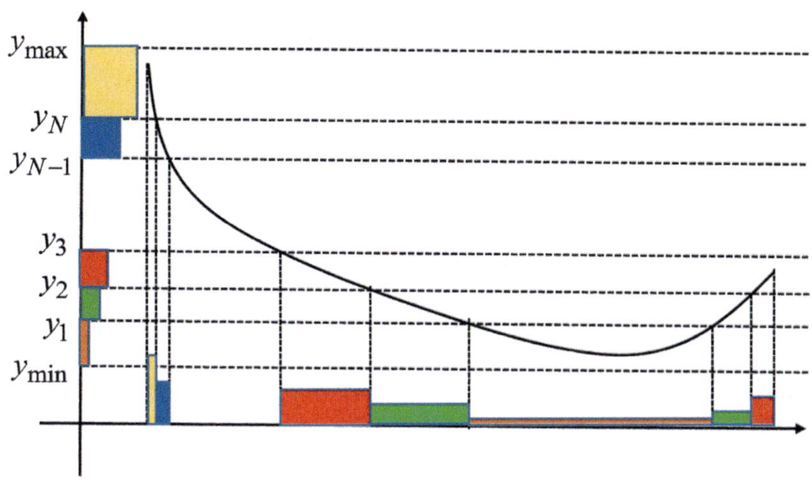

Figure 1.3. *Lebesgue's approach to integration: partition of the vertical axis*

We have:

$$I_{inf} = \sum_{i=1}^{M} y_i\, \ell_i \leq I \leq I_{isup} = \sum_{i=1}^{M} y_{i+1}\, \ell_i$$

Let us introduce:

$$\mu_{inf}(y) = \ell\left(S_{inf}(y)\right)\ ,\ S_{inf}(y) = \{x \in \Omega : f(x) < y\}\,.$$

Then

$$\ell_i = \mu_{inf}(y_{i+1}) - \mu_{inf}(y_i)$$

and

$$\begin{aligned} I_{inf} &= \sum_{i=1}^{M} y_i\, \ell_i = \sum_{i=1}^{M} y_i \left(\mu_{inf}(y_{i+1}) - \mu_{inf}(y_i)\right) \\ &= \sum_{i=2}^{M+1} y_{i-1}\, \mu_{inf}(y_i) - \sum_{i=1}^{M} y_i\, \mu_{inf}(y_i) \\ &= y_{M+1}\mu_{inf}(y_{M+1}) - y_1\mu_{inf}(y_1) - \sum_{i=2}^{M+1}(y_i - y_{i-1})\mu_{inf}(y_i)\ . \end{aligned}$$

Thus,

$$I_{inf} \longrightarrow\ y_{M+1}\mu_{inf}(y_{M+1}) - y_1\mu_{inf}(y_1) - \int_{y_1}^{y_{M+1}} \mu_{inf}(y)\, dy\,,$$

where the integral in this formula is a Riemann integral. So,

$$I \geq y_{M+1}\mu_{inf}(y_{M+1}) - y_1\mu_{inf}(y_1) - \int_{y_1}^{y_{M+1}} \mu_{inf}(y)\, dy\,. \quad [1.2]$$

Analogously,

$$I_{sup} = \sum_{i=1}^{M} y_{i+1}\, \ell_i = \sum_{i=1}^{M} y_{i+1} \left(\mu_{inf}(y_{i+1}) - \mu_{inf}(y_i)\right) = \sum_{i=2}^{M+1} y_i\, \mu_{inf}(y_i) - \sum_{i=1}^{M} y_{i+1}\, \mu_{inf}(y_i) = y_{M+1}\mu_{inf}(y_{M+1}) - y_1\mu_{inf}(y_1) - \sum_{i=1}^{M}(y_{i+1} - y_i)\mu_{inf}(y_i)\ .$$

Thus, we have

$$I_{sup} \longrightarrow\ y_{M+1}\mu_{inf}(y_{M+1}) - y_1\mu_{inf}(y_1) - \int_{y_1}^{y_{M+1}} \mu_{inf}(y)\, dy\,,$$

and

$$I \leq y_{M+1}\mu_{inf}(y_{M+1}) - y_1\mu_{inf}(y_1) - \int_{y_1}^{y_{M+1}} \mu_{inf}(y)\, dy\,. \quad [1.3]$$

Equations [1.2] and [1.3] show that:

$$I = y_{M+1}\mu_{\inf}\left(y_{M+1}\right) - y_1\mu_{\inf}\left(y_1\right) - \int_{y_1}^{y_{M+1}} \mu_{\inf}\left(y\right)dy. \qquad [1.4]$$

Notice that, if $y_{min} \leq min\{f(x): x \in \Omega\}$ and $y_{max} \geq max\{f(x): x \in \Omega\}$, then $\mu_{inf}(y_1) = 0$, while $\mu_{inf}(y_{M+1}) = b - a$ and we have:

$$I = (b-a)y_{M+1} - \int_{y_1}^{y_{M+1}} \mu_{inf}(y)\, dy\,.$$

Let

$$\mu_{sup}(y) = \ell\left(S_{sup}(y)\right)\,,\ S_{sup}(y) = \{x \in \Omega : f(x) \geq y\}\,.$$

Then,

$$\ell_i = \mu_{sup}(y_i) - \mu_{sup}(y_{i+1})$$

and

$$I_{inf} = \sum_{i=1}^{M} y_i\,\ell_i = \sum_{i=1}^{M} y_i\left(\mu_{sup}(y_i) - \mu_{sup}(y_{i+1})\right) = \sum_{i=1}^{M} y_i\,\mu_{sup}(y_i) - \sum_{i=2}^{M+1} y_{i-1}\,\mu_{sup}(y_i) = y_1\mu_{sup}(y_1) - y_{M+1}\mu_{sup}(y_{M+1}) + \sum_{i=2}^{M+1}(y_i - y_{i-1})\mu_{sup}(y_i)\ .$$

Thus,

$$I_{inf} \longrightarrow\ y_1\mu_{sup}(y_1) - y_{M+1}\mu_{sup}(y_{M+1}) + \int_{y_1}^{y_{M+1}} \mu_{sup}(y)\, dy\,,$$

where the integral in this formula is a Riemann one. So,

$$I \geq y_1\mu_{sup}(y_1) - y_{M+1}\mu_{sup}(y_{M+1}) + \int_{y_1}^{y_{M+1}} \mu_{sup}(y)\, dy\,. \qquad [1.5]$$

Analogously,

$$I_{sup} = \sum_{i=1}^{M} y_{i+1}\,\ell_i = \sum_{i=1}^{M} y_{i+1}\left(\mu_{sup}(y_i) - \mu_{sup}(y_{i+1})\right),$$

so that

$$I_{sup} = y_1\mu_{sup}(y_1) - y_{M+1}\mu_{sup}(y_{M+1}) - \sum_{i=1}^{M}(y_{i+1} - y_i)\mu_{sup}(y_i) \ .$$

Thus, we have:

$$I_{sup} \longrightarrow y_1\mu_{sup}(y_1) - y_{M+1}\mu_{sup}(y_{M+1}) + \int_{y_1}^{y_{M+1}} \mu_{sup}(y)\, dy \, ,$$

and

$$I \leq y_1\mu_{sup}(y_1) - y_{M+1}\mu_{sup}(y_{M+1}) + \int_{y_1}^{y_{M+1}} \mu_{sup}(y)\, dy \, . \quad [1.6]$$

Equations [1.5] and [1.6] show that:

$$I = y_1\mu_{\text{sup}}(y_1) - y_{M+1}\mu_{\text{sup}}(y_{M+1}) + \int_{y_1}^{y_{M+1}} \mu_{\text{sup}}(y)\,dy. \quad [1.7]$$

Notice that, if $y_{min} \leq \min\{f(x): x \in \Omega\}$ and $y_{max} \geq \max\{f(x): x \in \Omega\}$, then $\mu_{sup}(y_1) = b - a$, while $\mu_{sup}(y_{M+1}) = 0$ and we have:

$$I = (b - a)y_1 + \int_{y_1}^{y_{M+1}} \mu_{sup}(y)\, dy \, .$$

1.3. Matlab® classes for a Riemann integral by trapezoidal integration

The one-dimensional (1D) trapezoidal rule is implemented in Matlab® in the intrinsic function `trapz` and may be extended to multidimensional situations. Assume that the structure `p` has fields `p.x, p.y, p.z` corresponding to the coordinates of the points x_i, y_j, z_k of the partitions of the intervals and `p.dim` corresponding to the dimension, while table `F` contains the values of f - $F_i = f(x_i)$, or $F_{ij} = f(x_i, y_j)$, or $F_{ijk} = f(x_i, y_j, z_k)$, according to the dimension of the integral. Let us summarize the data in a structure `data` such that `data.points = p` and `data.values = F`. Then we may use the class below (Program 1.1).

```
classdef riemann
    methods (Static, Access = private)
        function v = trapz2(x1,x2,F)
            sy = zeros(size(x1));
            for i = 1: length(sy)
                sy(i) = trapz(x2,F(i,:));
            end;
            v = trapz(x1,sy);
        end
        function v = trapz3(x1,x2,x3,F)
            sz = zeros(length(x1),length(x2));
            for i = 1: length(x1)
                for j = 1: length(x2)
                    sz(i,j) = trapz(x3,F(i,j,:));
                end;
            end;
            sy = zeros(size(x1));
            for i = 1: length(x1)
                sy(i) = trapz(x2,sz(i,:));
            end;
            v = trapz(x1,sy);
        end
    end
    methods(Static)
        function v = trpzd(data)
            p = data.points;
            F = data.values;
            n = p.dim;
            switch n
                case 1
                    v = trapz(p.x,F);
                case 2
                    v = riemann.trapz2(p.x,p.y,F);
                case 3
                    v = riemann.trapz3(p.x,p.y,p.
z,F);
```

```
                otherwise
                    v = [];
                    disp('error in arguments for
trpzd');
            end
        end
        function v = mean_value(data)
            p = data.points;
            F = data.values;
            n = p.dim;
            switch n
                case 1
                    v = mean(F)*p.measure;
                case 2
                    v = mean(mean(F))*p.measure;
                case 3
                    v =
mean(mean(mean(F)))*p.measure;
                otherwise
                    v = [];
                    disp('error in arguments for
mean value');
                end
        end
    end
end
```

Program 1.1. *A class for the evaluation of Riemann integrals*

This class contains only trapezoidal and mean value integration methods, but it can be enriched by the user with other methods of numerical integration.

EXAMPLE 1.1.– Let us evaluate:

$$I = \int_0^1 dx \int_0^2 (x^2 + y^2)\, dy$$

by using the points `x=0:0.01:1;` and `y=0:0.01:2;`. Assuming that `F(i,j)=x(i)^2+y(j)^2;`, the code:

```
p.x = x;
p.y = y;
p.dim  = 2;
p.measure = 2;
data.points = p;
data.values = F;
v1 = riemann.trpzd(data)
v2 = riemann.mean_value(data)
```

produces `v1 = 3.3334, v2 = 3.3433.` The exact value is $\frac{10}{3} \approx 3.3333$. ■

EXAMPLE 1.2.– Let us evaluate:

$$I = \int_0^1 dx \int_0^2 dy \int_0^3 (x^2 + y^2 + z^2)\, dz$$

by using the points `x=0:0.01:1;,y=0:0.01:2;, z=0:0.01:3.` Assuming that `F(i,j,k)=x(i)^2+y(j)^2+z(k)^2`, the code:

```
p.x = x;
p.y = y;
p.z = z;
p.dim  = 3;
p.measure = 6;
data.points = p;
data.values = F3;
v1 = riemann.trpzd(F3,p)
v2 = riemann.mean_value(F3,p)
```

produces `v1 = 28.0003, v2 = 28.0600.` The exact value is 28. ■

The creation of the tables `F` from a subprogram `f` evaluating a function f is made by the following class:

```
classdef spam
    properties
    end
    methods (Static, Access = private)
        function v = vspam1(x,f)
            v = zeros(size(x));
            for i = 1: length(x)
                v(i) = f(x(i));
            end;
        end
        function v = vspam2(x,y,f)
            v = zeros(length(x),length(y));
            for i = 1: length(x)
                for j = 1: length(y)
                    v(i,j) = f(x(i),y(j));
                end;
            end;
        end
        function v = vspam3(x,y,z,f)
            v = zeros(length(x),length(y),length(z));
            for i = 1: length(x)
                for j = 1: length(y)
                    for k = 1: length(z)
                        v(i,j,k) = f(x(i),y(j),z(k));
                    end;
                end;
            end;
        end
        function v = tspam1(x,f)
            v = zeros(size(x));
            for i = 1: length(x)
                v(i) = f(x(i));
            end;
        end
        function v = tspam2(x,y,f)
            v = zeros(size(x));
            for i = 1: size(x,1)
                for j = 1: size(x,2)
                    v(i,j) = f(x(i,j),y(i,j));
                end;
            end;
```

```
        end
        function v = tspam3(x,y,z,f)
            v = zeros(size(x));
            for i = 1: size(x,1)
                for j = 1: size(x,2)
                    for k = 1: size(x,3)
                        v(i,j,k) =
f(x(i,j,k),y(i,j,k),z(i,j,k));
                    end;
                end;
            end;
        end
    end
    methods (Static)
        function v = points(method,f,x,y,z)
            n = nargin() - 2;
            if isnumeric(f)
                ff = @(x) f;
            else
                ff = f;
            end;
            switch method
                case 'vector'
                    switch n
                        case 1
                            v = spam.vspam1(x,ff);
                        case 2
                            fr = @(x,y) ff([x, y]);
                            v = spam.vspam2(x,y,fr);
                        case 3
                            fr = @(x,y,z) ff([x, y, z]);
                            v = spam.vspam3(x,y,z,fr);
                    end;
                case 'table'
                    switch n
                        case 1
                            v = spam.tspam1(x,ff);
                        case 2
                            fr = @(x,y) ff([x, y]);
                            v = spam.tspam2(x,y,fr);
```

```
                        case 3
                            fr = @(x,y,z) ff([x, y, z]);
                            v = spam.tspam3(x,y,z,fr);
                    end;
                otherwise
                    v = [];
                    disp('invalid method for
spam.points');
            end
        end
        function v = partition(f,p)
            n = p.dim;
            switch n
                case 1
                    v = spam.points('vector',f,p.x);
                case 2
                    v = spam.points('vector',f,p.x,p.y);
                case 3
                    v = spam.points('vector',f,p.x,
p.y,p.z);
                otherwise
                    v = [];
            disp('invalid partition dim for
spam.partition');
            end;
        end
    end
end
```

Program 1.2. *A class for the creation of the tables F*

This class generates a table of values $F_i = f(x_i)$, $F_{ij} = F(x_i, y_j)$, $F_{ij} = F(x_{ij}, yi_j)$, $F_{ijk} = f(x_i, y_j, z_k)$, or $F_{ijk} = f(x_{ijk}, y_{ijj}, z_{ijk})$, according to the data furnished. Data is stored in the structure `p`: `p.dim` is the dimension, `p.x, p.y, p.z` are the coordinates. `Method` takes the value `"vector"` or `"array"` according to the dimensions of `p.x, p.y, p.z`. For instance, the code:

```
x = 0:0.1:1;
y = 0:0.1:2;
f = @(x) sum(x.^2);
p.x = x;
p.y = y;
p.dim  = 2;
[X,Y] = meshgrid(x,y);
xx = permute(X,[2 1]);
yy = permute(Y,[2 1]);
F1 = spam.partition(f,p);
F2 = spam.points('table',f,xx,yy);
v = sqrt(sum(sum((F1-F2).^2)));
```

produces as result `v=0`. Both the tables `F1` and `F2` coincide, but F1 is generated by using vectors `x` and `y`, while `F2` is generated using two-dimensional (2D) arrays. Adding the lines:

```
z = 0:0.1:3;
p.z = z;
p.dim = 3;
[X,Y,Z] = meshgrid(x,y,z);
xx =  permute(X,[2 1 3]);
yy  =  permute(Y,[2 1 3]);
zz =  permute(Z,[2 1 3]);
F1 = spam.partition(f,p);
F2 = spam.points('table',f,xx,yy,zz);
v3 = sqrt(sum(sum(sum((F1-F2).^2))));
```

produces as result `v3=0`: the resulting arrays are equal. Here, F1 is generated by using vectors `x`, `y` and `z`, while `F2` is generated using three-dimensional (3D) arrays.

Notice that the class `Riemann` requests data in the vector form. The extension to data defined in the array form will be useful for integration using the intrinsic functions of Matlab (see section 1.6).

1.4. Matlab® classes for Lebesgue's integral

Equations [1.4] and [1.7] provide a practical method for the evaluation of I: the Lebesgue integral is replaced by a Riemann one, which may be numerically evaluated by quadrature methods using the values $\mu_{inf}(y_1), \dots, \mu_{inf}(y_{M+1})$ (when using equation [1.4]) or $\mu_{sup}(y_1), \dots, \mu_{sup}(y_{M+1})$ (when using equation [1.7]). The construction of the values of μ_{inf} or μ_{sup} is performed by using the subdivision of Ω: for each i, the measure of the part of Ω_i belonging to $S_{inf}(y)$ (respectively, $S_{sup}(y)$) is estimated and added to $\mu_{inf}(y_i)$ (respectively, $\mu_{sup}(y_i)$). For instance, we may use the following class:

```
classdef lebesgue
    methods  (Static, Access = private)
        function v =  hexainds(i,j,k)
            v = [
                i j k
                i+1 j k
                i+1 j+1 k
                i j+1 k
                i j k+1
                i+1 j k+1
                i+1 j+1 k+1
                i j+1 k+1
                ];
        end
        function v = quadinds(i,j)
            v = [
                i j
                i+1 j
                i+1 j+1
                i j+1
                ];
        end
        function [m, flag]  = inf1(y,x,yt)
            y2  = max(y);
            y1  = min(y);
            dx  = x(2)  - x(1);
            flag  = 1;
            if yt < y1
                m = 0;
                flag = 0;
            elseif yt >= y2
                m = dx;
```

```
        else
            if y1 == -Inf
                m = 0.5*dx;
            elseif y2 == Inf
                m = 0.5*dx;
            else
                slope = (y2- y1)/dx;
                m = (yt - y1)/slope;
            end;
        end;
    end
    function [m, flag] = sup1(y,x,yt)
        y2 = max(y);
        y1 = min(y);
        dx = x(2) - x(1);
        flag = 1;
        if yt > y2
            m = 0;
            flag = 0;
        elseif yt <= y1
            m = dx;
        else
            if y1 == -Inf
                m = 0.5*dx;
            elseif y2 == Inf
                m = 0.5*dx;
            else
                slope = (y2- y1)/dx;
                m = (y2 - yt)/slope;
            end;
        end;
    end
    function [m, flag]  = m2(y,x,yt,m1)
        s = zeros(4,1);
        flg = zeros(4,1);
        [s(1),flg(1)] = m1(y(1:2),[0,1],yt);
        [s(2),flg(2)] = m1(y(2:3),[0,1],yt);
        [s(3),flg(3)] = m1(y(3:4),[0,1],yt);
        [s(4),flg(41)] = m1([y(1),y(4)],[0,1],yt);
        ta = x(2,:) - x(1,:);
        tb = x(4,:) - x(1,:);
```

```
            m = mean(s)*norm(ta)*norm(tb);
            flag = max(flg);
        end
        function [m,flag] = m3(y,x,yt,m1)
            s = zeros(6,1);
            flg = zeros(6,1);
            ind = [
                1 2 3 4
                2 3 7 6
                3 4 8 7
                1 4 8 5
                1 2 6 5
                5 6 7 8
                ];
            for i = 1: 6
[s(i),flg(i)] = lebesgue.m2(y(ind(i,:)),[0 0; 0 1; 1
1; 1 0],yt,m1);
            end;
            ta = x(2,:) - x(1,:);
            tb = x(3,:) - x(2,:);
            tc = x(5,:) - x(1,:);
            m = mean(s)*norm(ta)*norm(tb)*norm(tc);
            flag = max(flg);
        end
        function [m, flag] = inf_el(y,x,yt,dim)
            switch dim
                case 1
                    [m, flag] =  lebesgue.inf1(y,x,yt);
                case 2
                    [m, flag] = lebesgue.m2(y,x,yt,
 @lebesgue.inf1);
                case 3
                    [m, flag] = lebesgue.m3(y,x,yt,
 @lebesgue.inf1);
            end
        end
        function [m, flag] = sup_el(y,x,yt,dim)
            switch dim
                case 1
                    [m, flag] = lebesgue.sup1(y,x,yt);
```

```
                case 2
                    [m, flag] = lebesgue.m2(y,x,yt,
@lebesgue.sup1);
                case 3
                    [m, flag] = lebesgue.m3(y,x,yt,
@lebesgue.sup1);
            end
        end
        function mu = msre(mes,y,data,pas,hw)
            p = data.points;
            f = data.values;
            n = p.dim;
            ny = length(y);
            mu = zeros(ny,1);
            switch n
                case 1
                    x1 = p.x;
                    n1 = length(x1);
                    for i =  1:  n1 - 1
                        xe = [x1(i) , x1(i+1)];
                        ye = [f(i) , f(i+1)];
                        flag = 1;
                        if pas > 0
                            ind = 1;
                        else
                            ind = ny;
                        end;
  while flag > 0 && ((ind > 1 && pas < 0)  || (ind <
ny && pas > 0))
                            [m, flag] =
mes(ye,xe,y(ind),n);
                            mu(ind) = mu(ind) +  m;
                            ind = ind + pas;
                        end;
                        wx = i/n1;
         text = [sprintf(' %4.1f ',fix(1000*wx)/10),'
% generated'];
                        if pas > 0
                            waitbar(wx,hw,['\mu_{sup}
:',text]);
                        else
```

```
                              waitbar(wx,hw,['\mu_{inf}
:',text]);
                          end;
                      end;
                      close(hw)
                  case 2
                      x1 = p.x;
                      x2 = p.y;
                      n1 = length(x1);
                      n2 = length(x2);
                      nt = n1*n2;
                      for i =  1:  n1 - 1
                          for j = 1: n2 - 1
                              ind = lebesgue.quadinds
(i,j);
                              xe = [ x1(ind(:,1))'
x2(ind(:,2))'];
                              ye = zeros(4,1);
                              for ii =  1:  4
                                  ye(ii) = f(ind(ii,1),
ind(ii,2));
                              end;
                              flag = 1;
                              if pas > 0
                                  ind = 1;
                              else
                                  ind = ny;
                              end;
   while flag > 0 && ((ind > 1 && pas < 0) || (ind <
ny && pas > 0))
                                  [m, flag] = mes(ye,xe,
y(ind),n);
                                  mu(ind) = mu(ind) +
m;
                                  ind = ind + pas;
                              end;
                              wx = (j + (i-1)*n2)/nt;
        text = [sprintf(' %4.1f ',fix(1000*wx)/10),' %
generated'];
                              if pas > 0
```

```
                                    waitbar(wx,hw,
['\mu_{sup} :',text]);
                                else
                                    waitbar(wx,hw,
['\mu_{inf} :',text]);
                                end;
                            end;
                        end;
                        close(hw)
                    case 3
                        x1 = p.x;
                        x2 = p.y;
                        x3 = p.z;
                        n1 = length(x1);
                        n2 = length(x2);
                        n3 = length(x3);
                        nt = n1*n2*n3;
                        for i =  1:  n1 - 1
                            for j = 1: n2 - 1
                                for k = 1:n3 - 1
                                    ind =
lebesgue.hexainds(i,j,k);
                    xe = [ x1(ind(:,1))'  x2(ind(:,2))'
x3(ind(:,3))'];
                                    ye = zeros(8,1);
                                    for ii =  1:  8
                     ye(ii) = f(ind(ii,1), ind(ii,2),
ind(ii,3));
                                    end;
                                    flag = 1;
                                    if pas > 0
                                        ind = 1;
                                    else
                                        ind = ny;
                                    end;
   while flag > 0 && ((ind > 1 && pas < 0) || (ind <
ny && pas > 0))
                                        [m, flag] =
mes(ye,xe,y(ind),n);
                                        mu(ind) = mu(ind)
+  m;
                                        ind = ind + pas;
```

```
                                    end;
                         wx = (k + (j-1)*n3 + (i-1)*n2*n3)/nt;
                 text = [sprintf(' %4.1f ',fix(1000*wx)/10),' %
generated'];
                                    if pas > 0
                                    waitbar(wx,hw,
['\mu_{sup} :',text]);
                                    else
                                    waitbar(wx,hw,
['\mu_{inf} :',text]);
                                    end;
                                end;
                            end;
                        end;
                        close(hw)
                    otherwise
                        mu = [];
                        disp('error in args for measure');
                        close(hw)
                end
            end
            function mu = mesinf(y,data,hw)
                if nargin() < 3
 hw = waitbar(0,'0 % generated','Name','Lebesgue inf
measure');
                end;
                mu = lebesgue.msre(@lebesgue.inf_el,
                y,data,-1,hw);
            end
            function mu = messup(y,data,hw)
                if nargin() < 3
 hw = waitbar(0,'0 % generated','Name','Lebesgue sup
measure');
                end;
                mu = lebesgue.msre(@lebesgue.sup_el,
y,data,1,hw);
            end
            function v = intinf(y,data,interpmet,hw)
                if nargin() < 4
hw = waitbar(0,'0 % generated','Name','intinf:
Lebesgue inf meas');
```

```
                end
                mu1 = lebesgue.mesinf(y,data,hw);
                ff = @(t) interp1(y,mu1,t,interpmet);
       v = y(end)*mu1(end) - y(1)*mu1(1) -
integral(ff,y(1),y(end));
            end
            function v = intsup(y,data,interpmet,hw)
                if nargin() < 4
hw = waitbar(0,'0 % generated','Name','intsup:
Lebesgue sup meas');
                end
                mu2 = lebesgue.messup(y,data,hw);
                ff = @(t) interp1(y,mu2,t,interpmet);
     v = y(1)*mu2(1) - y(end)*mu2(end) +
integral(ff,y(1),y(end));
            end
        end
        methods (Static)
            function mu1 = measure_inf(y,data,interpmet)
                aux = lebesgue.mesinf(y,data);
                mu1 = @(t) interp1(y,aux,t,interpmet);
            end
            function mu2 = measure_sup(y,data,interpmet)
                aux = lebesgue.messup(y,data);
                mu2 = @(t) interp1(y,aux,t,interpmet);
            end
            function v = integrate(y,data, metod,
 interpmet)
                switch metod
                    case 'inferior'
                        switch interpmet
                            case 'none'
                                mu1 = lebesgue.mesinf
 (y,data);
                                v = trapz(y,mu1);
                            otherwise
                                v =
lebesgue.intinf(y,data,interpmet);
                        end
                    case 'superior'
                        switch interpmet
```

```
                        case 'none'
                            mu2 = lebesgue.messup
(y,data);
                            v = trapz(y,mu2);
                        otherwise
                            v =
lebesgue.intsup(y,data,interpmet);
                    end
                case 'mean'
                    switch interpmet
                        case 'none'
                            mu1 =
lebesgue.mesinf(y,data);
                            aux1 = trapz(y,mu1);
                            mu2 =
lebesgue.messup(y,data);
                            aux2 = trapz(y,mu2);
                        otherwise
hw = waitbar(0,'0 % generated','Name','intmean:
Lebesgue inf meas');
                            aux1 =
lebesgue.intinf(y,data,interpmet,hw);
hw = waitbar(0,'0 % generated','Name','intmean:
Lebesgue sup meas');
                            aux2 =
lebesgue.intsup(y,data,interpmet,hw);
                    end
                    v = 0.5*(aux1+aux2);
                otherwise
                    v = [];
                    disp('error in args for
integral');
            end
        end
    end
end
```

Program 1.3. *A class for the evaluation of Lebesgue integrals*

At a glance, we see that this class is more complex than the class Riemann. This is due to the fact that two partitions are requested and must be treated. Methods `inf_el` and `sup_el` determine the part of $(x_{j-1}, x_j), 1 \leq j \leq N,$ which belongs to $S_{inf}(y_i)$ and $S_{sup}(y_i)$, respectively. These subprograms return approximations of the Lebesgue measure of the part of (x_1, x_2) lying in the region of interest ($S_{inf}(y_t)$ for `mes_inf` or $S_{sup}(y_t)$ for `mes_sup`). The special cases where either $y_1 = -\infty$ or $y_2 = +\infty$ are treated by considering that a half of the interval (x_1, x_2) belongs to the region of interest – this choice is arbitrary and may be modified by the user. For the standard situation where both y_1 and y_2 are real numbers, a linear interpolation determines the part of the interval belonging to the region of interest – this choice may also be modified by the user. When using this class, the evaluation of the integral involves the choices, on the one hand, the choice of the evaluation points (vector `y`) and, on the other hand, the choice among the use of μ_{inf} (equation [1.4]), of μ_{sup} (equation [1.7]) or the arithmetic mean of these results. In addition, it is possible to reduce the time of computation by evaluating these measures in a limited number of points and using the interpolation function `interp1` for the evaluation of the integral. The functions μ_{inf}, μ_{sup} are evaluated, respectively, by the methods `measure inf` and `measure_sup`, both returning a function generated by interpolation of the evaluated values, using `interp1` and the interpolation method `interpmet`.

EXAMPLE 1.3.– Let us consider $\Omega = (-1,1)$ and $f(x) = x^2$. Then,

$$\mu_{inf}(y) = \begin{cases} 0, if\ y \leq 0 \\ 2\sqrt{y}, if\ 0 < y < 1 \\ 2, if\ y \geq 1 \end{cases} ; \quad \mu_{sup}(y) = \begin{cases} 2, if\ y \leq 0 \\ 2(1 - \sqrt{y}), if\ 0 < y < 1 \\ 0, if\ y \geq 1 \end{cases}$$

For $N = 1024$, $M = 128$, $y_{\min} = -0.5$, $y_{\max} = 1.5$, $a = -1, b = 1$, the code:

```
hx = (b-a)/N;
x = a + hx*(0:N);
f = x.^2;
p.x = x;
p.dim = 1;
```

```
data.points = p;
data.values = f;
mu1 = lebesgue.measure_inf(y,data,'linear');
mu2 = lebesgue.measure_sup(y,data,'linear');
```

furnishes

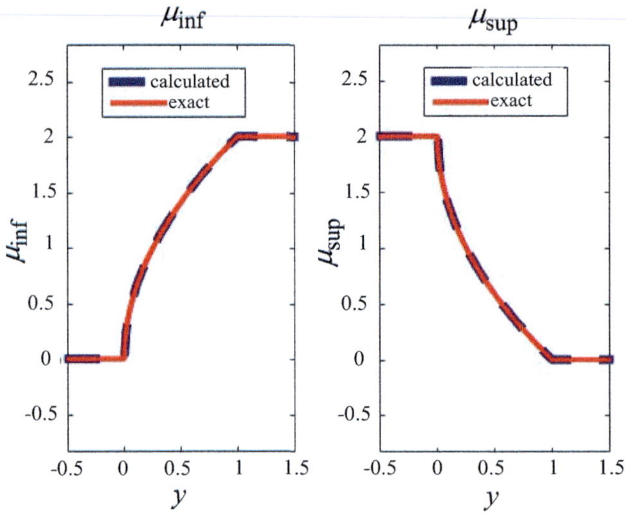

Figure 1.4. *Lebesgue measure obtained in example 1.3*

The code:

```
I1    =    lebesgue.integrate(y,data,'inferior',
'linear');
I2    =    lebesgue.integrate(y,data,'superior',
'linear');
I3 = lebesgue.integrate(y,data,'mean','linear')
```

furnishes `I1 = 0.66746`, `I2 = 0.66746` and `I3 = 0.66746`. The exact value is `2/3 ≈ 0.66667`. These results may be improved by refining the partitions, namely on the vertical axis. For instance, $N = 4096$, $M = 512$ lead to `I1 = 0.66677`, `I2 = 0.66677` and `I3 = 0.66677`. ■

When regular functions are considered – this is the case in example 1.3 – Lebesgue's approach is not the more efficient one. For instance, Riemann's approach furnishes more precise results. However, for functions having discontinuity points, Lebesgue's approach may be interesting (see examples below).

EXAMPLE 1.4.– Let us consider $\Omega = (0,2)$ and $f(x) = 1/\sqrt{x}$. Then,

$$\mu_{inf}(y) = \begin{cases} 0, if\ y \leq 1/\sqrt{2} \\ 2 - 1/y^2, otherwise \end{cases} \quad ; \quad \mu_{sup}(y) = \begin{cases} 2, if\ y \leq 1/\sqrt{2} \\ 1/y^2, otherwise \end{cases} .$$

Using $M = 256$, $y_{min} = 0$, $y_{max} = 100$, $N = 4096$, we obtain `I1 = 2.7905`, `I2 = 2.7906`, `I3 = 2.7906`, while the trapezoidal rule `trapz(x,f)` returns the value `Inf`. The exact value is `2*sqrt(2)` ≈ `2.8284`. For $M = 1024$, N = 4096, we have `I1 = I2 = I3 = 2.7959` and `trapz (x,f)` returns the value `Inf`. The measures μ_{inf} and μ_{sup} are in Figure 1.5. ■

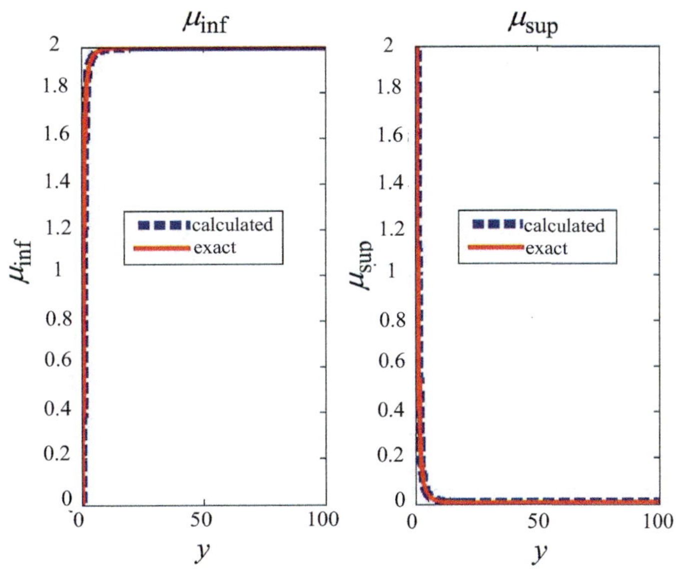

Figure 1.5. *Results for example 1.4*

EXAMPLE 1.5.– Let us consider $\Omega = (0,20)$ and $f(x) = log(|sin(\pi x)|)$. Then,

$$\mu_{inf}(y) = \begin{cases} (b-a), if\ y \geq 0 \\ \frac{2}{\pi}(b-a)\, sin^{-1}(e^y), otherwise \end{cases};$$

$$\mu_{sup}(y) = \begin{cases} 0, if\ y \geq 0 \\ (b-a)\left(1 - \frac{2}{\pi} sin^{-1}(e^y)\right), otherwise \end{cases}$$

In this case, when using N = 1024, $y_{min} = -2.5$, $y_{max} = 0.5$, M = 256, we obtain `I1 = I2 = I3 =-10.121`, while `trapz(x,f)` returns the value-`Inf`. For N = 4096, $y_{min} = -100$, $y_{max} = 0.5$, M = 1024, the values are `I1 = -14.3736, I2 = -14.3733, I3 = -14.3735` and `trapz(x,f)` returns the value-`Inf`. The measures μ_{inf} and μ_{sup} are in Figure 1.6. ■

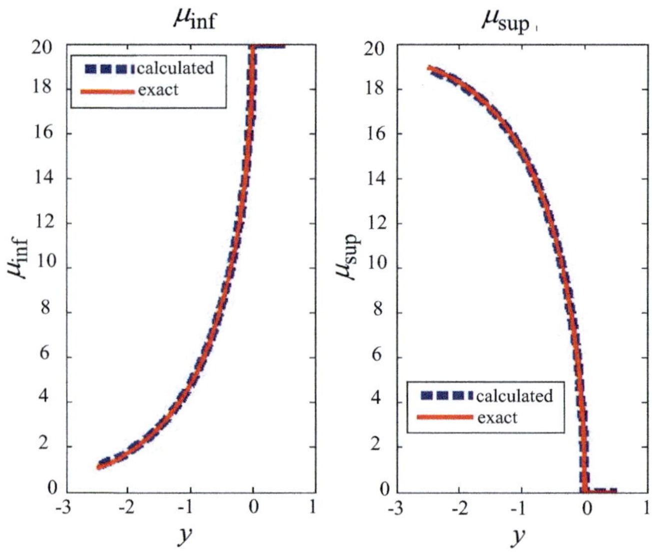

Figure 1.6. *Lebesgue measure in example 1.5*

EXAMPLE 1.6.– Let us consider $\Omega = (0,20)$ and $f(x) = sign(\sin(\pi x))$. Then,

$$\mu_{inf}(y) = \begin{cases} 0, if\ y \leq -1 \\ 1, if\ -1 < y < 1 \\ 2, if\ y \geq 1 \end{cases} ; \quad \mu_{sup}(y) = \begin{cases} 2, if\ y \leq -1 \\ 1, if\ -1 < y < 1 \\ 0, if\ y \geq 1 \end{cases}.$$

In this case, using N = 1024, $y_{min} = -1.5$, $y_{max} = 1.5$, $M = 256$, we obtain `I1= I2 = I3 =0.0097659` and `trapz(x,f)` returns the value `0.0098`. The exact value is `0`. The measures μ_{inf} and μ_{sup} are given in Figure 1.7. ■

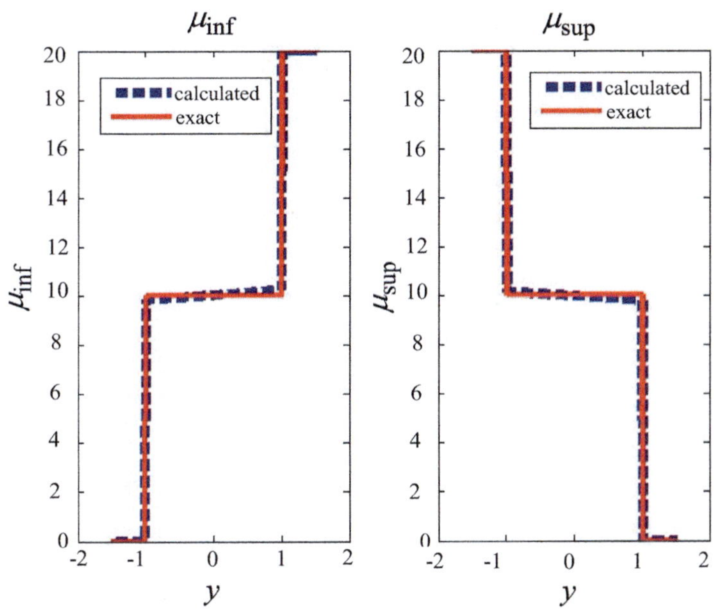

Figure 1.7. *Results for Lebesgue measure in example 1.6*

One of the interesting features of Lebesgue integrals is the fact that multidimensional integration reduces to 1D integration: indeed, only the evaluation of μ_{inf}, μ_{sup} involves multidimensional calculations.

Once these quantities are obtained, equation [1.4] or equation [1.7] is used in order to evaluate the integral – only 1D integration is involved in this calculation. Examples are given below.

EXAMPLE 1.7.– Let us evaluate:

$$I = \int_0^1 dx \int_0^2 (x^2 + y^2)\, dy$$

by using the points `x=0:step:1;` and `y=0:step:2;` with `step = 0.01`. Assuming that `F(i,j)=x(i)^2+y(j)^2`, the code:

```
p.x = x;
p.y = y;
p.dim  = 2;
yleb = 0:0.1:5;
data.points = p;
data.values = F;
I1 = lebesgue.integrate(yleb,data,'inferior','linear');
I2 = lebesgue.integrate(yleb,data,'superior','linear');
I3 = lebesgue.integrate(yleb,data,'mean','linear')
```

produces `I1=I2=I3=3.3334`. The exact value is $\frac{10}{3} \approx 3.3333$. ■

EXAMPLE 1.8.– Let us evaluate:

$$I = \int_0^1 dx \int_0^2 dy \int_0^3 (x^2 + y^2 + z^2)\, dz$$

by using the points `x=0:step:1;,y=0:step:2;,` `z=0:step:3;,` with `step = 0.05`. Assuming that `F(i,j,k)=x(i)^2+y(j)^2+z(k)^2`, the code:

```
p.x = x;
p.y = y;
p.z = z;
p.dim  = 3;
yleb = 0:0.2:14;
data.points = p;
data.values = F;
```

```
I1 = lebesgue.integrate(yleb,data,'inferior','linear');
I2 = lebesgue.integrate(yleb,data,'superior','linear');
I3 = lebesgue.integrate(yleb,data,'mean','linear')
```

produces `I1=I2=I3=28.0094`.

Using as interpolation method `'cubic'` (or `'pchip'`, for newer versions of Matlab®) produces the result `28.0090`. The exact value is 28. ■

1.5. Matlab® classes for evaluation of the integrals when *f* is defined by a subprogram

Let us assume that the values of $f(x)$ are furnished by a subprogram `f.m` which receives as argument a vector `x` and returns a value `f (x)`.

The internal functions for numerical integration are given in Table 1.1. Notice that `ay,by,az, bz` are function handles, i.e. anonymous functions or subprograms evaluating functions.

In order to use these intrinsic functions of Matlab®, we must transform `f`. Indeed, the intrinsic functions of Matlab® assume as arguments `(x,y)` or `(x,y,z)` and not a vector `x`. Moreover, they assume that the functions may receive multidimensional arrays of data and return a corresponding array of results. The adaptation is made by the subprograms of class `spam`. For instance, we may use the class given in program 1.4.

Intrinsic function	evaluates
`integral(f,a,b)` or `quadgk(f,a,b)`	$\int_a^b f(x)dx$
`integral2(f,ax,bx,ay,by)` or `quad2d(f,ax,bx,ay,by)`	$\int_{ax}^{bx} dx \int_{ay(x)}^{by(x)} f(x,y)dy$
`integral3(f,ax,bx,ay,by,az,bz)`	$\int_{ax}^{bx} dx \int_{ay(x)}^{by(x)} \int_{az(x,y)}^{bz(x,y)} f(x,y,z)dxdydz$

Table 1.1. *Intrinsic functions for the evaluation of integrals*

```
classdef subprogram_integration
    properties
        limits
        integrand
    end
    properties(Dependent)
        result
    end
    %
    methods
        function obj = subprogram_integration(f,xlim)
            if isnumeric(f)
                f = @(x) f;
            end;
            obj.integrand = f;
            obj.limits = xlim;
        end
        function v = get.result(obj)
            f = obj.integrand;
            xlim = obj.limits;
            a = xlim.lower;
            b = xlim.upper;
            n = xlim.dim;
            switch n
                case 1
                    fs = @(x) spam.points('table',
f,x);
                    v = integral(fs,a.x,b.x);
                case 2
                    fs = @(x,y) spam.points('table',
f,x,y);
                    if isnumeric(a.y)
                        a.y = @(x) a.y;
                    end;
                    ymin = @(x) spam.points('table',
a.y,x);
                    if isnumeric(b.y)
                        b.y = @(x) b.y;
                    end;
                    ymax = @(x) spam.points('table',
b.y,x);
```

```
                    v = integral2(fs,a.x,b.x,
ymin,ymax);
                    %   or quad2d(fs,a.x,b.x,
ymin,ymax);
                    %   or dblquad(fs,a.x,b.x,
a.y,b.y);
                case 3
                    fs = @(x,y,z) spam.points('table',
f,x,y,z);
                    if isnumeric(a.y)
a.y = @(x) a.y;
                    end;
                    ymin = @(x) spam.points('table',
a.y,x);
                    if isnumeric(b.y)
b.y = @(x) b.y;
                    end;
                    ymax = @(x) spam.points('table',
b.y,x);
                    if isnumeric(a.z)
                        a.z = @(x) a.z;
                    end;
                    zmin = @(x,y) spam.points('table',
a.z,x,y);
                    if isnumeric(b.z)
                        b.z = @(x) b.z;
                    end;
                    zmax = @(x,y)
spam.points('table',b.z,x,y);
                    v =
integral3(fs,a.x,b.x,ymin,ymax,zmin,zmax);
                    %  or
triplequad(fs,a.x,b.x,a.y,b.y,a.z,b.z);
            end;
        end
    end
end
```

Program 1.4. *A class for the integration of functions defined by subprograms*

In this program, `limits` (or `xlim`) is a data structure with the properties `dim, lower, upper`, which define dimension, lower and upper bounds for the integral, respectively.

EXAMPLE 1.9.– Let us consider the integration of:

$$f(x) = x_1 e^{-x_2} \quad on \ \Omega = \{x : 0 \le x_1 \le 1 \,, 0 \le x_2 \le x_1\}.$$

In this case,

```
xlim.lower.x = 0;
xlim.lower.y = 0;
xlim.upper.x = 1;
xlim.upper.y = @(x) x;
xlim.dim = 2;
f = @(x) x(1)*exp(-x(2));
sp = subprogram_integration(f,xlim);
v = sp.result
```

Program 1.5. *An example of evaluation*

This program produces `v = 0.2358`. The exact value is $\frac{2}{e} - \frac{1}{2} \approx 0.2358$. ■

EXAMPLE 1.10.– Let us consider the evaluation of the volume of a sphere. The code:

```
xlim.lower.x = -1;
xlim.lower.y = @(x) -sqrt(1- x(1)^2);
xlim.lower.z = @(x) -sqrt(1- x(1)^2  -x(2)^2);
xlim.upper.x = 1;
xlim.upper.y = @(x) sqrt(1- x(1)^2);
xlim.upper.z = @(x,y) sqrt(1- x(1)^2  - x(2)^2);
xlim.dim = 3;
f = 1;
sp = subprogram_integration(f,xlim);
v = sp.result
```

Program 1.6. *Evaluation of a 3D integral*

produces the result `v = 4.1888`. The exact result is $\frac{4\pi}{3} \approx 4.1888$. But the code:

```
xlim.lower.x = -1;
xlim.lower.y = -1;
xlim.lower.z = -1;
xlim.upper.x = 1;
xlim.upper.y = 1;
xlim.upper.z = 1;
xlim.dim = 3;
f = @(x) 1*(sum(x.^2)<1);
sp = subprogram_integration(f,xlim);
v = sp.result
```

Program 1.7. *Evaluation of a 3d integral with a discontinuous function*

produces an unsuccessful run: the result is `v = NaN`. ■

The intrinsic functions for older versions of Matlab® are given in Table 1.2. `triplequad` adapts to the same use as `integral3` by considering a box containing the region of integration and extending the integrand by zero outside the region of integration.

Notice that both `dblquad` *and* `triplequad` *assume that* f *accepts as input a vector* `x` *and scalars* `y` *and* `z` *and returns a vector of results, so that the preceding functions spam2 and spam3 have to be modified in order to satisfy these requirements.*

Intrinsic function	evaluates
`quad(f,a,b) orquadgk(f,a,b)`	$\int_a^b f(x)dx$
`quad2d(f,ax,bx,ay,by)`	$\int_{ax}^{bx} dx \int_{ay(x)}^{by(x)} f(x,y)dy$
`dblquad(f,ax,bx,ay,by)`	$\int_{ax}^{bx} dx \int_{ay}^{by} f(x,y)dy$
`triplequad(f,ax,bx,ay,by,az,bz)`	$\int_{a_x}^{b_x} dx \int_{a_y}^{b_y} \int_{a_z}^{b_z} f(x,y,z)dxdydz$

Table 1.2. *Intrinsic function for older versions of Matlab®0*

1.6. Matlab® classes for partitions including the evaluation of the integrals

Partitions may be assimilated to finite element meshes. For instance, in Matlab®, we may define classes having convenient properties and methods. For the situation where $\Omega = (a,b) \subset \mathbb{R}$, an example of such a class is given in program 1.8. Notice that the class includes a method for integration (it assumes finite values for all the elements of F).

```
classdef interval_partition
    properties
        x
    end
    properties(Constant)
        dim = 1
    end
    properties(Dependent)
        min
        max
        number_points
        number_intervals
        measure
    end
    methods
        function obj = interval_partition(points)
            obj.x = points;
        end
      function v =
integration(obj,method,F,np,metod,interpmetod)
            data.points = obj;
            data.values = F;
            switch method
                case 'riemann'
                    v = riemann.trpzd(data);
                case 'mean'
                    v = riemann.mean_value(data);
                case 'lebesgue'
                    m1 = min(F);
                    m2 = max(F);
                    pas = (m2 - m1)/np;
```

```
                    y = m1:pas:m2;
             v = lebesgue.integrate(y,data,metod,
interpmetod);
            end
        end
        function obj = set.x(obj,coordinates)
            xx = sort(coordinates);
            obj.x = xx;
        end
        function v = get.min(obj)
            v = min(obj.x);
        end
        function v = get.max(obj)
            v = max(obj.x);
        end
        function v = get.number_points(obj)
            v = length(obj.x);
        end
        function v = get.number_intervals(obj)
            v = length(obj.x)-1;
        end
        function v = get.measure(obj)
            v = max(obj.x) - min(obj.x);
        end
    end
end
```

Program 1.8. *Definition of a class for partition of intervals*

For instance, the code:

```
pas = 0.05;
x = 0:pas:1;
p = interval_partition(x);
```

creates the structure p of type interval_partition. Then,

```
f = @(x) exp(x);
F = spam.mapspam('vector',f,x);
```

```
v1 = p.integration('riemann',F)
v2 = p.integration('lebesgue',F,20,'inferior','pchip')
v3 = p.integration('mean',F)
```

evaluates the integral $I = \int_0^1 e^x dx = e - 1 \approx 1.7183$ using the trapezoidal rule or Lebesgue's method with 20 intervals on the vertical axis. For the points given, the result is `v1 = 1.7186, v2 = 1.7187, v3=1.7253`. It may be improved by considering a partition containing a larger number of points. For instance, if `pas = 0.01;` the result is `v1 = v2 = 1.7183, v3 = 1.7197.`

For $\Omega = (a_1, b_1) \times (a_2, b_2)$, we may define a class as follows:

```
classdef rectangle_partition
    properties
        x
        y
    end
    properties(Constant)
        dim = 2
    end
    properties(Dependent)
        number_points
        measure
    end
    methods(Access = private)
        function v = quadinds(obj,i,j)
            v = [
                i j
                i+1 j
                i+1 j+1
                i j+1
                ];
        end
        function [ind1, ind2] = tri24(obj,i,j)
            ind1 = [ i j; i+1 j; i j+1 ];
            ind2 = [ i+1 j; i+1 j+1; i j+1 ];
        end
```

```
function [ind1, ind2] = tri13(obj,i,j)
    ind1 = [ i j; i+1 j+1; i j+1 ];
    ind2 = [ i j; i+1 j; i+1 j+1 ];
end
function v = sumind(obj,F,ind)
    v = 0;
    for i = 1:size(ind,1)
        aux = F(ind(i,1), ind(i,2));
        v = v + aux;
    end;
end
function v = st24(obj,F)
    xx =  obj.x;
    yy = obj.y;
    s = 0;
    for i = 1: length(xx)-1
        hx = xx(i+1) - xx(i);
        for j = 1: length(yy)-1
            hy = yy(j+1) - yy(j);
            [ind1, ind2] = obj.tri24(i,j);
            surft = hx*hy/2;
            aux1 = obj.sumind(F,ind1);
            aux2 = obj.sumind(F,ind2);
            s = s + (aux1 + aux2)*surft;
        end
    end;
    v = s/3;
end
function v = st13(obj,F)
    xx =  obj.x;
    yy = obj.y;
    s = 0;
    for i = 1: length(xx)-1
        hx = xx(i+1) - xx(i);
        for j = 1: length(yy)-1
            hy = yy(j+1) - yy(j);
            surft = hx*hy/2;
            [ind1, ind2] = obj.tri13(i,j);
            aux1 = obj.sumind(F,ind1);
            aux2 = obj.sumind(F,ind2);
            s = s + (aux1 + aux2)*surft;
```

```
                end
            end;
            v = s/3;
        end
        function v = sq4(obj,F)
            xx = obj.x;
            yy = obj.y;
            s = 0;
            for i = 1: length(xx)-1
                hx = xx(i+1) - xx(i);
                for j = 1: length(yy)-1
                    hy = yy(j+1) - yy(j);
                    surfq = hx*hy;
                    ind = obj.quadinds(i,j);
                    saux = obj.sumind(F,ind);
                    aux = saux*surfq;
                    s = s + aux;
                end
            end;
            v = s/4;
        end
    end
    methods
        function obj = rectangle_partition(x1,x2)
            obj.x = x1;
            obj.y = x2;
        end
      function v =
integration(obj,method,F,np,metod,interpmetod)
            data.points = obj;
            data.values = F;
            switch method
                case 'trapezoid'
                    v = riemann.trpzd(data);
                case 'mean'
                    v = riemann.mean_value(data);
                case 'tri24'
                    v = obj.st24(F);
                case 'tri13'
                    v = obj.st13(F);
                case 'quad'
```

```
                    v = obj.sq4(F);
                case 'lebesgue'
                    m1 = min(min(F));
                    m2 = max(max(F));
                    pas =  (m2 - m1)/np;
                    yy = m1:pas:m2;
              v =
lebesgue.integrate(yy,data,metod,interpmetod);
                otherwise
                    v = [];
                    disp('invalid method');
            end;
        end
        function v = get.number_points(obj)
            v.x =length(obj.x);
            v.y =length(obj.y);
        end
        function v = get.measure(obj)
            dx = max(obj.x) - min(obj.x);
            dy = max(obj.y) - min(obj.y);
            v = dx*dy;
        end
    end
end
```

Program 1.9. *A class for partitions of rectangles*

For instance, the code:

```
pas = 0.05;
x = 0:pas:1;
y = 0:pas:2;
pp = rectangular_partition(x,y);
```

creates the structure pp of type rectangular_partition and the code:

```
f = @(x) exp(x(1) - x(2));
F = spam.mapspam('vector',f,pp);
v1 = pp.integration('trapezoid',F);
v2 = pp.integration('tri13',F);
```

```
v3 = pp.integration('tri24',F);
v4 = pp.integration('quad',F);
v5= pp.integration('lebesgue',F,40,'inferior','pchip');
v6= pp.integration('lebesgue',F,40,'superior','pchip');
v7 = pp.integration('lebesgue',F,40,'mean','pchip');
```

produces `v1 = v4 = 1.4864, v2 = 1.4860, v3 = 1.4867, v5 = v6 = v7 = 1.4861`. The exact value is $(e-1)(1-e^{-2}) \approx 1.4857$. When using `pas = 0.01`, the result is `v1 = v2 = v3 = v4 = 1.4858, v5=v6=v7=1.4856`.

EXAMPLE 1.11.– Let us consider the integration of $f(x) = x_1 e^{-x_2}$ on $\Omega = (0,1) \times (0,2)$. The exact value of the integral is $\frac{1}{2}(e^2 - 1)$. Using `pas = 0.05`, the results are `v1=v2=v4=v5=3.1946, v2=0.8597, v6=3.1942`. For `pas=0.01, v1=v2=v4=v5=3.1946, v3=3.1945, v6=3.1942, v7=3.1944`. ■

EXAMPLE 1.12.– Let us consider the integration of

$$f(x) = \begin{cases} x_1 e^{-x_2}, if\ x_2 \leq x_1 \\ 0, otherwise\ . \end{cases}$$

on $\boldsymbol{\Omega} = (\mathbf{0}, \mathbf{1}) \times (\mathbf{0}, \mathbf{1})$. This situation corresponds to the integration of

$$f(x) = x_1 e^{-x_2} \quad on\ \Omega = \{x : 0 \leq x_1 \leq 1\,, 0 \leq x_2 \leq x_1\}.$$

and the exact value of the integral is $\frac{2}{e} - \frac{1}{2} \approx 0.2358$. Using `pas = 0.01` and `np=100` intervals on the vertical axis, the results are `v1=v2=v4=0.2344, v3=0.2345, v5 = 0.2363, v6 = 0.2344, v7 = 0.2354`. Due to discontinuity, finer partitions are requested for a good precision. ■

In 3D situations, $\Omega = (a_1, b_1) \times (a_2, b_2) \times (a_3, b_3)$, we may define a class containing methods for integration by trapezoidal rule or integration using a tetrahedral or hexahedral mesh as follows:

```
classdef cobble_partition
    properties
        x
        y
        z
    end
    properties(Constant)
        dim = 3
    end
    properties(Dependent)
        number_points
        measure
    end
    %
    methods(Static, Access = Private)
        function v =  hexainds(i,j,k)
            v = [
                i j k
                i+1 j k
                i+1 j+1 k
                i j+1 k
                i j k+1
                i+1 j k+1
                i+1 j+1 k+1
                i j+1 k+1
                ];
        end
        function v = tetra5()
            v = [
                1 2 4 5
                2 3 4 7
                2 7 5 6
                4 5 7 8
                2 4 5 7
                ];
        end
        function v = tetra6()
            v = [
                3 8 4 2
                1 8 2 4
                7 2 8 3
```

```
                    1 2 8 5
                    2 5 6 8
                    2 8 6 7
                    ];
        end
        function v = sumind(obj,F,ind)
            v = 0;
            for i = 1:size(ind,1)
                aux = F(ind(i,1), ind(i,2), ind(i,3));
                v = v + aux;
            end;
        end
        function v = vol_tetra(obj,x,y,z,ind)
            xx = x(ind(:,1));
            yy = y(ind(:,2));
            zz = z(ind(:,3));
            v = abs(det([xx' yy' zz' ones(size
(xx'))]))/6;
        end
        function v = sumtetra(obj,x,y,z,F,
ind,tetras,volh)
            v = 0;
            sv = 0;
            for iaux = 1: size(tetras,1)
                iii = ind(tetras(iaux,:),:);
                saux = obj.sumind(F,iii);
                volt = obj.vol_tetra(x,y,z,iii);
                sv = sv + volt;
                v = v + saux*volt;
            end
            v = v*volh/sv;
        end
        function v = st5(obj,F)
            x = obj.x;
            y = obj.y;
            z = obj.z;
            tetras = obj.tetra5();
            s = 0;
            for i = 1: length(x)-1
                for j = 1: length(y)-1
```

```
                    for k = 1: length(z)-1
                        ind = obj.hexainds(i,j,k);
                        hx = x(i+1) - x(i);
                        hy = y(j+1) - y(j);
                        hz = z(k+1) - z(k);
                        volh = hx*hy*hz;
                      saux =
obj.sumtetra(x,y,z,F,ind,tetras,volh);
                        s = s + saux;
                    end;
                end
            end;
            v = 0.25*s;
        end
        function v = st6(obj,F)
            x = obj.x;
            y = obj.y;
            z = obj.z;
            tetras = obj.tetra6();
            s = 0;
            for i = 1: length(x)-1
                for j = 1: length(y)-1
                    for k = 1: length(z) - 1
                        ind = obj.hexainds(i,j,k);
                        hx = x(i+1) - x(i);
                        hy = y(j+1) - y(j);
                        hz = z(k+1) - z(k);
                        volh = hx*hy*hz;
                      saux =
obj.sumtetra(x,y,z,F,ind,tetras,volh);
                        s = s + saux;
                    end;
                end
            end;
            v = 0.25*s;
        end
        function v = sh8(obj,F)
            x = obj.x;
            y = obj.y;
            z = obj.z;
            s = 0;
```

```
                for i = 1: length(x)-1
                    hx = x(i+1) - x(i);
                    for j = 1: length(y)-1
                        hy = y(j+1) - y(j);
                        for k = 1: length(z) - 1
                            hz = z(k+1) - z(k);
                            volh = hx*hy*hz;
                            ind = obj.hexainds(i,j,k);
                            saux = obj.sumind(F,ind);
                            s = s + saux*volh;
                        end
                    end;
                end;
                v = s/8;
            end
        end
        methods
            function obj = cobble_partition(x1,x2,x3)
                obj.x = x1;
                obj.y = x2;
                obj.z = x3;
            end
            function v =
integration(obj,method,F,np,metod,interpmetod)
                data.points = obj;
                data.values = F;
                switch method
                    case 'trapezoid'
                        v = riemann.trpzd(data);
                    case 'mean'
                        v = riemann.mean_value(data);
                    case 'tetra5'
                        v = obj.st5(F);
                    case 'tetra6'
                        v = obj.st6(F);
                    case 'hexa'
                        v = obj.sh8(F);
                    case 'lebesgue'
                        m1 = min(min(min(F)));
                        m2 = max(max(max(F)));
                        pas =  (m2 - m1)/np;
```

```
                    yy = m1:pas:m2;
                v =
lebesgue.integrate(yy,data,metod,interpmetod);
                otherwise
                    v = [];
                    disp('invalid method');
            end;
        end
        function v = get.number_points(obj)
            v.x =length(obj.x);
            v.y =length(obj.y);
            v.z =length(obj.z);
        end
        function v = get.measure(obj)
            dx = max(obj.x) - min(obj.x);
            dy = max(obj.y) - min(obj.y);
            dz = max(obj.z) - min(obj.z);
            v = dx*dy*dz;
        end
    end
end
```

Program 1.10. *A class for three-dimensional partitions*

For instance, the code:

```
pas = 0.05;
x = 0:pas:1;
y = 0:pas:2;
z = 0:pas:3;
ppp = cobble_partition(x,y,z);
```

creates the structure pp of type `cobble partition`. The code:

```
v1 = ppp.integration('trapezoid',F);
v2 = ppp.integration('tetra5',F);
v3 = ppp.integration('tetra6',F);
v4 = ppp.integration('hexa',F);
v5=ppp.integration('lebesgue',F,40,'inferior','pchip');
v6=ppp.integration('lebesgue',F,40,'superior','pchip');
```

```
v7=ppp.integration('lebesgue',F,40,'mean','pchip');
```

produces `v1 = v4 = 13.3791`, `v2 = 13.3789`, `v3 = 13.3763`, `v4 = 13.3791`, `v5 = 13.4117`, `v6 =13.4086`, `v7 = 13.4102`. The exact value is $9(e-1)(1-e^{-2}) \approx 13.3716$.

EXAMPLE 1.13.– Let us consider the evaluation of the 3D integral:

$$I = \iiint_{\|x\|<1} 1 dx$$

We may consider $\Omega = (-1,1) \times (-1,1) \times (-1,1)$ and

$$f(x) = \begin{cases} 1, if \ \|x\| < 1, \\ 0, \ otherwise. \end{cases}$$

Then, we evaluate:

$$I = \iiint_{\Omega} f(x) dx \, .$$

Using `pas = 0.05`, the result obtained is `v1 = v2 = v3 = v4 = 4.1714`, `v5 = v6 = v7 = 4.1724`. The exact result is $\frac{4\pi}{3} \approx 4.1888$. ■

REMARK.– In some situations, the available data is not on a grid, but just on sparse points. This situation is considered in Chapter 3. ■

2

Variational Methods for Algebraic Equations

Let us begin our exploration of variational methods by a situation which may appear as simple, but contains the fundamental elements necessary to the implementation of variational methods in complex situations: the variational formulation of algebraic equations and its consequences.

Indeed, the fundamental variational standpoint consists of no more saying that $x = 0$, but that $x \in \mathbb{R}$ and $xy = 0, \forall y \in \mathbb{R}$.

The apparent simplicity of this fundamental change of viewpoint masks a radical transformation in the way of thinking, having profound consequences, which can be measured by the progress performed in engineering with the help of variational tools such as analytical mechanics, hamiltonian mechanics, control theory and modern numerical methods such as finite elements, finite volumes, spectral methods, particle methods and others.

Let us illustrate these consequences by adopting the variational point of view when studying a system of algebraic equations: for instance, let us consider;

$F: \mathbb{R}^n \longrightarrow \mathbb{R}^m,\ F = (F_1, \ \ldots, \ F_m)^t,\ F_i: \mathbb{R}^n \longrightarrow \mathbb{R}\ (i = 1, \ldots, m)$

and the algebraic equations;

$$x \in \mathbb{R}^n \ and \ F(x) = 0. \qquad [2.1]$$

The variational formulation of this system of algebraic equations reads as:

$$x \in \mathbb{R}^n \ and \ y^t F(x) = 0, \forall \, y \in \mathbb{R}^m. \qquad [2.2]$$

An alternative variational formulation is given by:

$$x \in \mathbb{R}^n \ and \ y_i F_i(x) = 0 \ for \ i = 1, \dots, m, \ \forall \, y \in \mathbb{R}^m. \qquad [2.3]$$

All these formulations are equivalent:

$$[2.1] \Leftrightarrow [2.2] \Leftrightarrow [2.3] .$$

For $m < n$, the equations are under-determined (less equations than unknowns); for $m > n$, the equations are over-determined (more equations than unknowns). This classification does not imply the existence or nonexistence of solutions.

2.1. Linear systems

Let us consider a full rank matrix A having dimensions $n \times n$ and a column matrix B having dimensions $n \times 1$. The linear system $AX = B$ corresponds to the particular situation where,

$$F(X) = AX - B ,$$

and we have $m = n$. The variational formulation is:

$$x \in \mathbb{R}^n \ and \ y^t A x = y^t B, \forall \, y \in \mathbb{R}^n .$$

Assume that we are interested in the single value of x_1: let us denote by $A(:,j)$ the j-th column of A and by $A(:,2:n)$ the matrix formed by the columns 2 to n of A:

$$A(:,j) = \begin{pmatrix} A_{1j} \\ \dots \\ A_{nj} \end{pmatrix} ;\ A(:,2:n) = (A(:,2) \quad \dots \quad A(:,n)) = \begin{pmatrix} A_{12} & \dots & A_{1n} \\ \dots & \dots & \dots \\ A_{n2} & \dots & A_{nn} \end{pmatrix}$$

When

$$y^t A(:,2:n) = 0 ,$$

we have:

$$y^t Ax = x_1 y^t A(:,1)$$

and

$$x_1 y^t A(:,1) = y^t B ,$$

which determines x_1. This idea can be directly generalized to the situation where the variables are separated in two disjoined lists $J = \{j_1, \dots, j_{n_j}\}$ and $K = \{1, \dots, n\} - \{j_1, \dots, j_{n_j}\} = \{k_1, \dots, k_{n-n_j}\}$. If we are interested in the single values of the n_j variables corresponding to the indexes J, we may consider:

$$A_J = \left(A(:,j_1) \quad \dots \quad A\left(:,j_{n_j}\right)\right) ;\ A_K = \left(A(:,k_1) \quad \dots \quad A\left(:,k_{n-n_j}\right)\right)$$

and

$$x_J = \left(x_{j_1} \quad \dots \quad x_{j_{n_j}}\right) ;\ x_K = \left(x_{k_1} \quad \dots \quad x_{k_{n-n_j}}\right).$$

For

$$y^t A_K = 0 ,$$

we have:

$$y^t Ax = y^t A_J x_J$$

and

$$y^t A_J x_J = y^t B .$$

This equality involves only the variables $\boldsymbol{x}_J$: for convenient choices of $\boldsymbol{y}$, it furnishes an $n_j \times n_j$ linear system for the n_j unknowns $\boldsymbol{x}_J$.

EXAMPLE 2.1.– Let us consider the linear system

$$\begin{pmatrix} 1 & 2 \\ 2 & 1 \end{pmatrix} \begin{pmatrix} x_1 \\ x_2 \end{pmatrix} = \begin{pmatrix} 1 \\ -1 \end{pmatrix} .$$

Taking $y = (2 \quad -1)^t$, we have:

$$y^t \begin{pmatrix} 1 & 2 \\ 2 & 1 \end{pmatrix} \begin{pmatrix} x_1 \\ x_2 \end{pmatrix} = y^t \begin{pmatrix} 1 \\ -1 \end{pmatrix} \Rightarrow (0 \quad 3) \begin{pmatrix} x_1 \\ x_2 \end{pmatrix} = 3 \Rightarrow x_2 = 1 .$$

Using $y = (-1 \quad 2)^t$, we have;

$$y^t \begin{pmatrix} 1 & 2 \\ 2 & 1 \end{pmatrix} \begin{pmatrix} x_1 \\ x_2 \end{pmatrix} = y^t \begin{pmatrix} 1 \\ -1 \end{pmatrix} \Rightarrow (3 \quad 0) \begin{pmatrix} x_1 \\ x_2 \end{pmatrix} = -3 \Rightarrow x_1 = -1 . \quad \blacksquare$$

EXAMPLE 2.2.– Let us consider the system shown in Figure 2.1.

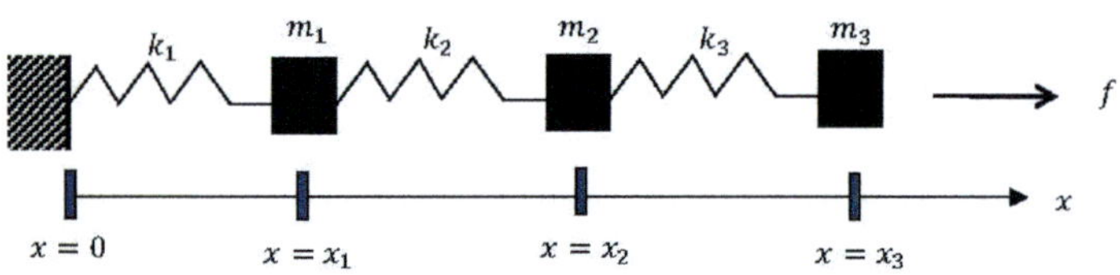

Figure 2.1. *A physical interpretation of the method*

The forces on the springs are, respectively, $k_1 u_1$, $k_2(u_2 - u_1)$, $k_3(u_3 - u_2)$, so that the equilibrium of the system reads as:

$$k_1 u_1 = k_2(u_2 - u_1)$$

$$k_2(u_2 - u_1) = k_3(u_3 - u_2)$$

$$k_3(u_3 - u_2) = f$$

what corresponds to the linear system $AU = B$, with

$$A = \begin{pmatrix} k_1 + k_2 & -k_2 & 0 \\ -k_2 & k_2 + k_3 & -k_3 \\ 0 & -k_3 & k_3 \end{pmatrix}, \ U = \begin{pmatrix} u_1 \\ u_2 \\ u_3 \end{pmatrix}, \ B = \begin{pmatrix} 0 \\ 0 \\ f \end{pmatrix}.$$

Assume that the force f is known and the displacements at the equilibrium u_i of each mass m_i have been measured. We are interested in determining the values of k_i. Taking $y = (1 \quad 1 \quad 1)^t$, we have:

$$y^{\mathrm{t}}\mathrm{AU} = y^{\mathrm{t}}\mathrm{B} \Longrightarrow k_1 u_1 = f.$$

Using $y = (0 \quad 1 \quad 1)^t$, we have:

$$y^{\mathrm{t}}\mathrm{AU} = y^{\mathrm{t}}\mathrm{B} \Longrightarrow k_2(u_2 - u_1) = f.$$

Using $y = (0 \quad 0 \quad 1)^t$, we have:

$$y^{\mathrm{t}}\mathrm{AU} = y^{\mathrm{t}}\mathrm{B} \Longrightarrow k_3(u_3 - u_2) = f.$$

Each one of the preceding choices leads to the determination of one of the values of k_i. These choices may be interpreted in terms of a physical quantity: the work of the springs' forces for a given displacement: for each displacement y, the work of two of the springs' forces is equal to zero, while the remaining one is different from zero. In terms of virtual works, each displacement makes work a single spring, what lead to the determination of its stiffness. For instance, the displacements $y = (1 \quad 1 \quad 1)^t$ correspond to an equal displacement of all the masses. In this case, only the first spring is under tension. Analogously, $y = (0 \quad 1 \quad 1)^t$ keeps the first and the third springs without tension. Finally, $y = (0 \quad 0 \quad 1)^t$ makes that the single third spring is under tension. This approach extends to more complex linear systems (see, for instance, [GRE 89]). ■

In practice, the vectors y have to be chosen in order to verify $y^t A_K = 0$. They may be determined by Gauss pivoting and operations on lines. For instance, we may use the program

```
classdef linear_system
    properties
        A
        B
    end
    methods(Static, Access = private)
        function [i_pivot, val_pivot] = lpivot(a,j_col)
            [val_pivot,ind] =
max(abs(a(j_col:size(a,1),j_col)));
            i_pivot = ind + j_col - 1;
        end
        function a_pivot = pmc(a,j_col,i_pivot)
            a_pivot = a;
            val_pivot = a_pivot(i_pivot, j_col);
            aux = a_pivot(j_col,:);
            a_pivot(j_col,:) = a_pivot(i_pivot, :);
            a_pivot(i_pivot,:) = aux;
            aux = a_pivot(j_col,:)/val_pivot;
            a_pivot(j_col, :) = aux;
            for i = j_col+1: size(a_pivot,1)
                a_pivot(i,:) = a_pivot(i,:)-
a_pivot(i,j_col)*aux;
            end;
        end
        function [error_flag,y, control] = nsgp(a,
tol_zero)
            ag = a;
            m = size(a,1);
            n = size(a,2);
            error_flag = 0;
            if n < m
       disp('error: number of equations >= number of
unknowns');
                error_flag = 1;
                return;
            end;
            cols_zero = [];
            for j_col = 1: m
            [i_pivot, val_pivot] =
linear_system.lpivot(ag,j_col);
                if val_pivot < tol_zero
```

```
                    cols_zero = [cols_zero j_col];
                else
                    ag = linear_system.pmc(ag,j_
col,i_pivot);
                end;
            end;
            if isempty(cols_zero)
                ns = n - m;
                y = zeros(n,ns);
                for j = 1: ns
                    y(m + j,j) = 1;
                    y(m,j) = -ag(m,m+j);
                    for i = 1: m-1
                        jj = m - i;
                        y(jj,j) = -
sum(ag(jj,jj+1:n)*y(jj+1:n,j));
                    end;
                end;
            elseif cols_zero(end) < m
                ns = n - m + length(cols_zero);
                y = zeros(n,ns);
                for i = 1: length(cols_zero)
                    y(cols_zero(i),i) = 1;
                end;
                for j = length(cols_zero)+1: ns
                    y(m + j-length(cols_zero),j) = 1;
                    y(m,j) = -ag(m,m+j-
length(cols_zero));
                    for i = 1: m-1
                        jj = m - i;
                        y(jj,j) = -
sum(ag(jj,jj+1:n)*y(jj+1:n,j));
                    end;
                end;
            else
                ns = n - m + length(cols_zero) - 1;
                y = zeros(n,ns);
                for i = 1: length(cols_zero)
                    y(cols_zero(i),i) = 1;
                end;
```

```
                ind = find(abs(ag(m,m+1:n)) >
tol_zero);
                for j = length(cols_zero)+1: ns
                    if abs(ag(m,m+j-length(cols_zero)))
< tol_zero
                        y(m + j-length(cols_zero),j) =
1;
                    else
                        inda = find(ind ~= m+j-
length(cols_zero));
                        if isempty(inda)
                            y(m + j-
length(cols_zero),j) = 0;
                        else
                            y(inda(1),j) = 1;
y(m+j-length(cols_zero),j)=-ag(m,inda(1))/ag(m,m+j-
length(cols_zero));
                        end;
                    end;
                    for i = 1: m-1
                        jj = m - i;
                        y(jj,j) = -
sum(ag(jj,jj+1:n)*y(jj+1:n,j));
                    end;
                end;
            end;
            control = norm(a*y)/sqrt(ns);
            return;
        end
        function [error_flag,y, control] = gpivot(a,
tol_zero)
            ag = a;
            m = size(a,1);
            n = size(a,2);
            error_flag = 0;
            if n < m
         disp('error: number of equations >= number of
unknowns');
                error_flag = 1;
                return;
            end;
```

```
            cols_zero = [];
            for j_col = 1: m
                [i_pivot, val_pivot] =
lpivot(ag,j_col);
                if val_pivot < tol_zero
                    cols_zero = [cols_zero j_col];
                else
                    ag = linear_system.pmc(ag,j_
col,i_pivot);
                end;
            end;
            if isempty(cols_zero)
                ns = n - m;
                y = zeros(n,ns);
                for j = 1: ns
                    y(m + j,j) = 1;
                    y(m,j) = -ag(m,m+j);
                    for i = 1: m-1
                        jj = m - i;
                        y(jj,j) = -
sum(ag(jj,jj+1:n)*y(jj+1:n,j));
                    end;
                end;
            elseif cols_zero(end) < m
                ns = n - m + length(cols_zero);
                y = zeros(n,ns);
                for i = 1: length(cols_zero)
                    y(cols_zero(i),i) = 1;
                end;
                for j = length(cols_zero)+1: ns
                    y(m + j-length(cols_zero),j) = 1;
                    y(m,j) = -ag(m,m+j-
length(cols_zero));
                    for i = 1: m-1
                        jj = m - i;
                        y(jj,j) = -
sum(ag(jj,jj+1:n)*y(jj+1:n,j));
                    end;
                end;
            else
                ns = n - m + length(cols_zero) - 1;
```

```
                y = zeros(n,ns);
                for i = 1: length(cols_zero)
                    y(cols_zero(i),i) = 1;
                end;
                ind = find(abs(ag(m,m+1:n)) >
tol_zero);
                for j = length(cols_zero)+1: ns
                    if abs(ag(m,m+j-length(cols_zero)))
< tol_zero
                        y(m + j-length(cols_zero),j) =
1;
                    else
                        inda = find(ind ~= m+j-
length(cols_zero));
                        if isempty(inda)
                            y(m + j-
length(cols_zero),j) = 0;
                        else
                            y(inda(1),j) = 1;
y(m+j-length(cols_zero),j)=-ag(m,inda(1))/ag(m,m+j-
length(cols_zero));
                        end;
                    end;
                    for i = 1: m-1
                        jj = m - i;
                        y(jj,j) = -
sum(ag(jj,jj+1:n)*y(jj+1:n,j));
                    end;
                end;
            end;
            control = norm(a*y)/sqrt(ns);
            return;
        end
    end
    methods
        function obj = linear_system(A,B)
            obj.A = A;
            obj.B = B;
        end
        function x = complete_solution(obj)
            x = obj.A \ obj.B;
```

```
        end
        function x = partial_solution(obj,
list_unknowns, tol_zero)
            x = zeros(size(obj.B));
            if isempty(list_unknowns)
                disp('empty list of unknowns');
                return;
            else
                list_u = sort(list_unknowns);
                list_t = (1:length(obj.B));
                list_t(list_u) = 0;
                ind = find(list_t > 0);
                list_e = list_t(ind);
            end;
            a_e = obj.A(:,list_e)';
      [error_flag,y, control] = linear_system.nsgp(a_e,
tol_zero);
            disp(['control = ', num2str(control)]);
            if error_flag > 0
                return;
            else
                aa = y'*obj.A;
                bb = y'*obj.B;
                aaa = aa(:,list_u);
                xxx = aaa \ bb;
                x(list_u) = xxx;
            end;
            x(list_e) = NaN;
            return;
        end
    end
end
```

Program 2.1. *A class for the determination of partial solutions of linear systems*

This class contains a method `partial_solution` which determines the values of the unknowns given in `list_unknowns` without determining the values of the other unknowns. For instance, the code:

```
n = 100;
% size of the system
a = randn(n,n);
% generates a gaussian distributed random matrix
b = rand(n,1);
% generates a uniformly distributed random second
member
ls = linear_system(a,b);
xsol = ls.complete_solution();
% find the exact solution
list_unknowns = [1 fix(n/10) fix(n/2)];
% unknowns to be determined
tol_zero = 1e-10;
% numerical value of zero
x = ls.partial_solution(list_unknowns, tol_zero);
% solution
err_abs = norm(x(list_unknowns)-
xsol(list_unknowns));
err_rel = err/norm(xsol(list_unknowns));
% comparison complete/partial
```

produces `err_abs =3.2601e-15, err_rel =4.8700e-15,` which corresponds to a relative error of 1e-11 %. Since the matrices are random, each run produces a different result. For instance, a second run produces `err_abs =5.6125e-15, err_rel =5.0186e-15,` which corresponds to a relative error of 1e-12 %. The user will have different results, due to randomness in the definition of the matrices.

2.2. Algebraic equations depending upon a parameter

Let us consider the system of algebraic equations:

$$t \in (a,b), x \in \mathbb{R}^n \; and \; F(x,t) = 0.$$

where $F: \mathbb{R}^n \times (a,b) \longrightarrow \mathbb{R}^m$, $F = (F_1, \; \dots, \; F_m)^t$, $F_i: \mathbb{R}^n \times (a,b) \longrightarrow \mathbb{R}$, for $1 \leq i \leq m$ and t is a real parameter. Since the equations depend upon t, the solution x depends on t: $x = x(t)$ and it is more appropriate to write

$$t \in (a,b), x(t) \in \mathbb{R}^n \; and \; F(x(t),t) = 0 \quad [2.4]$$

The variational formulation of this system of algebraic equations may be formulated as:

$$t \in (a,b), x(t) \in \mathbb{R}^n \ and \ y^t F(x(t),t) = 0, \forall\, y \in \mathbb{R}^m.$$

A practical way in order to eliminate the parameter t is to consider $= y(t)$:

$$t \in (a,b), x(t) \in \mathbb{R}^n \ and \ y(t)^t F(x(t),t) = 0, \forall\, y(t) \in \mathbb{R}^m$$

and integrate this last equation on (a,b):

$$x: (a,b) \longrightarrow \mathbb{R}^n \ and \ \int_a^b y(t)^t F(x(t),t)dt = 0, \forall\, y: (a,b) \longrightarrow \mathbb{R}^m \quad [2.5]$$

As in the standard situation, an alternative variational formulation is given by:

$$t \in (a,b), x(t) \in \mathbb{R}^n \ and \ y_i F_i(x(t),t) = 0 \ for\ i = 1, \dots, m, \forall\, y \in \mathbb{R}^m.$$

Here yet, a practical way in order to eliminate the parameter t is to consider $y_i = y_i(t)$:

$$t \in (a,b), x(t) \in \mathbb{R}^n \ and \ y_i(t) F_i(x(t),t) = 0 \ for\ i = 1, \dots, m, \forall\, y \in \mathbb{R}^m$$

and integrate this equation on (a,b):

$$\begin{gathered} x: (a,b) \longrightarrow \mathbb{R}^n, \\ \int_a^b y_i(t) F_i(x(t),t)dt = 0 \ \ for\ i = 1, \dots, m, \forall\, y: (a,b) \longrightarrow \mathbb{R}^m. \end{gathered} \quad [2.6]$$

As in the standard situation, all these formulations are equivalent:

$$[2.4] \Leftrightarrow [2.5] \Leftrightarrow [2.6].$$

2.2.1. *Approximation of the solution by collocation*

For engineers, a practical question concerns the determination of approximate solutions. For instance, the solution may be determined at some particular values of t – for instance, $t_1, \dots, t_{ns}$ – leading to particular values of y– for instance, $x_1, \dots, x_{ns}$ – and an interpolation may be performed in order to generate values for arbitrary t. This approach is usually referred to as *collocation*. In practice, it involves a family of interpolation functions

$$\mathcal{B} = \{\varphi_i\}_{1 \le i \le k}, \varphi_i: (a, b) \longrightarrow \mathbb{R},$$

such as, for instance, $\varphi_i(t) = t^{i-1}$ (polynomial interpolation). Then, an approximation:

$$x(t) \approx \sum_{p=1}^{k} u_p \varphi_p(t), u_p \in \mathbb{R}^n \; for \; p = 1, \dots, k.$$

is determined by solving the linear system of algebraic equations:

$$x(t_r) \approx \sum_{p=1}^{k} u_p \varphi_p(t_r) = x_r, 1 \le r \le ns,$$

The unknowns to be determined are the coefficients $u_p \in \mathbb{R}^n$, for $p = 1, \dots, k$. They can be arranged in a matrix $U = (u_{pj})_{\substack{1 \le p \le k \\ 1 \le j \le n}}$, such that $u_{pj} = (u_p)_j$ However, the data $x_r \in \mathbb{R}^n$ may be arranged in a matrix $X = (x_{ri})_{\substack{1 \le r \le ns \\ 1 \le i \le n}}$ such that $x_{ri} = (x_r)_i$, so that these equations read as:

$$AU = X,$$

where, for $1 \le i, j \le n$, $1 \le p \le k$, $1 \le r \le ns$,

$$A_{ripj} = \varphi_p(t_r).$$

These equations may be transformed in a standard linear system involving two indexes by using the transformation.

$$index(k_1, k_2, n_1) = k_1 + (k_2 - 1)n_1 \;\; (1 \le k_1 \le n_1).$$

By setting $I = index(r, i, ns)$, $J = index(p, j, k)$, $\mathcal{A}_{IJ} = A_{ripj}$, $\mathcal{B}_I = X_{ri}$, $\mathcal{U}_J = U_{pj}$, we have $\mathcal{AU} = \mathcal{B}$. This linear system has dimension $n.ns \times k.n$. For $ns > k$, it is overdetermined and least squares solution may be used – under Matlab®, this is made automatically. Notice that it may also be solved component by component: let us introduce:

$$M_{pr} = \varphi_p(t_r),\ N_i = X(:,i)^t, U_i = U(:,i)^t$$

Then, for $1 \leq i \leq n$,

$$MU_i = N_i \,.$$

Thus, the solution may be sequentially determined by solving n linear systems of size $ns \times k$.

2.2.2. *Variational approximation of the solution*

Collocation may be sensitive to errors in the numerical determination of $x_1, \ldots, x_{ns}$. The variational formulations presented furnish a more robust method for the determination of the solution.

By denoting $\Phi = (\varphi_1, \ldots, \varphi_k)^t$, we have $x \approx U^t\Phi$. By taking $y_i(t) = \varphi_j(t)$ in equation [2.6], we obtain, for $1 \leq i \leq m; 1 \leq p \leq k$.

$$\mathcal{F}_{pi}(U) = \int_a^b \varphi_p(t) F_i(U^t\Phi, t)dt = 0 \,, \qquad [2.7]$$

so that;

$$\mathcal{F}(U) = 0;\ \mathcal{F}(U) = \left(\mathcal{F}_{pi}(U)\right)_{\substack{1 \leq p \leq k \\ 1 \leq i \leq m}} \,. \qquad [2.8]$$

This system contains $k \times m$ equations for $k \times n$ unknowns. When $m = n$, the numbers of equations and unknowns coincide For $m < n$, the equations are *under-determined* (less equations than unknowns); for $m > n$, the equations are *over-determined* (more equations than unknowns). As in the previous situation, this classification does not imply the existence or nonexistence of solutions. Equation [2.7] must be solved by an appropriated method, such as, for instance,

Newton–Raphson, quasi-newton or fixed-point iterations. As an alternative, we may look for:

$$U = (u_{pj})_{\substack{1\le p\le k \\ 1\le j\le n}} \text{ such that } J(U) = \textstyle\sum_{p,i}\left(\mathcal{F}_{pi}(U)\right)^2 \text{ is minimal}. \quad [2.9]$$

We observe that;

$$\frac{\partial}{\partial U_{qj}} J(U) = \textstyle\sum_{p,i}\left(\mathcal{F}_{pi}(U)\frac{\partial \mathcal{F}_{pi}}{\partial U_{qj}}(U)\right), \quad [2.10]$$

with

$$\frac{\partial \mathcal{F}_{pi}}{\partial U_{qj}}(U) = \int_a^b \varphi_p(t)\varphi_j(t)\frac{\partial F_i}{\partial x_q}(U^t\Phi, t)dt. \quad [2.11]$$

These equations may be invoked for the evaluation of the gradients when using optimization methods.

2.2.3. *Linear equations*

The particular situation where

$$F(x,t) = A(t)x - B(t)$$

leads to a linear system. In this case,

$$\mathcal{F}(U) = \int_a^b \Phi^{\mathrm{t}}(AU^t\Phi - B)dt = 0\ ,$$

so that;

$$\mathcal{F}_{pi}(U) = \textstyle\sum_{j,q} U_{qj}\int_a^b \varphi_p(t)A_{ij}(t)\varphi_q(t)dt - \int_a^b \varphi_p(t)B_i(t)dt = 0$$

and

$$\mathcal{A}U = \mathcal{B}\,,$$

where

$$\mathcal{A}_{piqj} = \int_a^b \varphi_p(t)A_{ij}(t)\varphi_q(t)dt \quad ; \quad \mathcal{B}_{pi} = \int_a^b \varphi_p(t)B_i(t)dt \,.$$

These equations may be transformed in a standard linear system involving two indexes by using the transformation previously introduced: $I = index(p,i,k)$, $J = index(q,j,k)$, $\mathbb{A}_{IJ} = \mathcal{A}_{piqj}$, $\mathbb{B}_I = \mathcal{B}_{pi}$, $\mathbb{U}_J = U_{qj}$. Then, $\mathbb{A}\mathbb{U} = \mathbb{B}$.

2.2.4. Connection to orthogonal projections

The variational approach may be interpreted in terms of *orthogonal projection*: let V be the set of all the functions defined on (a,b) and taking their values on $\mathbb{R}$.

$$V = \{v: (a,b) \longrightarrow \mathbb{R}\} \,.$$

V is a linear space (i.e. a vector space) and $\mathcal{B} = \{\varphi_i\}_{1 \le i \le k}$ generates a finite dimensional subspace $S \subset V$:

$$S = [\mathcal{B}] = \left\{ s^t\Phi = \sum_{p=1}^k s_p\varphi_p \; : \; s = (s_1, \dots, s_k)^t \in \mathbb{R}^k \right\}.$$

Since $\dim(S) = k$, S may be assimilated to $\mathbb{R}^k$: the element $s = s^t\Phi \in \mathrm{S}$ is entirely defined by the vector $s = (s_1, \dots, s_k)^t \in \mathbb{R}^k$. Let us consider a particular scalar product for vectors $u, v \in \mathbb{R}^k$:

$$(u, v) = u^t W v \;, \; W_{ij} = \int_a^b \varphi_i(t)\varphi_j(t)dt \,.$$

We have:

$$(u, v) = \int_a^b (u^t\Phi)(v^t\Phi)dt = \int_a^b u(t)v(t)dt$$

where $= u^t\Phi$, $v = v^t\Phi \in \mathrm{S} \subset \mathrm{V}$. Let us extend this definition as being a scalar product on the whole space V:

$$(u, v) = \int_a^b u(t)v(t)dt \;, for \; u, v \; \in V.$$

Recall that $u \perp v \Leftrightarrow (u, v) = 0$, i.e. u and v are orthogonal if and only if their scalar product is null. Let $P_S: V \rightarrow S$ denote the orthogonal projection onto S. For $u \in V$, $P_S(u)$ is characterized by (see section 3.7)

$$P_S(u) \in S \; and \; u - P_S(u) \perp S \,,$$

or, equivalently

$$P_S(u) \in S \; and \; u - P_S(u) \perp \varphi_j, for \; j = 1, \dots, k.$$

Equation [2.7] shows that, for $1 \leq i \leq m$,

$$F_i(U^t\Phi(t), t) \;\perp\; \varphi_j, for \; j = 1, \dots, k.$$

So, for $1 \leq i \leq m$,

$$F_i(U^t\Phi(t), t) \;\perp\; S.$$

Thus,

$$0 \in S \;\; and \;\; F_i(U^t\Phi(t), t) - 0 \;\perp\; S.$$

So, the orthogonal projection of $F_i(U^t\Phi(t), t)$ onto S is null for $1 \leq i \leq m$:

$$P_S\big(F_i(U^t\Phi(t), t)\big) = 0, \;\; for \; 1 \leq i \leq m \,.$$

Let us consider the Cartesian product $S^m = S \times \dots \times S \;\; (m \; times)$. We have also;

$$0 \in S^m \;\; and \;\; F(U^t\Phi(t), t) - 0 \;\perp\; S^m \;.$$

Thus, the orthogonal projection of $F(U^t\Phi(t), t)$ onto S^m is null:

$$P_{S^m}\big(F(U^t\Phi(t), t)\big) = 0 \,.$$

In the particular situation where

$$F(x, t) = x(t) - f(t),$$

we have, on the one hand, $m = n$ and, on the other hand,

$$U^t\Phi \in S^m and \;\; f - U^t\Phi \perp S^m .$$

Thus, $P_{S^m}(f) = U^t\Phi$ and, on the one hand,

$$\sum_{p=1}^{k} u_{pi}\varphi_p \;\; is\ the\ orthogonal\ projection\ of\ f_i\ onto\ S$$

while, on the other hand (recall that $m = n$),

$$\sum_{p=1}^{k} \boldsymbol{u_p}\varphi_p \;\; is\ the\ orthogonal\ projection\ of\ \boldsymbol{f}\ onto\ S^m .$$

2.2.5. *Numerical determination of the orthogonal projections*

Let us consider an element $u \in V^n$ and its orthogonal projection $P_{S^n}(u)$ onto S^n. We have:

$$P_{S^n}(u) \in S^n \;\; and\ u - P_{S^n}(u) \perp S^n.$$

Thus, $P_{S^n}(u) = (Pu_1, \ldots, Pu_n)$, where;

$$Pu_i = \sum_{p=1}^{k} u_{pi}\varphi_p \quad and\ (u_i - Pu_i, \varphi_i) = 0, for\ i = 1, \ldots, k.$$

So,

$$\sum_{p=1}^{k} (\varphi_q, \varphi_p) u_{pi} = (u_i, \varphi_q), for\ q = 1, \ldots, k.$$

Analogously to collocation, these equations form a linear system for the coefficients $U = (u_{pj})_{\substack{1\le p\le k \\ 1\le j\le n}}$. By setting, for $1 \le p, q \le k$, $1 \le i, j \le n$,

$$A_{qipj} = (\varphi_q, \varphi_p)\delta_{ij}\, , B_{qi} = (u_i, \varphi_q)\, ,$$

we have:

$$AU = B\, .$$

These equations may be transformed in a standard linear system involving two indexes by setting $I = index(q, i, k)$,

$J = index(p,j,k)$, $\mathcal{A}_{IJ} = A_{qipj}$, $\mathcal{B}_I = B_{qi}$, $\mathcal{U}_J = U_{pj}$: then $\mathcal{A}\mathcal{U} = \mathcal{B}$. This linear system has dimension $kn \times kn$. It may also be solved component by component: let us introduce:

$$M_{pq} = \left(\varphi_p, \varphi_q\right), N_i = \left(B_{1i}, \dots, B_{ki}\right)^t, U_i = \left(U_{1i}, \dots, U_{ki}\right)^t$$

Then, for $1 \leq i \leq n$,

$$MU_i = N_i \, .$$

Thus, the solution may be sequentially determined by solving n linear systems of size $\times k$.

2.2.6. *Matlab® classes for a numerical solution*

In order to determine numerical solutions, we need to define the family $\mathcal{B} = \{\varphi_i\}_{1 \leq i \leq k}$. For instance, in one-dimensional (1D) situations, we may use a polynomial family defined by the following class:

```
classdef polynomial_basis
    properties
        zero
        xmin
        xmax
        nel
    end
    properties(Dependent)
        spmatrix
        intmatrix
    end
    methods
        function obj =
polynomial_basis(zero,xmin,xmax,degree)
            obj.xmin = xmin;
            obj.xmax = xmax;
            obj.zero = zero;
            obj.nel = degree + 1;
        end
```

```
        function A = get.spmatrix(obj)
            k = obj.nel;
            a = obj.xmin;
            b = obj.xmax;
            A = zeros(k, k);
            for i = 1: k
                A(i,i) = (b-a)/(2*i-1);
                for j = i+1:k
                    aux = (b-a)/(i+j-1);
                    A(i,j) = aux;
                    A(j,i) = aux;
                end;
            end;
        end
        function B = get.intmatrix(obj)
            k = obj.nel;
            a = obj.xmin;
            b = obj.xmax;
            B = zeros(k,1);
            for i = 1: k
                B(i) = (b-a)/i;
            end;
        end
        function v = value(obj,i,t)
            if i == 1
                v = 1;
            else
                s = (t-obj.xmin)/(obj.xmax -
obj.xmin);
                if abs(s) < obj.zero
                    v = 0;
                elseif abs(s-1) < obj.zero
                    v = 1;
                elseif i == 2
                    v = s;
                else
                    v = s^(i-1);
                end;
```

```
            end
        end
        function v = d1(obj,i,t)
            if i == 1
                v = 0;
            elseif i == 2
                v = 1;
            else
                s = (t-obj.xmin)/(obj.xmax -
obj.xmin);
                if abs(s) < obj.zero
                    v = 0;
                elseif abs(s-1) < obj.zero
                    v = i-1;
                elseif i == 3
                    v = 2*s;
                else
                    v = (i-1)*s^(i-2);
                end;
            end
            v = v/(obj.xmax - obj.xmin);
        end
        function v = d2(obj,i,t)
            if i < 3
                v = 0;
            elseif i == 3
                v = 2;
            else
                s = (t-obj.xmin)/(obj.xmax -
obj.xmin);
                if abs(s) < obj.zero
                    v = 0;
                elseif abs(s-1) < obj.zero
                    v = (i-1)*(i-2);
                elseif i == 4
                    v = 6*s;
                else
                    v = (i-1)*(i-2)*s^(i-3);
                end;
            end
```

```
                v = v/(obj.xmax - obj.xmin)^2;
            end
        end
    end
```

Program 2.2. *Polynomial basis*

In this code, the polynomials are $\varphi_i(t) = \left(\frac{t-a}{b-a}\right)^{i-1}$, with `xmin=a` and `xmin= b`. Property `zero` defines numbers which are treated as zero. `degree` defines the degree of the polynomial to be used. Property `spmatrix` is a square matrix containing the integrals of the products of the elements of the basis, while property `intmatrix` is a vector containing the integrals of the elements themselves.

Let us assume that the data is given by a structure `approxdata` having as properties: `points`, `values`, `subprogram`, which contain, respectively, the points `x`, the corresponding values `f(x)` and a subprogram evaluating `f(x)`. In this case, the coefficients of an approximation and, namely, orthogonal projections may be evaluated by using the following class:

```
classdef basis
    properties
        phi
        nel
    end
    methods (Access = private)
        function v = select_index_f(obj,i,f,t)
            aux = f(t);
            v = aux(i);
            return;
        end
        function A = collocmat(obj,data)
            k = obj.nel;
            x = data.points.x;
            y = data.values;
            n = length(y);
            A = zeros(n,k);
```

```
            for i = 1: n
                for j = 1:k
                    A(i,j) = obj.phi.value(j,x(i));
                end;
            end;
        end
        function A = varmat_mean(obj,data)
            k = obj.nel;
            p = data.points;
            y = data.values;
            A = zeros(k, k);
            for i = 1: k
                phi_i = @(t) obj.phi.value(i,t);
                fi = spam.partition(phi_i,p);
                A(i,i) = mean(fi.^2);
                for j = i+1:k
                    phi_j = @(t) obj.phi.value(j,t);
                    fj = spam.partition(phi_j,p);
                    aux = mean(fi.*fj);
                    A(i,j) = aux;
                    A(j,i) = aux;
                end;
            end;
        end
        function A = varmat_int(obj,data)
            k = obj.nel;
            p = data.points;
            y = data.values;
            A = zeros(k, k);
            for i = 1: k
                phi_i = @(t) obj.phi.value(i,t);
                ff = @(t) phi_i(t)^2;
                fs = @(x) spam.points('table',ff,x);
                A(i,i) =
integral(fs,obj.phi.xmin,obj.phi.xmax);
                for j = i+1:k
                    phi_j = @(t) obj.phi.value(j,t);
                    ff = @(t) phi_i(t)*phi_j(t);
                    fs = @(x)
spam.points('table',ff,x);
```

```
                    aux =
integral(fs,obj.phi.xmin,obj.phi.xmax);
                    A(i,j) = aux;
                    A(j,i) = aux;
                end;
            end;
        end
        function B = sm_mean(obj,data,j)
            k = obj.nel;
            p = data.points;
            y = data.values(:,j);
            B = zeros(k, 1);
            for i = 1: k
                phi_i = @(t) obj.phi.value(i,t);
                fi = spam.partition(phi_i,p);
                B(i) = mean(fi.*y);
            end;
        end
        function B = sm_int(obj,data,j)
            k = obj.nel;
            p = data.points;
            f = @(t)
obj.select_index_f(j,data.subprogram,t);
            B = zeros(k, 1);
            for i = 1: k
                phi_i = @(t) obj.phi.value(i,t);
                ff = @(t) phi_i(t)*f(t);
                fs = @(t) spam.points('table',ff,t);
                B(i) =
integral(fs,obj.phi.xmin,obj.phi.xmax);
            end;
        end
    end
    methods
        function obj = basis(phi)
            obj.phi = phi;
            obj.nel = phi.nel;
        end
        function U = approx_coeffs(obj,metod,data)
            switch metod
                case 'collocation'
                    A = obj.collocmat(data);
```

```
                    U = zeros(obj.nel,data.dim);
                    for j = 1: data.dim
                        U(:,j) = A \ data.values(:,j);
                    end;
                case 'variational_mean'
                    A = obj.varmat_mean(data);
                    U = zeros(obj.nel,data.dim);
                    for j = 1: data.dim
                        B = obj.sm_mean(data,j);
                        U(:,j) = A \B;
                    end;
                case 'variational_int'
                    A = obj.varmat_int(data);
                    U = zeros(obj.nel,data.dim);
                    for j = 1: data.dim
                        B = obj.sm_int(data,j);
                        U(:,j) = A \B;
                    end;
                case 'variational_sp'
                    A = obj.phi.spmatrix;
                    U = zeros(obj.nel,data.dim);
                    for j = 1: data.dim
                        B = obj.sm_int(data,j);
                        U(:,j) = A \B;
                    end;
                otherwise
                    U = [];
                    disp('error in args for
coeeficients');
            end
        end
        function PX = projection(obj,u,xs)
            ns = size(xs,2);
            k = size(u,1);
            if k ~= obj.nel
                PX = [];
                disp('first dim of coeffs differs from
basis nel');
                return;
            end
            nd = size(u,2);
```

```
            PX = zeros(nd,ns);
            for p = 1: k
                u_p = u(p,:)';
                for j = 1: ns
                  PX(:,j) = PX(:,j) +
u_p*obj.phi.value(p,xs(:,j));
                end;
            end
        end
        function v = integrate(obj,u)
            k = size(u,1);
            if k ~= obj.nel
                v = [];
                disp('first dim of coeffs differs from
basis nel');
                return;
            end
            n = size(u,2);
            v = zeros(n,1);
            B = obj.phi.intmatrix;
            for i = 1: n
                v(i) = u(:,i)'*B;
            end
        end
        function dPX = derive(obj,u,xs)
            ns = size(xs,2);
            k = size(u,1);
            if k ~= obj.nel
                dPX = [];
                disp('first dim of coeffs differs from
basis nel');
                return;
            end
            nd = size(u,2);
            dPX = zeros(nd,ns);
            for p = 1: k
                u_p = u(p,:)';
                for j = 1: ns
              dPX(:,j) = dPX(:,j) +
u_p*obj.phi.d1(p,xs(:,j));
                end;
            end
        end
    end
end
```

Program 2.3. *Approximation*

This class contains a method `approx_coefficients` which determines the coefficients of the approximation according to the method chosen: `'collocation'` uses the collocation approach, while the others use the variational approach, but differ in the manner where the integrals are evaluated: `'variational_mean'` uses the data in order to evaluate a mean value approximation, `'variational_int'` uses the quadrature by Matlab®, `'variational_sp'` uses the matrix of integrals `spmatrix` furnished in the definition of the polynomial basis. The class also contains methods `integrate` and `derive` for the evaluation of an integral on the interval `(xmin, xmax)` and the derivative at a vector of points, respectively.

Let us illustrate the use of these classes. For instance, let us consider the noisy data generated as follows:

```
f = @(t) exp(t);
t = (-1:0.2:1);
p.x = t';
p.dim = 1;
x = spam.partition(f,p);
noise = 0.1;
td = (-1:0.05:1);
xd = spam.points('vector',f,td);
vd = (1+ noise*(2*rand(size(xd))-1)).*xd;
fnoise = @(t) interp1(td,vd,t,'linear');
data.points = p;
data.values = (1+ noise*(2*rand(size(x))-1)).*x;
data.subprogram = @(t) fnoise(t);
data.dim = 1;
```

These commands generate noisy data representing the exponential function on the interval $(-1,1)$. The level of noise is 10%. Notice that the subprogram evaluating the function is not the exact exponential, but a noisy version generated by linear interpolation of noisy data. The commands

```
xmin = -1;
xmax = 1;
degree = 4;
zero = 1e-3;
```

```
phi = polynomial_basis(zero,xmin,xmax,degree);
b = basis(phi);
u1 = b.approx_coeffs('collocation',data);
```

generate the coefficients of the approximation by collocation by using a polynomial basis involving a maximal degree 4. The command

```
u2 = b.approx_coeffs('variational_mean',data);
```

generates the coefficients corresponding to the variational approximation and evaluate the integrals by using the means. The command

```
u3 = b.approx_coeffs('variational_int',data);
```

generates the variational approximation and evaluates the integrals by using subprogram integration.

The results are shown in Figure 2.2 (recall that data is given on $(-1,1)$ and points outside this interval are extrapolated ones):

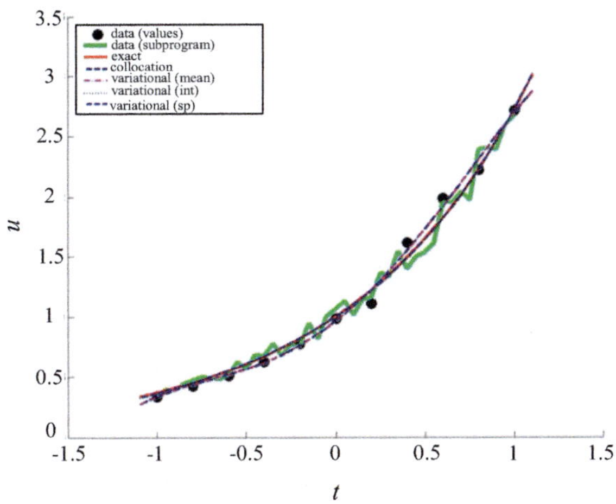

Figure 2.2. *Orthogonal projection in a simple noisy situation. For a color version of the figure, see www.iste.co.uk/souzadecursi/variational.zip*

The derivative is shown below.

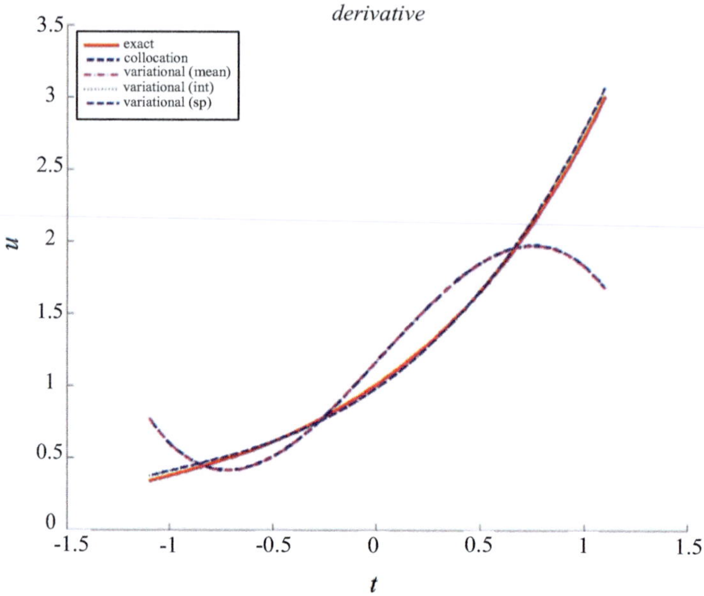

Figure 2.3. *Derivative in a simple noisy situation. For a color version of the figure, see www.iste.co.uk/souzadecursi/variational.zip*

We observe that, on the one hand, the results by collocation and variational approach with integrals evaluated by the mean are close and, on the other hand, the results furnished by the variational approach with more sophisticated integration are close to exact ones. Concerning the evaluation of the integral of f on the interval $(-1,1)$, the commands

```
v1 = b.integrate(u1);
v2 = b.integrate(u2);
v3 = b.integrate(u3);
v4 = b.integrate(u4);
ff = @(t) b.projection(u4,t);
v5 = integral(f,-1,1);
```

furnish `v1 = v2 =` 2.3535, `v3 = v4 = v5 =` 2.3442, while the exact value is $e - e^{-1} \approx 2.3504$.

Results for the vector $(t) = (\sin(\pi t) \quad \exp(t))^t$ are given in Figures 2.4 and 2.5. We use a degree 5. In this situation, we obtain `v1 = v2 = (-0.0139 2.3507),v3 = 0.0027 2.3658)`.

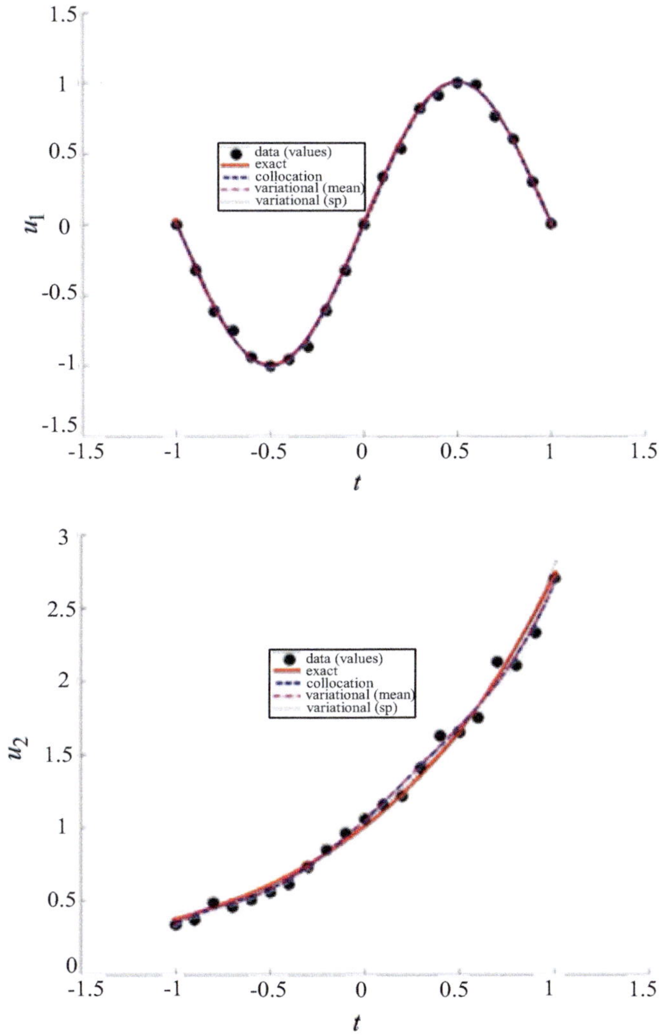

Figure 2.4. *Orthogonal projection in a simple noisy situation. For a color version of the figure, see www.iste.co.uk/souzadecursi/variational.zip*

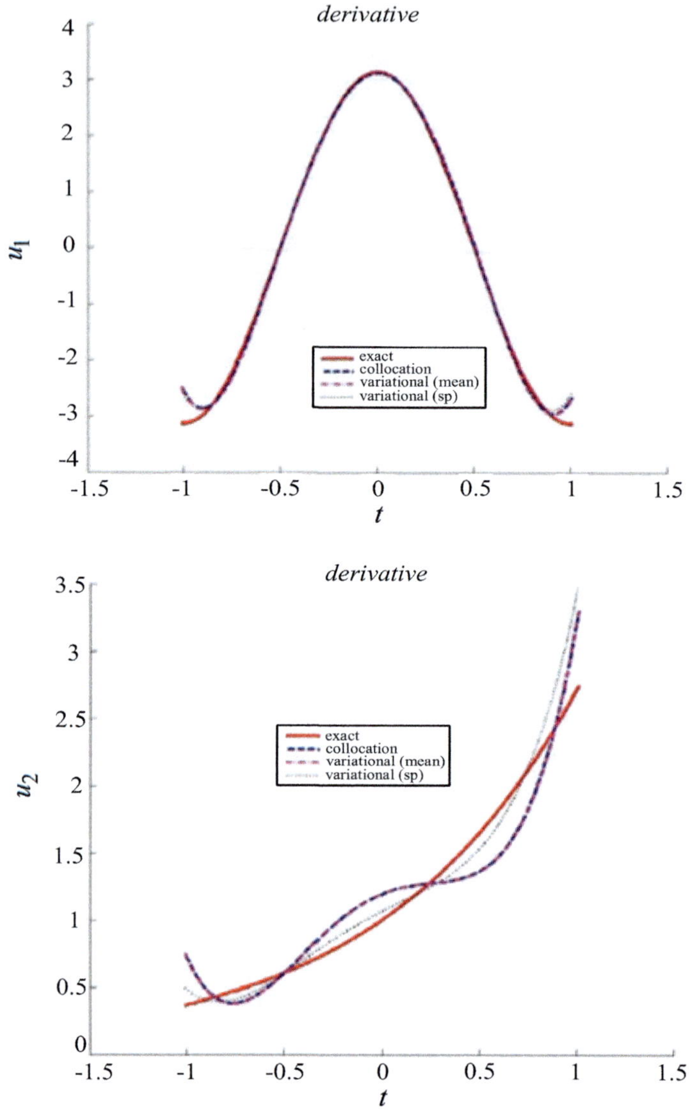

Figure 2.5. *Derivative in a simple noisy situation. For a color version of the figure, see www.iste.co.uk/souzadecursi/variational.zip*

When considering nonlinear parametric equations involving a single parameter, we may use the following class

```
classdef parametric_algebraic
    properties
        eqs
        basis
        ns
    end
    methods (Static, Access = private)
        function v = select_index(i,f)
            v = f(i);
            return;
        end
        function v = to1d(U,n1,n2)
            v = zeros(n1*n2,1);
            ind = 0;
            for i = 1: n1
                for j= 1: n2
                    ind = ind + 1;
                    v(ind) = U(i,j);
                end
            end
        end
        function v = to2d(u,n1,n2)
            v = zeros(n1,n2);
            ind = 0;
            for i = 1: n1
                for j= 1: n2
                    ind = ind + 1;
                    v(i,j) = u(ind);
                end
            end
        end
        function v = mean_Fcal(obj,U)
            F = obj.eqs.subprogram;
            m = obj.eqs.neq;
            k = obj.basis.nel;
            a = obj.basis.phi.xmin;
```

```
                b = obj.basis.phi.xmax;
                v = zeros(m,k);
                ts = a:(b-a)/obj.ns:b;
                p.x = ts;
                p.dim = 1;
                px = @(t) obj.basis.projection(U,t);
                fx = @(t) F(px(t), t);
                for i = 1: m
                    f_i = @(t)
obj.select_index(i,fx(t));
                    fi = spam.partition(f_i,p);
                    for j = 1: k
                        phi_j = @(t)
obj.basis.phi.value(j,t);
                        fj = spam.partition(phi_j,p);
                        v(i,j) = mean(fi.*fj);
                    end;
                end;
            end
            function v = int_Fcal(obj,U)
                F = obj.eqs.subprogram;
                m = obj.eqs.neq;
                k = obj.basis.nel;
                a = obj.basis.phi.xmin;
                b = obj.basis.phi.xmax;
                v = zeros(m,k);
                px = @(t) obj.basis.projection(U,t);
                fx = @(t) F(px(t), t);
                for i = 1: m
                    f_i = @(t)
obj.select_index(i,fx(t));
                    for j = 1: k
                        f_j = @(t)
obj.basis.phi.value(j,t);
                        ff = @(t) f_i(t)*f_j(t);
                        fff = @(t)
spam.points('table',ff,t);
                        v(i,j) = integral(fff,a,b);
                    end;
```

```
            end;
        end
        function v = objectif(obj,metod,u,n1,n2)
            U = obj.to2d(u,n1,n2);
            switch metod
                case 'mean'
                    aux = obj.mean_Fcal(obj,U);
                case 'integral'
                    aux = obj.int_Fcal(obj,U);
                otherwise
                    v = [];
                    return;
            end
            m = obj.eqs.neq;
            k = obj.basis.nel;
            v = norm(aux)/sqrt(m*k);
        end
        function v =
objectif_system(obj,metod,u,n1,n2)
            U = obj.to2d(u,n1,n2);
            switch metod
                case 'mean'
                    aux = obj.mean_Fcal(obj,U);
                case 'integral'
                    aux = obj.int_Fcal(obj,U);
                otherwise
                    v = [];
                    return;
            end
            v =
obj.to1d(aux,size(aux,1),size(aux,2));
        end
    end
    methods
        function obj =
parametric_algebraic(equations,basis,ns)
            obj.eqs = equations;
            obj.basis = basis;
            if nargin() > 2
```

```
                    obj.ns = ns;
                end
            end
            function U = solve_min(obj,metod,U0,nitmax)
                n1 = obj.basis.nel;
                n2 = obj.eqs.dim;
                fobj = @(u)
obj.objectif(obj,metod,u,n1,n2);
  options =
optimset('MaxIter',nitmax,'MaxFunEvals',10*nitmax);
                u0 = obj.to1d(U0,n1,n2);
                u = fminsearch(fobj,u0,options);
                U = obj.to2d(u,n1,n2);
            end;
            function U = solve_zero(obj,metod,U0,nitmax)
                n1 = obj.basis.nel;
                n2 = obj.eqs.dim;
      fobj = @(u)
obj.objectif_system(obj,metod,u,n1,n2);
      options =
optimset('MaxIter',nitmax,'MaxFunEvals',10*nitmax);
                u0 = obj.to1d(U0,n1,n2);
                u = fsolve(fobj,u0,options);
                U = obj.to2d(u,n1,n2);
            end;
        end
end
```

Program 2.4. *Variational solution of parametric algebraic equations*

In this class, the solution is obtained either by minimizing the quadratic mean square of the elements of $\mathcal{F}(U)$or by solving the equations $\mathcal{F}(U) = 0$. The optimization step is performed by using the intrinsic Matlab® function `fminsearch`., while the determination of a zero is performed by the intrinsic function `fzero`. Parameters such as number of iterations, function evaluations, precision are defined by using `optimset`. Choice `'mean'` evaluates the integrals by using the mean value approximation with `ns` equal subintervals, while `'integral'` uses the intrinsic quadrature of Matlab®. The equations are given by structured data in `equations`, which has as fields `subprogram`, `neq`, `dim`, which correspond, respectively, to the subprogram which furnishes the value of $F(x,t)$, the number of equations and the number of unknowns (dimension of the unknown).

EXAMPLE 2.3.–Let us consider the simple situation where,

$$F(x,t) = x - sin(t)\,, t \in (0,\pi).$$

We use the polynomial basis with a degree 6. The code is:

```
xmin = 0;
xmax = pi;
ns = 500;
degree = 6;
zero = 1e-3;
phi = polynomial_basis(zero,xmin,xmax,degree);
b = basis(phi);
f = @(x,t) x - sin(t);
eqs.subprogram = f;
eqs.dim = 1;
eqs.neq = 1;
```

The obvious solution is $x = sin(t)$. The results furnished by using:

```
pa = parametric_algebraic(eqs,b,ns);
nitmax = 10000;
U1 = pa.solve_zero('integral',U0,nitmax);
U2 = pa.solve_min('mean',U0,nitmax)
```

are shown in Figure 2.6 below (left: methods `integral` and `solve_zero`, right: methods `mean` and `solve_min`). ■

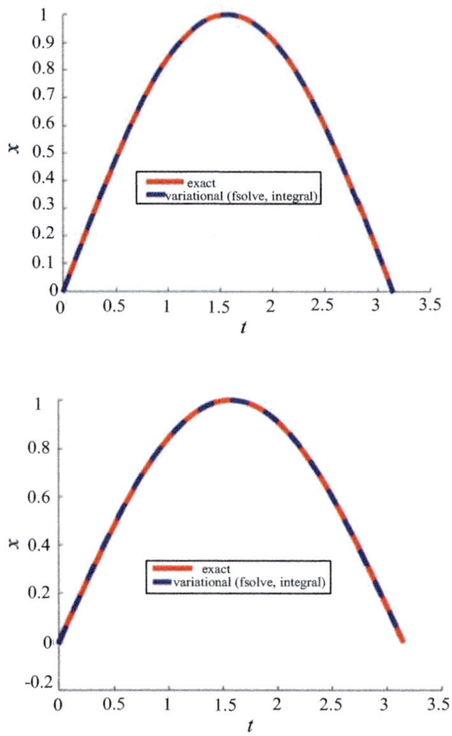

Figure 2.6. *Solutions obtained for a polynomial of degree 6. For a color version of the figure, see www.iste.co.uk/souzadecursi/variational.zip*

EXAMPLE 2.4.– Let us consider the simple situation where,

$$F(x,t) = \begin{pmatrix} x_1 + x_2 - sin(t) \\ x_1 - x_2 - cos(t) \end{pmatrix}, t \in (0, \pi).$$

We use the polynomial basis with a degree 5. The code is:

```
xmin = 0;
xmax = pi;
ns = 500;
degree = 5;
zero = 1e-3;
phi = polynomial_basis(zero,xmin,xmax,degree);
b = basis(phi);
```

```
f = @(x,t) [x(1) + x(2) - sin(t) ; x(1) - x(2) -
cos(t)];
eqs.subprogram = f;
eqs.dim = 2;
eqs.neq = 2;
pa = parametric_algebraic(eqs,b,ns);
U1 = pa.solve_zero('mean',U0,nitmax);
U2 = pa.solve_min('integral',U0,nitmax);
```

The obvious solution is $x_1 = (sin(t) + \cos(t))/2$, $x_1 = (sin(t) - \cos(t))/2$. The results are shown in Figures 2.7 (using method `solve_zero`, integrals evaluated by the mean) and 2.8 (using method `solve_min`, integrals evaluated by quadrature) below. ■

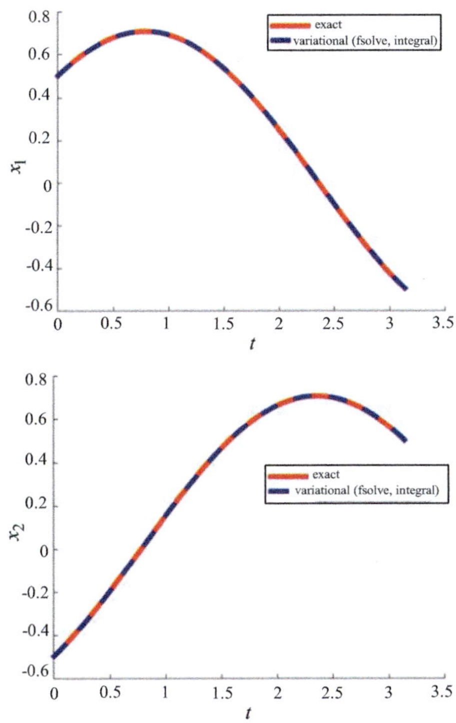

Figure 2.7. *Solutions obtained for a polynomial of degree 5 and "mean" method. For a color version of the figure, see www.iste.co.uk/souzadecursi/variational.zip*

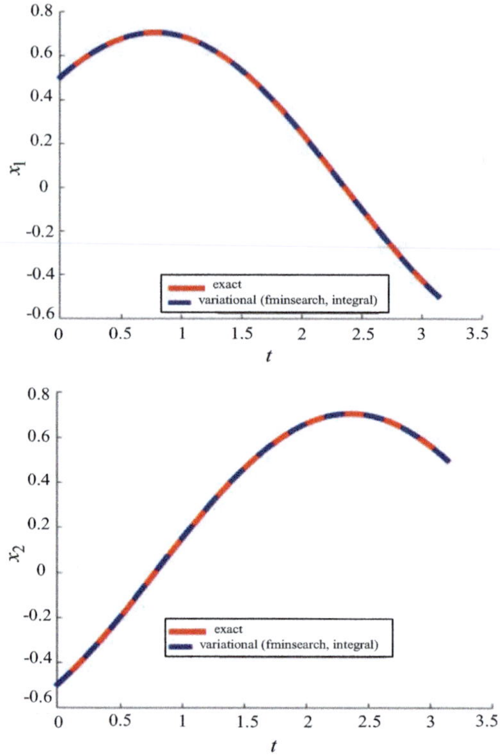

Figure 2.8. *Solutions obtained for a polynomial of degree 5 and "integral" method. For a color version of the figure, see www.iste.co.uk/souzadecursi/variational.zip*

EXAMPLE 2.5.– Let us consider the simple situation where

$$F(x,t) = x^2 - sin(t)\,, t \in (0,\pi).$$

We use the polynomial basis with a degree 6. The results furnished by using method `solve_zero` are shown in Figure 2.9. ■

EXAMPLE 2.6.– Let us consider the simple situation where

$$F(x,t) = \begin{pmatrix} x_1^3 + x_2^3 - sin(t) \\ x_1^3 - x_2^3 - cos(t) \end{pmatrix}, t \in (0,\pi).$$

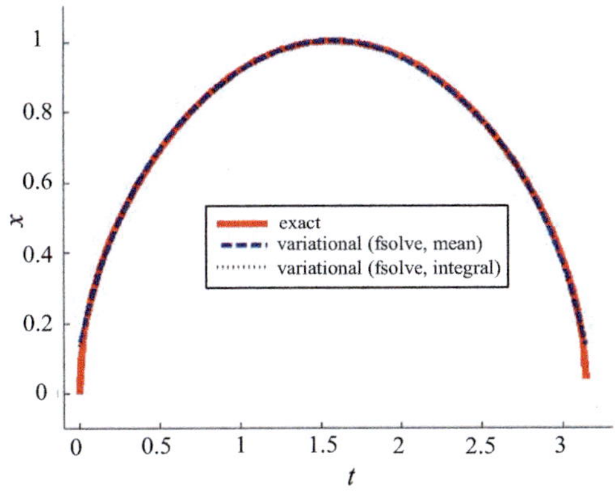

Figure 2.9. *Solutions obtained for a polynomial of degree 6. For a color version of the figure, see www.iste.co.uk/souzadecursi/variational.zip*

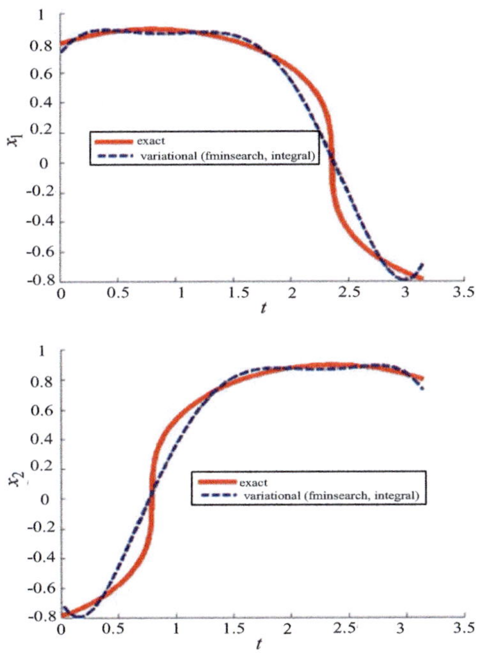

Figure 2.10. *Solutions obtained for a polynomial of degree 5. For a color version of the figure, see www.iste.co.uk/souzadecursi/variational.zip*

We use the polynomial basis with a degree 5. The results are shown in Figure 2.10 (using method `solve_zero`). ■

Other families may be used, such as, for instance, the trigonometric family: $\varphi_1(t) = 1$, $\varphi_{2i}(t) = cos(it)$, $\varphi_{2i+1}(t) = sin(it)$, or the family associated with finite elements P1. These families are defined by the classes below.

```
classdef trigonometrical_basis
    properties
        zero
        xmin
        xmax
        nel
    end
    properties(Dependent)
        spmatrix
    end
    methods
        function obj =
trigonometrical_basis(zero,xmin,xmax,order)
            obj.xmin = xmin;
            obj.xmax = xmax;
            obj.zero = zero;
            obj.nel = 2*order + 1;
        end
        function A = get.spmatrix(obj)
            k = obj.nel;
            a = obj.xmin;
            b = obj.xmax;
            A = zeros(k, k);
            A(1,1) = 1;
            for i = 1: (k-1)/2
                A(2*i,1) = sin(i)/i;
                A(2*i+1,1) = (1-cos(i))/i;
                A(1,2*i) = sin(i)/i;
                A(1,2*i+1) = (1-cos(i))/i;
                A(2*i,2*i) = 0.5 + 0.25*sin(2*i)/i;
                A(2*i+1,2*i+1) = 0.5 -
0.25*sin(2*i)/i;
```

```
                A(2*i+1,2*i) = 0.5*sin(i)^2/i;
                A(2*i,2*i+1) = 0.5*sin(i)^2/i;
                for j = i+1:(k-1)/2
            aux = (i*cos(i)*sin(j) -
j*cos(j)*sin(i))/(j^2 - i^2);
                    A(2*i,2*j) = aux;
                    A(2*j,2*i) = aux;
         aux = (j - i*sin(i)*sin(j) -
j*cos(j)*cos(i))/(j^2 - i^2);
                    A(2*i,2*j+1) = aux;
                    A(2*j+1,2*i) = aux;
          aux = (j*cos(i)*sin(j) -
i*cos(j)*sin(i))/(j^2 - i^2);
                    A(2*i+1,2*j+1) = aux;
                    A(2*j+1,2*i+1) = aux;
         aux = (j*sin(i)*sin(j) + i*cos(j)*cos(i) -
i)/(j^2 - i^2);
                    A(2*i+1,2*j) = aux;
                    A(2*j,2*i+1) = aux;
                end;
            end;
            A = (b-a)*A;
        end
        function B = get.intmatrix(obj)
            k = obj.nel;
            a = obj.xmin;
            b = obj.xmax;
            B = zeros(k,1);
            B(1) = 1;
            for i = 1: (k-1)/2
                B(2*i) = sin(i)/i;
                B(2*i+1) = (1-cos(i))/i;
            end;
            B = (b-a)*B;
        end
        function v = value(obj,i,t)
            if i == 1
                v = 1;
            else
```

```
                s = (t-obj.xmin)/(obj.xmax -
obj.xmin);
                if mod(i,2) == 0
                    if abs(s) < obj.zero
                        v = 1;
                    else
                        aux = 0.5*i*s;
                        v = cos(aux);
                    end;
                else
                    if abs(s) < obj.zero
                        v = 0;
                    else
                        aux = 0.5*(i-1)*s;
                        v = sin(aux);
                    end;
                end;
            end;
        end
        function v = d1(obj,i,t)
            if i == 1
                v = 0;
            else
                s = (t-obj.xmin)/(obj.xmax -
obj.xmin);
                if mod(i,2) == 0
                    if abs(s) < obj.zero
                        v = 0;
                    else
                        aux = 0.5*i*s;
                        v = -0.5*i*ds*sin(aux);
                    end;
                else
                    if abs(s) < obj.zero
                        v = 0.5*(i-1)*ds;
                    else
                        aux = 0.5*(i-1)*s;
                        v = 0.5*(i-1)*ds*cos(aux);
                    end;
                end;
                v = v/(obj.xmax - obj.xmin);
```

```
            end;
        end
        function v = d2(obj,i,t)
            if i == 1
                v = 0;
            else
                s = (t-obj.xmin)/(obj.xmax -
obj.xmin);
                if mod(i,2) == 0
                    if abs(s) < obj.zero
                        v = -1;
                    else
                        aux = 0.5*i*s;
                        v = -0.5*i^2*cos(aux);
                    end;
                else
                    if abs(s) < obj.zero
                        v = 0;
                    else
                        aux = 0.5*(i-1)*s;
                        v = -0.25*(i-1)^2*sin(aux);
                    end;
                end;
                v = v/(obj.xmax - obj.xmin)^2;
            end;
        end
    end
end
```

Program 2.5. *Trigonometrical basis*

```
classdef p1_basis
    properties
        zero
        xmin
        xmax
        nel
        step
        nodes
```

```
    end
    properties(Dependent)
        spmatrix
    end
    methods
        function obj =
p1_basis(zero,xmin,xmax,ns)
            obj.xmin = xmin;
            obj.xmax = xmax;
            obj.zero = zero;
            obj.nel = ns + 1;
            obj.step = (xmax - xmin)/ns;
            obj.nodes = xmin:obj.step:xmax;
        end
        function A = get.spmatrix(obj)
            k = obj.nel;
            h = obj.step;
            A = zeros(k, k);
            A(1,1) = h/3;
            A(1,2) = h/6;
            for i = 2: k-1
                A(i,i) = 2*h/3;
                A(i,i-1) = h/6;
                A(i,i+1) = h/6;
            end;
            A(k,k) = h/3;
            A(k-1,k) = h/6;
        end
        function B = get.intmatrix(obj)
            k = obj.nel;
            h = obj.step;
            B = zeros(k,1);
            B(1) = h/2;
            for i = 2: k-1
                B(i) = h;
            end;
            B(k) = h/2;
        end
        function v = value(obj,i,t)
            t_i = obj.nodes(i);
            h = obj.step;
```

```
                aux = abs((t - t_i)/h);
                if aux > 1
                    v = 0;
                else
                    v = 1 - aux;
                end;
            end
            function v = d1(obj,i,t)
                t_i = obj.nodes(i);
                h = obj.step;
                aux = (t - t_i)/h;
                if aux > 1
                    v = 0;
                elseif aux < -1
                    v = 0;
                elseif aux > obj.zero
                    v = -1/h;
                elseif aux < -obj.zero
                    v = 1/h;
                else
                    v = 0;
                end;
            end
            function v = d2(obj,i,t)
                v = [];
                disp('no d2 for P1 family');
            end
        end
end
```

Program 2.6. *P1 Finite Element basis*

Let us illustrate the use of these classes: consider again $f(x,t) = x - \sin(t)$, what corresponds to the orthogonal projection of $x = \sin(t)$ onto the linear subspaces generated by the basis. The results are shown in the figures below. P1 Finite Elements have a piecewise constant derivative.

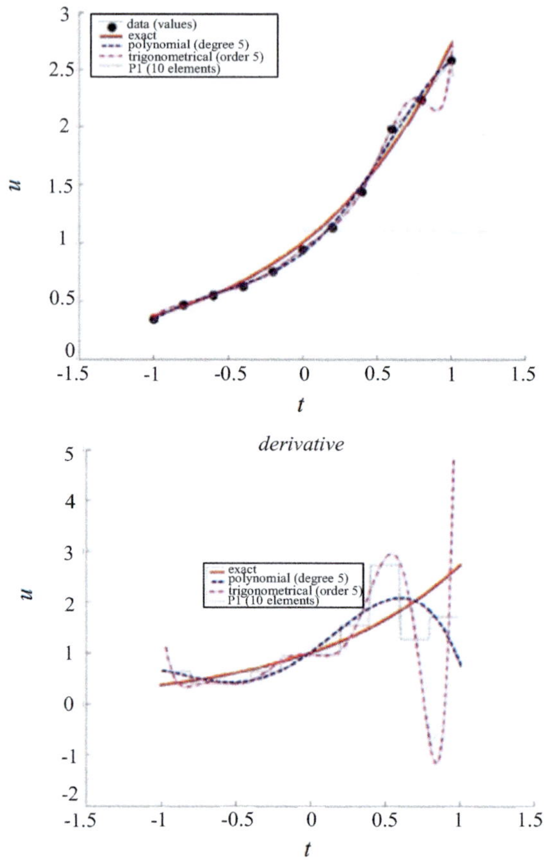

Figure 2.11. *Solutions obtained by collocation using different basis. For a color version of the figure, see www.iste.co.uk/souzadecursi/variational.zip*

2.3. Exercises

EXERCISE 2.1.– Consider the linear system:

$$\begin{pmatrix} 1 & 2 & 3 & 4 \\ 1 & -1 & -1 & 1 \\ 0 & 2 & 1 & 3 \\ 4 & 3 & 2 & 0 \end{pmatrix} \begin{pmatrix} x_1 \\ x_2 \\ x_3 \\ x_4 \end{pmatrix} = \begin{pmatrix} 9 \\ 1 \\ 3 \\ 5 \end{pmatrix}$$

Find a vector y such that:

$$y^t \begin{pmatrix} 2 \\ -1 \\ 2 \\ 3 \end{pmatrix} = y^t \begin{pmatrix} 3 \\ -1 \\ 1 \\ 2 \end{pmatrix} = 0$$

and determine x_1 and x_4 by the variational method. ■

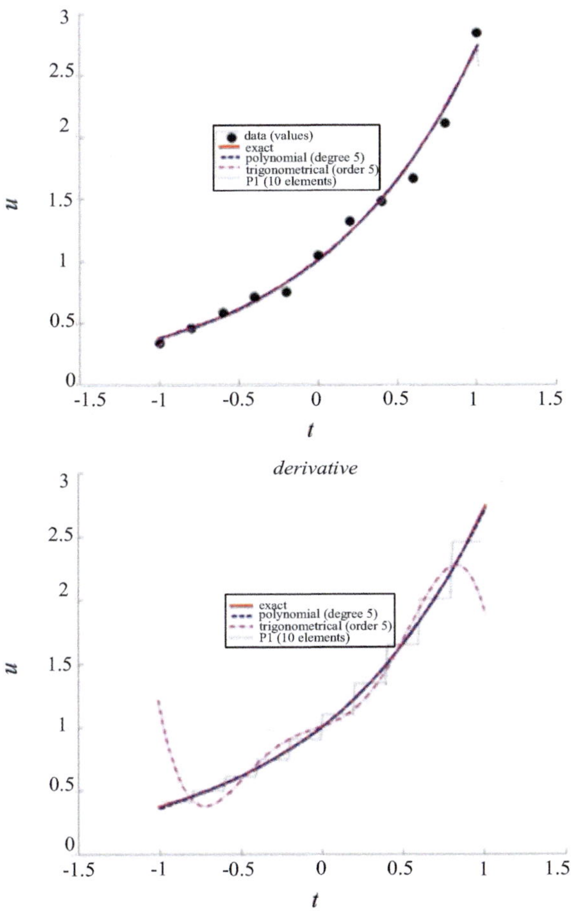

Figure 2.12. *Solutions obtained by variational approach (mean) using different basis. For a color version of the figure, see www.iste.co.uk/souzadecursi/variational.zip*

EXERCISE 2.2.– Consider the linear system:

$$\begin{pmatrix} 1 & 2 & 3 & 4 & 5 \\ 5 & 1 & 2 & 3 & 4 \\ 2 & 1 & -1 & -1 & 1 \\ 3 & 0 & 2 & 1 & 3 \\ 4 & 4 & 3 & 2 & 0 \end{pmatrix} \begin{pmatrix} x_1 \\ x_2 \\ x_3 \\ x_4 \\ x_5 \end{pmatrix} = \begin{pmatrix} 18 \\ 9 \\ 1 \\ 9 \\ 4 \end{pmatrix}$$

Find a vector y such that:

$$y^t \begin{pmatrix} 2 \\ 1 \\ 1 \\ 0 \\ 4 \end{pmatrix} = y^t \begin{pmatrix} 3 \\ 2 \\ -1 \\ 2 \\ 3 \end{pmatrix} = y^t \begin{pmatrix} 5 \\ 4 \\ 1 \\ 3 \\ 0 \end{pmatrix} = 0$$

and determine x_1 and x_4 by the variational method. ■

EXERCISE 2.3.– Consider the family $\psi_n(s) = s^{n-1}$, defined on $\Omega = (0,1)$. Denote

$$(u, v) = \int_0^1 u(s)v(s)ds .$$

and the family $\varphi_n(s)$ given by:

$$\varphi_1 = \psi_1 \, ; \, \varphi_n = \psi_n - \sum_{i=1}^{n-1} \frac{(\psi_n, \varphi_i)}{(\varphi_i, \varphi_i)} \varphi_i \; (n > 1)$$

Show that $(\varphi_n, \varphi_m) = 0 \; for \; m < n$. Conclude that $(\varphi_n, \varphi_m) = 0 \; for \; m \neq n$, so that the matrix defined by the property `spmatrix` is diagonal. Show that the orthogonal projection $P_{S^m}(f) = U^t\Phi$ satisfies:

$$u_i = \frac{(f, \varphi_i)}{(\varphi_i, \varphi_i)} .$$

■

EXERCISE 2.4.– Consider the family $\varphi_1(t) = 1$, $\varphi_{2n}(t) = cos(2n\pi t)$, $\varphi_{2n+1}(t) = sin(2n\pi t)$, defined on $\Omega = (0,1)$. Show that $(\varphi_n, \varphi_m) = 0 \; for \; m \neq n$. Conclude that the matrix defined by the property spmatrix is diagonal and show that the orthogonal projection $P_{S^m}(f) = U^t \Phi$ satisfies:

$$u_i = \frac{(f, \varphi_i)}{(\varphi_i, \varphi_i)} \ .$$

Verify that $(\varphi_1, \varphi_1) = 1$ and $(\varphi_i, \varphi_i) = 1/2$, for $i > 1$. Determine the coefficients u_i of the expansion of $f(t) = sin\ (t)$.

3

Hilbert Spaces for Engineers

David Hilbert is considered to be one of the most influential mathematicians of the 20th Century. He was a German mathematician, educated at Konigsberg, where he completed his PhD thesis under the direction of Ferdinand Lindemann. At the same time, Hermann Minkowski was also a PhD student of Lindemann – they presented their PhD's in the same year (1885) and thereafter became good friends. Hilbert was acquainted with some of the best mathematicians of his time and contributed significantly to important mathematical problems. He also supported the candidacy of Emmy Noether to teach at the University of Gottingen – at these times, this university was considered one of the best mathematical centers of the world and women had difficulty in finding academic positions. The history of mathematics teems with anecdotes and quotes of Hilbert. At the 2nd International Congress of Mathematics held in Paris, in 1900, Hilbert presented a speech where he evoked unsolved problems. His speech was transcribed in a long article, first published in German, but which was quickly translated into English [WIN 02] and is known globally as the *23 Hilbert problems* or the *Hilbert program* – this text has widely influenced Mathematics along the 20th Century. *The Hilbert program* addressed fundamental difficulties of Mathematics – some of them remain unresolved to this day – and their study led to a large development of new ideas and methods. For instance, the research connected to the second problem (the consistency of arithmetic) have led to the denial of any divine status to Mathematics and brought it to the dimension of a human activity based on human choices, biases and beliefs. The particularity of Mathematics remains in its quest for consistency and its extremely codified way to build results by logical deductions and demonstrations based on previous results. Mathematicians make choices by establishing axioms, which are the foundations of their buildings. Then, they patiently construct the whole structure by exploring the consequences of the axioms and generating results which are new bricks that make the building higher. The question of the compatibility of all the bricks among themselves is difficult and leads to the analysis of quality of the foundations, which is connected to the coherence of the results obtained, which may be weakened by paradoxes found in the development of the theory. The works of Kurt Godel [GOD 31] and Alfred Tarski [TAR 36] tend to show that we cannot be sure of the global compatibility of all the bricks and, so, of the coherence of the structure. But this apparent weakness transforms into force, since it gives power to imagination and total freedom for the construction of alternate theories, based on new axioms, which makes the possible developments of Mathematics virtually inexhaustible. This has led, for instance, to new geometries which have found applications in Physics and new function theories which have found their application in Physics and Engineering.

Ironically, Hilbert spaces were not formalized by Hilbert himself, but by other people. The three major contributions came from Maurice Fréchet [FRE 06], Stefan Banach [BAN 22] and John Von Neumann [VON 32]. Fréchet established a formalism based on the idea of distance in his PhD Thesis, prepared under the direction of Jacques Hadamard and presented in Paris in 1905. Banach presented the formalism of normed spaces in his PhD (1920 at Lvov), under the direction of Hugo Steinhaus – a former student of Hilbert. Banach also published an article presenting the formalization of normed spaces. Von Neumann worked with Hilbert and stated the axiomatic presentation of Hilbert spaces in his book. The original paper by Hilbert (the fourth of the series of six works on integral equations published between 1904 and 1910 [HIL 04a, HIL 04b, HIL 05, HIL 06a, HIL 06b, HIL 10, HIL 12] contains all the papers) did not present a formal theory and was limited to square summable functions [WEY 44].

In order to take the full measure of the importance of this work and the extent of its applications, the reader must recall that one of the main reasons for the existence of Mathematics is the need for the determination of the solutions of equations and the prediction of the evolution of systems or the determination of some of their parameters (namely, for safety or obtaining a desired result). Human activities often involve the generation of data, its analysis and conclusions about the future from observations or the determination of some critical parameters from data. This work is based on models representing the behavior of the system under analysis. However, the models involve equations that must be solved, either to generate the model or to determine the evolution of the system. For instance, Engineering and Physics use such an approach. Many of the interesting situations involve the solution of various types of equations (algebraic, differential, partial differential, integral, etc.). When studying integral, differential or partial differential equations, a major problem has to be treated, connected to the infinity: while the equation $x \in \mathbb{R}\ and\ f(x) = 0$ concerns a single value of x, the equation $x(t) \in \mathbb{R}$ and $f(x(t), t) = 0, t \in (a, b)$ concerns infinite values of x (one for each t). Analogously, integral, differential or partial differential equations concern the determination of infinitely many values. Since the determination of infinite actions takes an infinite time, procedures like limits and approximations have been introduced. Instead of achieving an infinite amount of solutions, it seems more convenient to determine close approximations of the real solution by achieving a finite number of solutions.

The practical implementation of this simple idea begs answering some questions. For instance: what is the sense of "close", i.e. who are the neighbors at a given distance of an element? How do we ensure that the approximations are close to "the real solution", i.e. that there are neighbors of the real solution at an arbitrarily small distance? How do we obtain a sequence of solutions which are "closer and closer (i.e. convergent) to the real solution", i.e. belonging to arbitrarily small neighborhoods of the real solution? Moreover, what is a solution? Mathematically, these questions may be formulated as follows: what is the set $(x; \varepsilon) = \{y: distance\ between\ y\ and\ x \leq \varepsilon\}$? How do we ensure that an approximation $\hat{x}$ verifies $\hat{x} \in V(x; \varepsilon)$? How do we generate an element $x_\varepsilon \in V(x;\varepsilon)$ for an arbitrarily small $\varepsilon > 0$? By taking $\varepsilon = 1/n$ and denoting $x_n = x_\varepsilon$, this last question reads as: how to generate an element x_n such that $distance\ between\ x_n\ and\ x \leq 1/n$, i.e. how do we generate a sequence $\{x_n : n \in \mathbb{N}\}$ which converges to x? How do we define x into a convenient way?

In the mathematical construction, elements such as the neighborhoods, the convergent sequences and approximations are defined by the *topology* of the space.

It is quite intuitive that we are interested in the construction of a theory and methods leading to the solution of the largest possible number of equations: for practical reasons, we want to be able to solve as many equations as possible. This means that we want to be able to generate the largest number of converging sequences $\{x_n : n \in \mathbb{N}\}$ or, equivalently, that our main interest is the construction of spaces tending to include in the neighborhoods of given functions as many elements as possible (i.e. $V(x;\varepsilon)$ is as large as possible).

Increasing the members of a neighborhood, of solvable equations or of convergent sequences consists of *weakening* a topology: weaker topologies have more convergent sequences, more solvable equations and more elements in a neighborhood. But weaker topologies also introduce weaker regularity and a more complex behavior of the functions, so that a convenient topology establishes a balance between regularity and solvability.

In order to generate convenient topologies, we observe that Mathematics is constructed into a hierarchical way: the starting level is Set Theory, which defines objects that are mere collections and only involves operations such as intersection, union and difference.

In the next level, we consider operations involving the elements of the set itself: groups, rings, vectorial spaces. At this level, we have no concept of "neighbour", "convergence" or "approximation". In order to give a mathematical sense to these expressions, we need a topology.

A convenient way to define a topology consists of the introduction of a tool measuring the distance between two objects – i.e. a metric

$d(u,v)$giving the distance between two elements u,v of the set (this was the work of Fréchet): then, we may define neighborhoods of u by considering the elements v such that $d(u,v) \leq \varepsilon$, where ε is a small parameter: $V(u;\varepsilon) = \{v: d(u,v) \leq \varepsilon\}$.

We may use a simple way to measure distances, which consists of using a norm $\|\blacksquare\|$ measuring the "length" of the elements of V: $d(u,v) = \|u - v\|$, i.e. the distance between u and $v$is measured as the length of the difference between the elements u,v (this was the work of Banach).

The next level consists of introducing a measure of the angle between the elements u,v – this is made by using the scalar product (u,v) and was the work of Von Neumann. We will see that such a structure has more complex neighbourhoods and leads to a larger number of converging sequences. Thus, this last level leads to more converging sequences, more complete neighborhoods and gives at the same time measures of angles, lengths and distances. It corresponds to Hilbert Spaces and furnishes an interesting framework for the definition of convenient topologies.

3.1. Vector spaces

Vector spaces are a fundamental structure for variational methods. The definition of a vector space assumes that two sets were defined: a set Λ of *scalars* and a set V of *vectors*. The set of scalars is usually the set of real numbers $\mathbb{R}$ or the set of complex numbers $\mathbb{C}$ – even if, in general, we may consider any *field,* i.e. any algebraical structure having operations of addition and multiplication with their inverses (subtraction and division). In the sequel, we consider *vector spaces on* $\mathbb{R}$, i.e. we assume that $\Lambda = \mathbb{R}$. However, in a general approach, V may appear as arbitrary, but in our applications V will be a functional space, i.e. the elements of V will be considered as functions. The formal definition of a vector space on $\mathbb{R}$ (resp. Λ) assumes that V possesses two operations: addition $u + v$ of two elements $u,v \in V$ and multiplication λu of an element u of V by a scalar $\lambda \in \mathbb{R}$ (respectively, Λ). Then,

DEFINITION 3.1.– Let $V \neq \emptyset$ be such that:

$$\lambda u + v \in V, \ \forall \lambda \in \mathbb{R}, u, v \in V \, ;$$

$$1v = v, \forall \, v \in V \, .$$

$$u + (v + w) = (u + v) + w, \ \forall u, v, w \in V \, ;$$

$$\alpha u + \beta u = (\alpha + \beta)u, \forall u \in V, \alpha, \beta \in \mathbb{R};$$

$$\alpha(\beta u) = (\alpha\beta)u, \forall u \in V, \alpha, \beta \in \mathbb{R};$$

$$\alpha(u + v) = \alpha u + \alpha v, \ \forall u, v \in V, \alpha, \in \mathbb{R};$$

$$u + v = v + u \, , \ \forall u, v \in V;$$

$$\exists! \ 0 \in V \colon u + 0 = u, \ \forall u \in V \, .$$

$$\forall u \in V \colon \exists! - u \in V \, : \, u + (-u) = 0 \, .$$

Then, V is a vector space on $\mathbb{R}$. ■

We have:

THEOREM 3.1.– Let V be a vector space space on $\mathbb{R}$ and $S \subset V$. S is a vector space on $\mathbb{R}$ if and only if $0 \in S$ and $\lambda u + v \in V, \forall \lambda \in \mathbb{R}, u, v \in S$. In this case, S is said to be a vector subspace of V.

PROOF.– We have$(-1)u + 0 = -u$ (see exercises). Thus,$-u \in S, \forall \, u \in S$. Thus, all the conditions of definition 3.1 are satisfied by S.

EXAMPLE 3.1.– Let us consider $\Omega \subset \mathbb{R}^n$ and $V = \{v \colon \Omega \to \mathbb{R}^k\}$. V is a vector space on $\mathbb{R}$ (see exercises). Thus any subset of V such that $0 \in S$ and $\lambda u + v \in V, \forall \lambda \in \mathbb{R}, u, v \in S$ is also a vector space on $\mathbb{R}$. For instance:

$$S = \left\{ v \in V : \int_\Omega |v(x)|^2 dx < \infty \right\} \; ;$$

$$S = \left\{ v \in V : \int_\Omega \left(|v(x)|^2 + |\nabla v(x)|^2\right) dx < \infty \right\}$$

are both vector subspaces of V. ■

DEFINITION 3.2.– Let $F = \{\varphi_\lambda : \lambda \in \Lambda\} \subset \mathrm{V},\ \Lambda \neq \emptyset,\ \ F \neq \emptyset$. F is linearly *independent* (or *free*) if and only if a finite linear combination $a_1\varphi_{\lambda_1} + a_n\varphi_{\lambda_n}$ of its elements is null if and only if all the coefficients a_i are null:

$$\sum_{i=1}^{n} a_i\varphi_{\lambda_i} = 0 \Leftrightarrow a_i = 0, for\ 1 \leq i \leq n\ .$$

We denote by $[F\,]$ the set of the finite linear combinations of elements of F:

$$[F] = \left\{v = \sum_{i=1}^{n} v_i\varphi_{\lambda_i} : n \in \mathbb{N}^*;\ v_i \in \mathbb{R}\ and\ \lambda_i \in \Lambda, 1 \leq i \leq n\right\}.$$

We say that F is a *generator* (or that F *spans* V) if and only if $[F] = V$. We say that F is a *basis* of V if and only if F is free and Fspans V. The number of elements of the basis is the dimension of V: if the basis contains n elements, we say that V is n-dimensional (ordim $(V)\ =\ n$); if V contains infinitely many elements, we say that V is infinite dimensional (or $\dim(V) =\ +\infty$). ■

Any vector space containing at least one element non null has a basis (see [SOU 10]). In the following, we are particularly interested in countable families F, i.e. in the situations where $\Lambda = \ \mathbb{N}^*$. We will extend the notion of "basis" to the notion of "Hilbert basis", where we *have not* if $[F] = V$, but V is the limit of elements of $[F]$.

3.2. Distance, norm and scalar product

The structure of a vector space does not contain elements for the definition of neighbors, i.e. does not allow us to give a sense to the expressions "close to" and "convergent". In order to give a practical content to these ideas, we must introduce new concepts: the distance between elements of V, the norm of an element of V, the scalar product of elements of V.

3.2.1. *Distance*

The notion of distance allows the definition of neighbors and quantifies as far or as close are two elements.

DEFINITION 3.3.– *Let* $d: V \times V \rightarrow \mathbb{R}$ *be a function such that*

$$d(u,v) \geq 0 \; and \; d(u,v) = d(v,u), \;\; \forall \, u, v \in V \, ;$$

$$d(u,u) = 0 \; andd(u,v) = 0 \;\; \Longrightarrow u = v, \;\; \forall \, u, v \in V \, ;$$

$$d(u,v) \leq d(u,w) + d(w,u), \;\; \forall u, v, w \in V \, ;$$

Then, d is a distance on V. ■

As previously observed, we are not interested in the numeric value of a distance, but in the definition of families of neighbors in order to approximate solutions and study convergence of sequences. For such a purpose, the numeric value of a distance is not useful as information. For instance, if d is a distance on V and $\eta: \mathbb{R} \rightarrow \mathbb{R}$ is a strictly increasing application such that $\eta(0) = 0$ and $\eta(a+b) \leq \eta(a) + \eta(b)$, then $d_\eta(u,v) = \eta\big(d(u,v)\big)$ is a distance on V (see exercises). Thus, we may give an arbitrary value to the distance between two elements of V. Moreover, distances may be pathological: for instance, the definition

$$d(u,v) = \begin{cases} 0, if \; u = v; \\ 1, if \; u \neq v\,. \end{cases}$$

corresponds to a distance. This distance is not useful, since the small neigborhoods of u contain a single element, which is u itself: $d(u,v) \leq \varepsilon < 1 \Longrightarrow v = u$. Thus, $V(u;\varepsilon) = \{u\}$ and the associated topology contains just a few converging sequences and neighborhoods are poor in elements.

Distances allow the definition of limits and continuity. Since our presentation focuses on Hilbert spaces, these notions are introduced later, when considering norms.

3.2.2. *Norm*

The notion of norm allows the definition of the length of a vector and, at the same time, may be used in order to define a distance. In Mechanics, norms may also be interpreted in terms of internal energy of a system.

DEFINITION 3.4.– Let $\| \blacksquare \|: V \to \mathbb{R}$ be a function such that

$$\| v \| \geq 0\,,\ \forall\, v \in V\,;$$

$$\| v \| = 0 \Rightarrow v = 0\,, \forall\, v \in V\,;$$

$$\| \lambda v \| = |\lambda| \| v \|, \lambda \in \mathbb{R},\ v \in V\,;$$

$$\| u + v \| \leq \| u\| + \| v \|,\ \forall u, v \in V\,;$$

Then, $\| \blacksquare \|$ is a norm on V. ■

Classical examples of norms on $V = \mathbb{R}^n$ are $\big(x = (x_1, \dots, x_n)\big)$:

$$\|x\|_p = (\textstyle\sum_{i=1}^{n} |x_i|^p)^{\frac{1}{p}}\ ,\ \|x\|_\infty = \max\{|x_i| : 1 \leq i \leq n\}\,.$$

THEOREM 3.2.– Let $\| \blacksquare \|$ be a norm on V. Then $d(u, v) = \| u - v \|$is a distance on V. ■

PROOF.– Immediate. ■

When considering functional spaces, such as subspaces of $V = \{v: \Omega \subset \mathbb{R}^n \to \mathbb{R}^k\}$ continuity may be an essential assumption in order to define norms. For instance, let us consider $p \in \mathbb{N}^*$ and

$$\|v\|_p = \left(\int_\Omega |v(x)|^p\, dx\right)^{\frac{1}{p}}.$$

Let $S = \{v \in V: v\ is\ continuous\ on\ \Omega$ and $\|v\| < \infty\}$, then $\|\blacksquare\|$ is a norm on S. The assumption of continuity is used in order to show that

$$\|v\|_p = 0 \Rightarrow v = 0\,.$$

Indeed, for v continuous on Ω:

$$\int_\Omega |v(x)|^p \, dx = 0 \underbrace{\Rightarrow} v(x) = 0, \forall x \in \Omega .$$

If $x \in \Omega$ is such that $|v(x)|^p > 0$, the, by continuity, there is $\varepsilon > 0$ such that $|v(y)|^p > 0$ on $B_\varepsilon(x) = \{y \in \Omega : \|y - x\| \leq \varepsilon\}$ (Corollary 3.1). Then,

$$0 = \int_\Omega |v(x)|^p \, dx \geq \int_{B_\varepsilon} |v(x)|^p \, dx > 0$$

and we have $0 > 0$, what is a contradiction. The classical counter-example establishing that continuity must be assumed is furnished by a function v such that $v(x) = 0$ everywhere on Ω, except on a finite or enumerable number of points. For instance,

$$\Omega = (0,1), v(x) = 0, if\ x \notin \mathbb{Q}\ ; v(x) = 1, if\ x \in \mathbb{Q} .$$

Here, $\| v \| = 0$, but $v \neq 0$. In the following, we show how to we lift this restriction by using equivalence relations that generate the functional space $L^p(\Omega)$ – there will be a price to pay: it will be the loss of the concept of value at a point: the individual value $v\ (x)$ of v in a point x has no meaning for elements of $L^p(\Omega)$.

As an alternative norm, we may also consider

$$\|v\|_\infty = sup\{|v(x)| : x \in \Omega\} .$$

In this case, the assumption of continuity is not necessary. Indeed,

$$\|v\|_\infty = 0 \Longrightarrow |v(x)| \leq 0, \forall x \in \Omega \Longrightarrow v(x) = 0, \forall x \in \Omega.$$

3.2.3. *Scalar product*

The notion of scalar product allows the definition of angles between vectors and, at same time, may be used in order to define a norm. In Mechanics, scalar products may also be interpreted as the virtual work of internal efforts.

DEFINITION 3.5.– Let $(\blacksquare, \blacksquare): V \times V \to \mathbb{R}$ be a function such that

$$(v, v) \geq 0\,,\ \forall\, v \in V\,;$$

$$(v, v) = 0 \Longrightarrow v = 0\,, \forall\, v \in V\,;$$

$$(u, v) = (v, u),\ \forall\, u, v \in V\,;$$

$$(u + \lambda v, w) = (u, w) + \lambda(v, w),\ \forall\, u, v, w \in V\,, \boldsymbol{\lambda} \in \mathbb{R}\,;$$

Then, $(\blacksquare, \blacksquare)$ is a scalar product on V. ■

A classical example of scalar product on $V = \mathbb{R}^n$ is$\big(x = (x_1, \dots, x_n), y = (y_1, \dots, y_n)\big)$:

$$x.y = \textstyle\sum_{i=1}^{n} x_i y_i\,.$$

THEOREM 3.3.– Let $(\blacksquare, \blacksquare): V \times V \to \mathbb{R}$ be a scalar product on V. Then, $\forall u, v \in V, \lambda \in \mathbb{R}$,

i) $(u + \lambda v, u + \lambda v\,) = (u, u) + 2\lambda(u, v) + \lambda^2(v, v)$;

ii) $(u + v, u + v) + \ (u - v, u - v) = 2[(u, u) + (v, v)]$;

iii) $|(u, v)| \leq \sqrt{(u, u)}\sqrt{(v, v)}$

iv) $\sqrt{(u - v, u - v)} \leq \sqrt{(u, u)} + \sqrt{(v, v)}$

v) $\sqrt{(u + v, u + v)} \leq \sqrt{(u, u)} + \sqrt{(v, v)}$.

vi) $\left|\sqrt{(u, u)} - \sqrt{(v, v)}\right| \leq \sqrt{(u - v, u - v)}$ ■

PROOF.– We have

$$(u + \lambda v, u + \lambda v\,) = \ (u, u + \lambda v) + \lambda(v, u + \lambda v),$$

so that,

$$(u + \lambda v, u + \lambda v\,) = \ (u, u) + \lambda(u, v) + \lambda(v, u) + \lambda^2(v, v)$$

and we have (i). Thus:

$$\lambda = 1 : (u+v, u+v) = (u,u) + 2(u,v) + (v,v)$$

$$\lambda = -1 : (u-v, u-v) = (u,u) - 2(u,v) + (v,v)$$

By adding these two equalities, we have (ii).

Let $f(\lambda) = (u + \lambda v, u + \lambda v)$. We have $f(\lambda) \geq 0, \forall \lambda \geq 0$. Thus,

$$[2(u,v)]^2 \leq 4(u,u)(v,v)$$

and we have (iii).

Since

$$(u+v, u+v) = (u,u) + 2(u,v) + (v,v) \leq (u,u) + 2\sqrt{(u,u)}\sqrt{(v,v)} + (v,v),$$

we have

$$(u+v, u+v) \leq \left(\sqrt{(u,u)} + \sqrt{(v,v)}\right)^2,$$

what implies (iv). Replacing v by $-v$, we obtain (v).

Finally, we observe that

$$\left(\sqrt{(u,u)} - \sqrt{(v,v)}\right)^2 = (u,u) - 2\sqrt{(u,u)}\sqrt{(v,v)} + (v,v),$$

so that (from (iii))

$$\left(\sqrt{(u,u)} - \sqrt{(v,v)}\right)^2 \leq (u,u) - 2(u,v) + (v,v) = (u-v, u-v)$$

and we have (vi). ■

THEOREM 3.4.– Let $(\blacksquare, \blacksquare): V \times V \to \mathbb{R}$ be a scalar product on V. Then, $\|u\| = \sqrt{(u,u)}$ is a norm on V. ■

PROOF.– We have:

$$\|u\| = \sqrt{(u,u)} \geq 0 \quad and \quad \|u\| = 0 \Longrightarrow (u,u) = 0 \Longrightarrow u = 0\,;$$

$$\|\lambda u\| = \sqrt{(\lambda u, \lambda u)} = \sqrt{\lambda^2 (u,u)} = |\lambda| \|u\|\,.$$

Finally, 3.2.5. (v) implies that $\|u + v\| \leq \|u\| + \|v\|$.

Functional spaces, such as subspaces of $V = \{\boldsymbol{v}: \Omega \subset \mathbb{R}^n \to \mathbb{R}^k\}$ generally involve continuity assumptions in order to define scalar products. Analogously to norms, the assumption of continuity is necessary in order to verify that

$$(v,v) = 0 \Longrightarrow v = 0\,.$$

For instance, if

$$(u,v) = \int_\Omega u(x).v(x)dx$$

and $S = \{v \in V: v \; is \; continuous \; on \; \Omega$ and $(v,v) < \infty\}$, then $(\blacksquare, \blacksquare)$ defines a scalar product on S. Analagously to the case of norms (use $p = 2$),

$$(v,v) = 0 \Longrightarrow v(x) = 0, \forall x \in \Omega\,.$$

In the following, continuity assumption is lifted by using functional space $L^2(\Omega)$. As previously observed, the consequence is the loss of the meaning of point values $v(x)$.

This limitation may be avoided in particular situations where the scalar product implies the continuity. For instance, let us consider,

$$(u,v) = \int_\Omega [u(x).v(x) + \nabla u(x): \nabla v(x)]dx,$$

where $A: B = \sum_{i,j} A_{ij} B_{ij}$. In this case, the continuity of the elements u, v yields from the existence of the gradients ∇u, ∇v and we may give a meaning to the punctual values $u(x)$ and (x) .

DEFINITION 3.6.– The angle $\theta(u,v)$ between the vectors u and v is

$$\theta = \cos^{-1}\left(\frac{(u,v)}{\|u\|\|v\|}\right).$$

u and v are *orthogonal* if and only if $(u,v) = 0$ Notation: $u \perp v$. For $S \subset V$, we denote by $S^{\perp}$ the set of the elements of V which are orthogonal to all the elements of S:

$$S^{\perp} = \{u \in V : (u,v) = 0, \forall v \in S\}.$$ ■

DEFINITION 3.7.– Let $F = \{\varphi_n : n \in \Lambda \subset \mathbb{N}^*\}$ be a countable family. We say that F is an *orthogonal family* if and only if

$$(\varphi_i, \varphi_j) = 0\,, for\ i \neq j\,, i, j \in \Lambda.$$

We say that F is an *orthonormed family* if and only if it is orthogonal and, *moreover*,

$$(\varphi_i, \varphi_i) = 1,\ \forall i \in \Lambda\ .$$ ■

A countable free family $G = \{\phi_n : n \in \Lambda \subset \mathbb{N}^*\}$ may be transformed into an orthonomal family $F = \{\varphi_n : n \in \Lambda \subset \mathbb{N}^*\}$ by the *Gram–Schmidt procedure*, which has been created for the solution of least-square linear problems. It has been proposed by Jorgen Pedersen Gram [GRA 83] and formalized by Ehrlich Schmidt [SCM 07] (although early uses by Pierre Simon de Laplace have been noticed [LEO 13]). The Gram–Schmidt procedure reads as:

1) Set

$$\varphi_0 = \frac{\phi_0}{\|\phi_0\|},$$

2) For $k > 0$, set

$$\varphi_k = \frac{\psi_k}{\|\psi_k\|},\quad \psi_k = \phi_k - \sum_{i=0}^{k-1}(\phi_k, \varphi_i)\varphi_i.$$

In this case,

$$(\varphi_i, \varphi_j) = 0\,, for\ i \neq j\,, i, j \in \Lambda;\ (\varphi_i, \varphi_i) = 1,\ \forall i \in \Lambda\,.$$

The practical implementation requests the evaluation of scalar products and norms.

As an example, let us consider $\Omega = (-1,1)$ and

$$(u, v) = \int_\Omega u(x)v(x)dx$$

The family $\phi_n(x) = x^n$ is not orthonormal, since (*mod(p,q)* denotes the remainder of the integer division *p/q*)

$$(\phi_i, \phi_j) = \int_{-1}^{1} x^{i+j} dx = \frac{1 + (-1)^{mod(i+j,2)}}{i+j+1} = \begin{cases} 0, i+j\ impair \\ \frac{1}{i+j+1}, i+j\ pair \end{cases}.$$

If the Gram–Schmidt procedure is applied, we have:

$$\varphi_0 = \frac{1}{\sqrt{2}}\,, \varphi_1 = \sqrt{\frac{3}{2}}\,x\,, \varphi_2 = \sqrt{\frac{45}{8}}\left(x^2 - \frac{1}{3}\right)\,, \ldots$$

3.2.4. *Cartesian products of vector spaces*

In practice, we often face situations where $V = V_1 \times \ldots V_d$ is the Cartesian product of vector spaces. In this case, elements $u, v \in \mathbb{V}$ are given by:

$$u = (u_1, \ldots, u_d), v = (v_1, \ldots, v_d),\ u_i, v_i \in V_i\ .$$

In such a situation:

– If the distance on V_i is d_i, then the distance d_V on V may be defined as:

$$d_V(u, v) = \sum_{i=1}^{d} d_i(u_i, v_i)$$

– Analogously, if the norm on V_i is $\|\blacksquare\|_i$, then the norm $\|\blacksquare\|_V$ on V may be defined as:

$$\|u\|_V = \sum_{i=1}^{d}\|u_i\|_i$$

– Finally, if the scalar product on V_i is $(\blacksquare,\blacksquare)_i$, then the scalar product $(\blacksquare,\blacksquare)_V$ on V may be defined as:

$$(u,v)_V = \sum_{i=1}^{d}(u_i,v_i)_i$$

When $V_1 = \cdots = V_d = \mathbb{V}$, we have $V = \mathbb{V}^d$ and

$$d_V(u,v) = \sum_{i=1}^{d} d_{\mathbb{V}}(u_i,v_i)\ ,\|u\|_V = \sum_{i=1}^{d}\|u_i\|_{\mathbb{V}}\ ,(u,v)_V = \sum_{i=1}^{d}(u_i,v_i)_{\mathbb{V}}$$

3.2.5. *A Matlab® class for scalar products and norms*

Below we give an example of a Matlab® class for the evaluation of scalar products and norms. The class contains two scalar products:

$$(u,v) = \int_\Omega u.vdx \quad and \quad (u,v) = \int_\Omega (u.v + \nabla u.\nabla v)dx\ ,$$

Evaluated by methods `sp0` and `sp1`, respectively. The associated norms are evaluated by methods `n0` and `n1`, respectively. For each method, the functions are furnished by structured data. In both the cases, the structure has as properties `dim`, `dimx` and `component`. `dim` and `dimx` are integer values defining the dimension of the function and the dimension of its argument x. `component` is a cell array of length `dim` containing the values of the components and their gradients: each element of `component` is a structure having properties `value` and `grad`. When using `'subprogram'`, `value` and `grad` are Matlab subprograms (*i.e.* anonymous functions). For `'table'`, `value` is an array of values and `grad` is a cell arrays containing tables of values: `u.component{i}.value` is a table of values of component u_i and `u.component{i}.grad{j}` is a table of values of $\partial u_i/\partial x_j$.

For `'subprogram'`, parameter `xl` contains the limits of the region, analogous to those used in class `subprogram_integration`. For `'table'`, `xl` is a structure containing the points of integration, analogously to those used in class `riemann`.

```
classdef scalar_product
%       methods (Static, Access = private)
%       end
    methods (Static)
        function val = sp0(u,v, xl, met)
            switch met
                case 'subprogram'
                    val = 0;
                    for i = 1: u.dim
   f = @(x) dot(u.component{i}.value(x),
v.component{i}.value(x));
                val = val +
subprogram_integration(f,xl).result;
                    end;
                case 'table'
                    data.points = xl;
                    val = 0;
                    for i = 1: u.dim
                        uu = u.component{i}.value;
                        vv = v.component{i}.value;
                        data.values = uu.*vv;
                        val = val + riemann.trpzd(data);
                    end;
                otherwise
                    val = [];
            end
        end
        function val = n0(u,xl, met)
            v = scalar_product.sp0(u,u, xl, met);
            val = sqrt(v);
        end
        function val = sp1(u,v, xl, met)
            switch met
                case 'subprogram'
                    v0 = 0;
                    v1 = 0;
                    for i = 1: u.dim
  f = @(x) dot(u.component{i}.value(x),
v.component{i}.value(x));
```

```
                        v0 = v0 +
subprogram_integration(f,xl).result;
    f = @(x) dot(u.component{i}.grad(x),
v.component{i}.grad(x));
                        v1 = v1 +
subprogram_integration(f,xl).result;
                    end;
                    val = v0 + v1;
                case 'table'
                    data.points = xl;
                    val = 0;
                    for i = 1: u.dim
                        uu = u.component{i}.value;
                        vv = v.component{i}.value;
                        data.values = uu.*vv;
                        val = val + riemann.trpzd(data);
                    end;
                    for i = 1: length(u)
                        for j = 1: u.dimx
                            uu = u.component{i}.grad{j};
                            vv = v.component{i}.grad{j};
                            data.values = uu.*vv;
                            val = val +
riemann.trpzd(data);
                        end;
                    end;
                otherwise
                    val - [];
            end
        end
        function val = n1(u,xl, met)
            v = scalar_product.sp1(u,u, xl, met);
            val = sqrt(v);
        end
    end
end
```

Program 3.1. *A class for the determination of partial solutions of linear systems*

EXAMPLE 3.2.– Let us evaluate both the scalar products and norms for the functions $u, v: (0,1) \to \mathbb{R}$ given by $u(x) = x, v(x) = x^2$. The functions may be defined as follows:

```
function v = u1()
v.dim = 1;
v.dimx = 1;
v. component =
cell(v.dim,1);
v.component{1}.value =
@(x) val(x);
v.component{1}.grad = @(x)
gradval(x);
end
%
function v = val(x)
v = x;
return;
end
%
function gv = gradval(x)
gv = 1;
return;
end
```

```
function v = v1()
v.dim = 1;
v.dimx = 1;
v. component =
cell(v.dim,1);
v.component{1}.value =
@(x) val(x);
v.component{1}.grad =
@(x) gradval(x);
end
%
function v = val(x)
v = x^2;
return;
end
%
function gv = gradval(x)
gv = 2*x;
return;
end
```

Program 3.2. *Definition of u and v in example 3.2*

The code:

```
xlim.lower.x = 0;
xlim.upper.x = 1;
xlim.dim = 1;
u = u1();
v = v1();
sp = scalar_product();
sp0uv = sp.sp0(u,v, xlim,'subprogram');
n0u = sp.n0(u,xlim,'subprogram');
n0v = sp.n0(v,xlim,'subprogram');
sp1uv = sp.sp1(u,v, xlim,'subprogram');
n1u = sp.n1(u,xlim,'subprogram');
n1v = sp.n1(v,xlim,'subprogram');
```

produces the results `sp0uv = 0.25`, `n0u = 0.57735`, `n0v = 0.44721`, `sp1uv = 1.25`, `n1u = 1.1547`, `n1v = 1.2383`.

The exact values are, respectively, $\frac{1}{4} = 0.25$, $\frac{1}{\sqrt{3}} \approx 0.5774$, $\frac{1}{\sqrt{5}} \approx 0.4472$, $\frac{5}{4} = 1.25$, $\sqrt{\frac{4}{3}} \approx 1.1547$, $\sqrt{\frac{23}{15}} \approx 1.2383$.

We generate cell arrays containing tables of values by using:

```
x = 0:0.01:1;
p.x = x;
p.dim = 1;
U.dim = 1;
U.dimx = 1;
U.component = cell(u.dim,1);
U.component{1}.value =
spam.partition(u.component{1}.value,p);
U.component{1}.grad = cell(u.dim,1);
U.component{1}.grad{1} =
spam.partition(u.component{1}.grad,p);
V.dim = 1;
V.dimx = 1;
V.component = cell(v.dim,1);
V.component{1}.value =
spam.partition(v.component{1}.value,p);
V.component{1}.grad = cell(v.dim,1);
V.component{1}.grad{1} =
spam.partition(v.component{1}.grad,p);
```

Then, the code:

```
sp0UV = sp.sp0(U,V,p,'table');
n0U = sp.n0(U,p,'table');
n0V = sp.n0(V,p,'table');
sp1UV = sp.sp1(U,V,p,'table');
n1U = sp.n1(U,p,'table');
n1V = sp.n1(V,p,'table');
```

produces the results `sp0UV = 0.25003`, `n0U = 0.57736`, `n0V = 0.44725`, `sp1UV = 1.25`, `n1U = 1.1547`, `n1V = 1.2383`. ■

EXAMPLE 3.3.– Let us evaluate both the scalar products and norms for the functions $u, v: (0,1) \times (0,2) \to \mathbb{R}$ given by $u(x) = x_1 x_2, v(x) = x_1^2 + x_2$. The functions may be defined as follows:

``function v = u2()``
``v.dim = 1;``
``v.dimx = 2;``
``v. component =``
``cell(v.dim,1);``
``v.component{1}.value =``
``@(x) val(x);``
``v.component{1}.grad =``
``@(x) gradval(x);``
``end``
``%``
``function v = val(x)``
``v = x(1)*x(2);``
``return;``
``end``
``%``
``function gv = gradval(x)``
``gv = [x(2) x(1)];``
``return;``
``end``
	``function v = v2()``
	``v.dim = 1;``
	``v.dimx = 2;``
	``v. component =``
	``cell(v.dim,1);``
	``v.component{1}.value =``
	``@(x) val(x);``
	``v.component{1}.grad = @(x)``
	``gradval(x);``
	``end``
	``%``
	``function v = val(x)``
	``v = x(1)^2 + x(2);``
	``return;``
	``end``
	``%``
	``function gv = gradval(x)``
	``gv = [2*x(1) 1];``
	``return;``
	``end``

Program 3.3. *Definition of u and v in example 3.3*

The code:

```
xlim.lower.x = 0;
xlim.upper.x = 1;
xlim.lower.y = 0;
xlim.upper.y = 2;
xlim.dim = 2;
u = u2();
v = v2();
sp = scalar_product();
sp0uv = sp.sp0(u,v, xlim,'subprogram');
n0u = sp.n0(u,xlim,'subprogram');
n0v = sp.n0(v,xlim,'subprogram');
sp1uv = sp.sp1(u,v, xlim,'subprogram');
n1u = sp.n1(u,xlim,'subprogram');
n1v = sp.n1(v,xlim,'subprogram');
```

produces the results `sp0uv = 1.8333, n0u = 0.94281, n0v = 2.0976, sp1uv = 4.8333, n1u = 2.0548, n1v = 3.0111.` The exact values are, respectively, $\frac{11}{6} \approx 1.8333$, $\frac{2\sqrt{2}}{3} \approx 0.9428$, $\sqrt{\frac{22}{5}} \approx 2.0976$, $\frac{29}{6} \approx 4.8333$, $\frac{\sqrt{38}}{3} \approx 2.0548$, $2\sqrt{\frac{34}{15}} \approx 3.0111$.

We use the function

```
function v = select_index(ind,u,x)
aux = u(x);
v = aux(ind);
return;
end
```

and we generate cell arrays containing tables of values by the code below:

```
x = 0:0.01:1;
y = 0:0.01:2;
p.x = x;
p.y = y;
p.dim  = 2;
U.dim = 1;
U.dimx = 2;
U.component = cell(u.dim,1);
U.component{1}.value =
spam.partition(u.component{1}.value,p);
U.component{1}.grad = cell(u.dim,1);
V.dim = 1;
V.dimx = 2;
V.component = cell(u.dim,1);
V.component{1}.value =
spam.partition(v.component{1}.value,p);
V.component{1}.grad = cell(u.dim,1);
for i = 1: u.dim
    for j = 1: p.dim
        du = @(x) select_index(j,u.component{i}.grad,x);
        U.component{i}.grad{j} = spam.partition(du,p);
        dv = @(x) select_index(j,v.component{i}.grad,x);
        V.component{i}.grad{j} = spam.partition(dv,p);
    end;
end
sp0UV = sp.sp0(U,V,p,'table');
n0U = sp.n0(U,p,'table');
n0V = sp.n0(V,p,'table');
sp1UV = sp.sp1(U,V,p,'table');
n1U = sp.n1(U,p,'table');
n1V = sp.n1(V,p,'table');
```

It produces the results `sp0UV = 1.8334, n0U = 0.94284, n0V = 2.0977, sp1UV = 4.8334, n1U = 2.0548, n1V = 3.0111.` ■

3.2.6. *A Matlab® class for Gram–Schmidt orthonormalization*

The class defined in the preceding section may be used in order to transform a general family $G = \{\phi_i: 1 \leq i \leq n\}$ into an orthonomal family $F = \{\varphi_i: 1 \leq i \leq n\}$ by the Gram-Schmidt procedure.

Let us assume that the family G is defined by a cell array `basis` such that `basis{i}` defines ϕ_i according to the standard defined in section 3.2.5 : `basis{i}` is a structure having properties `dim`, `dimx` and `component`. `component` is a structure having fields `value` and `grad`. The class given in program 3.4 generates a cell array `hilbert_basis` defining the family F, resulting from the application of the Gram-Schmidt's procedure to G. The class uses subprograms `sp_space` and `n_space`, which are assumed to evaluate the scalar product and the norm.

```
classdef gram_schmidt
    methods (Static)
     function hilbert_basis =
subprograms(basis,sp_space,n_space)
            n = length(basis);
            hilbert_basis = cell(n,1);
            nf = n_space(basis{1});
            nd = basis{1}.dim;
            nx = basis{1}.dimx;
            hilbert_basis{1}.dim = nd;
            hilbert_basis{1}.dimx = nx;
            hilbert_basis{1}.component = cell(nd,1);
            for k = 1: nd
                hilbert_basis{1}.component{k}.value =
@(x) basis{1}.component{k}.value(x)/nf;
                hilbert_basis{1}.component{k}.grad =
@(x) basis{k}.component{k}.grad(x)/nf;
            end;
            for i = 2: n
                f = cell(nd,1);
                for k = 1: nd
                    f{k}.value =
basis{i}.component{k}.value;
```

```
                    f{k}.grad =
basis{i}.component{k}.grad;
                end;
                for j = 1: i-1
                    sp =
sp_space(basis{i},hilbert_basis{j});
                    for k = 1: nd
                        f{k}.value = @(x) f{k}.value(x)
- sp*hilbert_basis{j}.component{k}.value(x);
                        f{k}.grad = @(x) f{k}.grad(x) -
sp*hilbert_basis{j}.component{k}.grad(x);
                    end;
                end;
                ff.dim = nd;
                ff.dimx = nx;
                ff.component = f;
                nf = n_space(ff);
                hilbert_basis{i}.dim = basis{i}.dim;
                hilbert_basis{i}.dimx = basis{i}.dimx;
                hilbert_basis{i}.component =
cell(nd,1);
                for k = 1: nd
                    hilbert_basis{i}.component{k}.value
= @(x) ff.component{k}.value(x)/nf;
                    hilbert_basis{i}.component{k}.grad
= @(x) ff.component{k}.grad(x)/nf;
                end;
            end;
            return;
        end
        function hilbert_basis =
tables(basis,sp_space,n_space)
            n = length(basis);
            hilbert_basis = cell(n,1);
            nf = n_space(basis{1});
            nd = basis{1}.dim;
            nx = basis{1}.dimx;
            hilbert_basis{1}.dim = nd;
            hilbert_basis{1}.dimx = nx;
            hilbert_basis{1}.component = cell(nd,1);
            for i = 1: nd
                hilbert_basis{1}.component{i}.value =
basis{1}.component{i}.value/nf;
                hilbert_basis{1}.component{i}.grad =
cell(nx,1);
                for j = 1: nx

hilbert_basis{1}.component{i}.grad{j} =
basis{1}.component{i}.grad{j}/nf;
                end;
            end;
            for k1 = 2: n
                f = cell(nd,1);
                for k = 1: nd
                    f{k}.value =
```

```
basis{k1}.component{k}.value;
                    for j = 1: nx
                        f{k}.grad{j} =
basis{k1}.component{k}.grad{j};
                    end;
                end;
                for k2 = 1: k1-1
                    sp =
sp_space(basis{k1},hilbert_basis{k2});
                    for k = 1: nd
                        f{k}.value = f{k}.value -
sp*hilbert_basis{k2}.component{k}.value;
                        for j = 1: nx
                            f{k}.grad{j} = f{k}.grad{j}
- sp*hilbert_basis{k2}.component{k}.grad{j};
                        end;
                    end;
                end;
                ff.dim = nd;
                ff.dimx = nx;
                ff.component = cell(nd,1);
                for i = 1: nd
                    ff.component{i} = f{i};
                end;
                nf = n_space(ff);
                hilbert_basis{k1}.dim = nd;
                hilbert_basis{k1}.dimx = nx;
                hilbert_basis{k1}.component =
cell(nd,1);
                for i = 1: nd
                   hilbert_basis{k1}.component{i}.value
                     = ff.component{i}.value/nf;
                    hilbert_basis{k1}.component{i}.grad
                     = cell(nx,1);
                    for j = 1: nx
                        hilbert_basis{k1}.component{i}.
                        grad{j} = ff.component{i}.
                        grad{j}/nf;
                    end;
                end;
            end;
            return;
        end
    end
end
```

Program 3.4. *A class for the Gram-Schmidt orthogonalization*

EXAMPLE 3.4.– Let us consider the family $\phi_i(x) = x^{i-1}$.

```
function v = u3(i)
v.dim = 1;
v.dimx = 1;
v. component = cell(v.dim,1);
v.component{1}.value = @(x)
val(x,i);
v.component{1}.grad = @(x)
gradval(x,i);
end
%
function v = val(x,i)
if i == 1
    v = 1;
elseif i == 2
    v = x;
else
    v = x^(i-1);
end;
return;
end
%
function gv = gradval(x,i)
if i == 1
    gv = 0;
elseif i == 2
    gv = 1;
elseif i == 3
    gv = 2*x;
else
    gv = (i-1)*x^(i-2);
end;
return;
end
```

Program 3.5. *Definition of the family G in example 3.4*

The family is created by the code

```
basis = cell(n,1);
for i = 1: n
    basis{i} = u3(i);
end;
```

We apply the procedure on $(-1, 1)$ with the scalar product sp0:

```
xlim.lower.x = -1;
```

```
xlim.upper.x = 1;
xlim.dim = 1;
sp_space = @(u,v)
scalar_product.sp0(u,v,xlim,'subprogram');
n_space = @(u)
scalar_product.n0(u,xlim,'subprogram');
hb =
gram_schmidt.subprograms(basis,sp_space,n_space);
```

The results are shown in Figure 3.1 below:

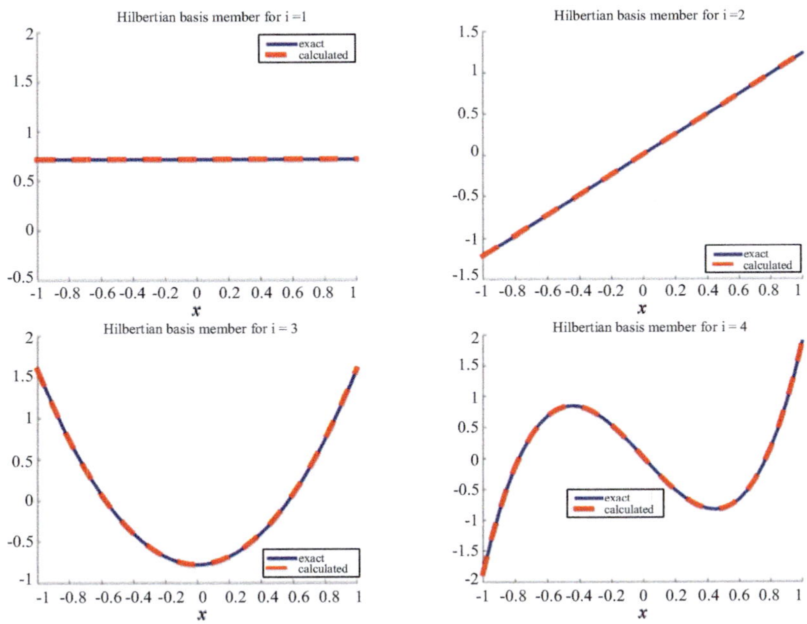

Figure 3.1. *Results for example 3.4 (using sp0 and 'subprograms'). For a color version of the figure, see www.iste.co.uk/souzadecursi/variational.zip*

We generate discrete data associated with these functions by using the code:

```
x = -1:0.02:1;
p.x = x;
p.dim = 1;
tb = cell(n,1);
```

```
for i = 1: n
      u = u3(i);
      U.dim = 1;
U.dimx = 1;
      U.component = cell(u.dim,1);
      U.component{1}.value =
spam.partition(u.component{1}.value,p);
      U.component{1}.grad = cell(u.dim,1);
      U.component{1}.grad{1} =
spam.partition(u.component{1}.grad,p);
      tb{i} = U;
end;
```

Then, the commands

```
sp_space = @(u,v)
scalar_product.sp0(u,v,p,'table');
n_space = @(u) scalar_product.n0(u,p,'table');
hb = gram_schmidt.tables(tb,sp_space,n_space);
```

furnish the result shown in Figure 3.2 below:

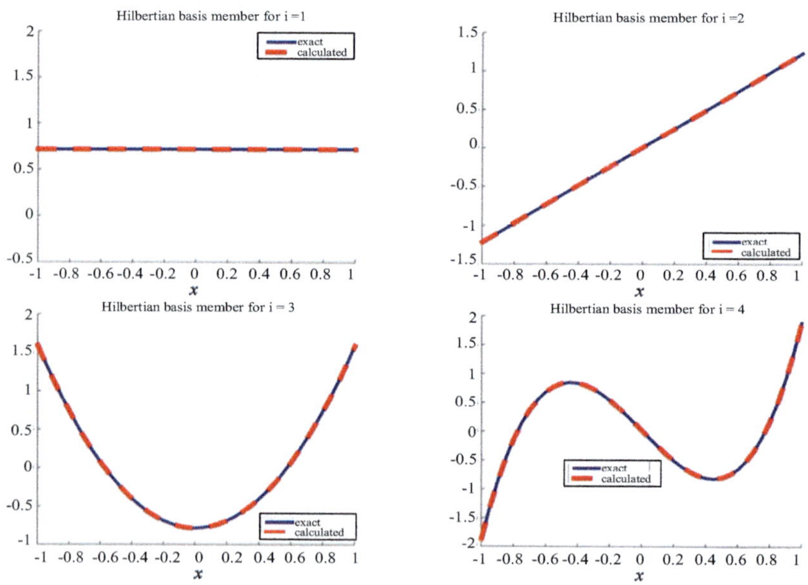

Figure 3.2. *Results for example 3.4 (using sp0 and 'tables'). For a color version of the figure, see www.iste.co.uk/souzadecursi/variational.zip*

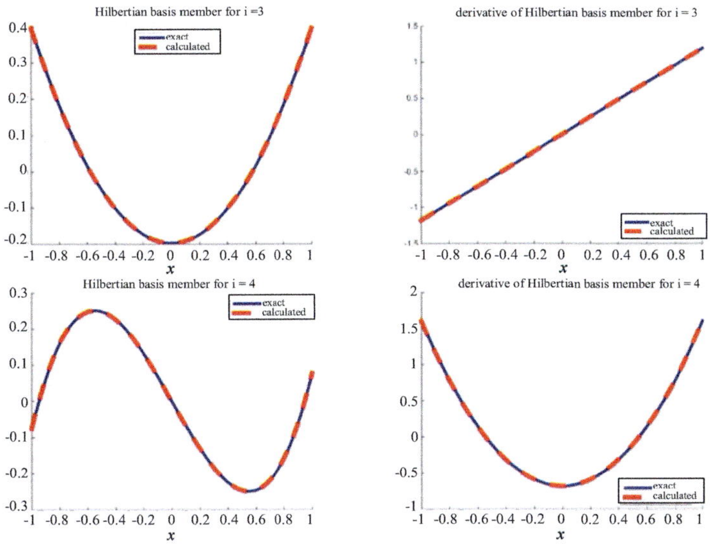

Figure 3.3. *Results for example 3.4 using sp1 and 'subprograms'. For a color version of the figure, see www.iste.co.uk/souzadecursi/variational.zip*

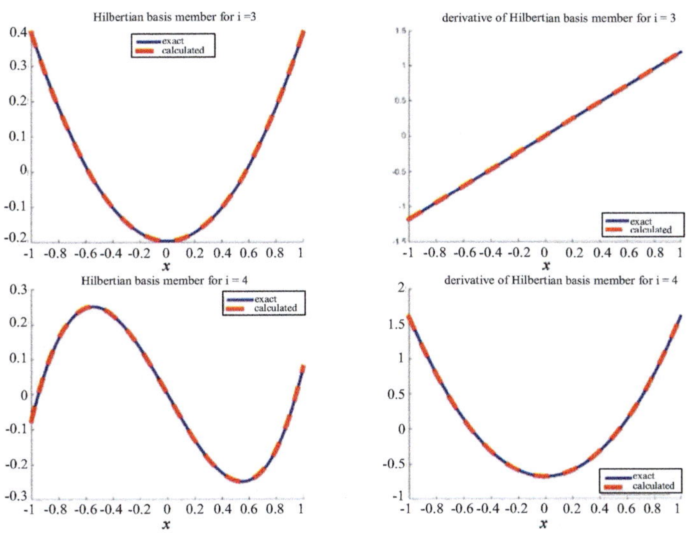

Figure 3.4. *Results for example 3.4 using sp1 and 'tables'. For a color version of the figure, see www.iste.co.uk/souzadecursi/variational.zip*

When the procedure is applied with the scalar product sp1, the results differ. For instance, we obtain the results exhibited in Figures 3.3 and 3.4:

As an alternative, we may define the basis by using the classes previously introduced (section 2.2.6). For instance,

```
zero = 1e-3;
xmin = -1;
xmax = 1;
degree = 3;
bp = polynomial_basis(zero,xmin,xmax,degree);
%
basis = cell(n,1);
for i = 1: n
    basis{i}.dim = 1;
    basis{i}.dimx = 1;
    basis{i}.component = cell(1,1);
    basis{i}.component{1}.value = @(x) bp.value(i,x);
    basis{i}.component{1}.grad = @(x) bp.d1(i,x);
end;
```

This definition leads to the same results. ■

3.3. Continuous maps

As previously indicated, distances allow the introduction of the concept of continuity of a function. For a distance defined by a norm (as in theorem 3.2),

DEFINITION 3.8.– Let V, W be two vector spaces having norms$\|\blacksquare\|_v$, $\|\blacksquare\|_w$, respectively. Let $T: V \to W$ be a function. We say that T is continuous at the point $u \in V$ if and only if

$$\forall \varepsilon > 0 : \exists \delta(u, \varepsilon) \text{ } such \text{ } that \text{ } \|u - v\|_V \leq \delta(u, \varepsilon) \Longrightarrow \|T(u) - T(v)\|_W \leq \varepsilon .$$

We say that T is continuous on $A \subset V$ if and only if $\forall u \in A : T$ is continuous at u.

We say that T is uniformly continuous on $A \subset V$ if and only if δ is independent of u on A: $\delta(u, \varepsilon) = \delta(\varepsilon), \ \forall u \in A$. ■

An example of uniformly continuous function is $T: V \to \mathbb{R}$ given by $T(u) = \|u\|$: we have $|\|u\| - \|v\|| \leq \|u - v\|$, so that $\delta(u, \varepsilon) = \varepsilon$.

Continuous functions have remarkable and useful properties. For instance,

PROPOSITION 3.1.– Let V be a vector space having norm $\|\blacksquare\|$. Let $T: V \to \mathbb{R}$ be continuous at the point $u \in V$. If $T(u) > 0$ (resp. $(u) < 0$), then there exists $\delta > 0$ such that

$$\|u - v\| \leq \delta \Longrightarrow T(v) > 0 \ (resp. \ T(v) < 0). \qquad \blacksquare$$

PROOF.– Take $\varepsilon = |T(u)|/2$ and $\delta = \delta(u, \varepsilon)$. From definition 3.5,

$$\|u - v\| \leq \delta \Longrightarrow T(u) - \frac{|T(u)|}{2} \leq T(v) \leq T(u) + \frac{|T(u)|}{2}$$

and we have the result stated. ■

COROLLARY 3.1.– Let V, W be two vector spaces having norms $\|\blacksquare\|_v$, $\|\blacksquare\|_w$, respectively. Let $F: V \to W$ be continuous at the point $u \in V$. If $F(u) \neq 0$ then there exists $\delta > 0$ such that

$$\|u - v\|_V \leq \delta \Longrightarrow F(v) \neq 0 \ and \ \|F(v)\|_W > 0\,. \qquad \blacksquare$$

PROOF.– Since F is continuous at u,

$$\forall \varepsilon > 0: \exists \delta(u, \varepsilon) \ such \ that \ \|u - v\|_V \leq \delta(u, \varepsilon) \Longrightarrow \|F(u) - F(v)\|_W \leq \varepsilon\,.$$

Let $T(v) = \|F(v)\|_W$. We have

$$|T(v) - T(u)| = |\|F(v)\|_W - \|F(u)\|_W| \leq \|F(v) - F(u)\|_W\,,$$

therefore, that T is continuous at u and the result follows from proposition 3.1. ■

3.4. Sequences and convergence

Our main objective is the determination of numerical approximations to equations (algebraic, differential, partial differential, etc.). Approximations are closely connected to converging sequences.

3.4.1. *Sequences*

DEFINITION 3.9.– A *sequence* of elements of V (or a sequence on V) is an application $\mathbb{N}^* \ni n \longrightarrow u_n \in V$. Such a sequence is denoted $\{u_n : n \in \mathbb{N}^*\} \subset V$.

Let $\mathbb{N}^* \ni k \longrightarrow n_k \in \mathbb{N}^*$ be a strictly increasing map. Then $\mathbb{N}^* \ni k \longrightarrow u_{n_k} \in V$ is a subsequence $\{u_{n_k} : k \in \mathbb{N}^*\} \subset \{u_n : n \in \mathbb{N}^*\}$. ■

We observe that $n_k \to +\infty$ when $k \to +\infty$: we have $n_{k+1} > n_k$ so that $n_{k+1} \geq n_k + 1$. Thus, $n_{k+1} \geq n_1 + k \geq k$ and $n_k \to +\infty$.

In the context of engineering, the main interest of sequences the main interest lies in their use in the construction of approximations, whether for functions (for instance, signals) or solutions of equations. Approximations are connected to the notion of convergence. In the following, we present this concept and we distinguish strong and weak convergence.

3.4.2. *Convergence (or strong convergence)*

DEFINITION 3.10.– Let V be a vector space having norm $\| \blacksquare \|$. $\{u_n : n \in \mathbb{N}^*\} \subset V$ converges (or strongly converges) to $u \in V$ if and only if $\|u_n - u\| \longrightarrow 0$, when $n \longrightarrow +\infty$. Notation: $u_n \to u$or $u = \lim_{n \to +\infty} u_n$. ■

EXAMPLE 3.5.– Let us consider $u_n : (0,1) \longrightarrow \mathbb{R}$ given by

$$u_n(x) = \begin{cases} 1, \ x \in \left(0, \frac{1}{2}\right) \\ 1 - nx, x \in \left(\frac{1}{2}, \frac{1}{2} + \frac{1}{n}\right) \\ 0, x \in \left(\frac{1}{2} + \frac{1}{n}, 1\right). \end{cases}$$

and

$$u(x) = \begin{cases} 1, & x \in \left(0, \frac{1}{2}\right) \\ 0, & x \in \left(\frac{1}{2}, 1\right). \end{cases}$$

We have

$$\|u_n - u\|_p = \left(\int_{\frac{1}{2}}^{\frac{1}{2}+\frac{1}{n}} |u_n(x)|^p \, dx \right)^{\frac{1}{p}} \leq \left(\int_{\frac{1}{2}}^{\frac{1}{2}+\frac{1}{n}} 1 \, dx \right)^{\frac{1}{p}} = \left(\frac{1}{n}\right)^{\frac{1}{p}} \longrightarrow 0, for\ n \longrightarrow +\infty.$$

Thus, $u_n \rightarrow u$ on $V = \{v: \Omega \rightarrow \mathbb{R}: v\ is\ continuous\ on\ \Omega \text{ and } \|v\|_p < \infty\}$. Notice that the limit $u \notin V$, since it is discontinuous at $x = 1/2$, while $u_n \in V$ is continuous on Ω, $\forall n$. ■

THEOREM 3.5.– $u_n \rightarrow u$ if and only if $u_{n_k} \rightarrow u$, for any subsequence. ■

PROOF.– Assume that $\|u_n - u\| \longrightarrow 0$. Then:

$\forall \varepsilon > 0: \exists n_0(\varepsilon) > 0$ such that $n \geq n_0(\varepsilon) \Longrightarrow \|u_n - u\| \leq \varepsilon.$

Let $k_0(\varepsilon)$ be such that $n_{k_{0(\varepsilon)}} \geq n_0(\varepsilon)$. Then

$$k > k_0(\varepsilon) \Longrightarrow n_k > n_{k_{0(\varepsilon)}} \geq n_0(\varepsilon) \Longrightarrow \left\|u_{n_k} - u\right\| \leq \varepsilon.$$

Thus, $u_{n_k} \rightarrow u$. However, assume that $u_{n_k} \rightarrow u$, for any subsequence. Let us suppose that $u_n \nrightarrow u$. Then, there exists $\varepsilon > 0$ such that,

$$\forall m > 0: \exists n(m) > m \text{ such that } \left\|u_{n(m)} - u\right\| > \varepsilon.$$

Let us generate a subsequence as follows: $n_1 = n(1)$ and $n_{k+1} = n(n_k)$, $\forall k \geq 1$. Then

$$\varepsilon < \left\| u_{n_k} - u \right\| \longrightarrow 0 \Rightarrow \varepsilon \leq 0 \ .$$

Thus, $0 < \varepsilon \leq 0$, what is a contradiction. So, $u_n \to u$. ■

We have also

THEOREM 3.6.– LetV, W be two vector spaces having norms $\|\blacksquare\|_V, \|\blacksquare\|_W$, respectively. $F: V \to W$ is continuous at $u \in V$ if and only if $u_n \longrightarrow u \Rightarrow F(u_n) \longrightarrow F(u)$. ■

PROOF.– See [SOU 10].

One of the main differences between vector spaces formed by functions (*i.e.* functional spaces) and standard finite dimensional spaces such as $\mathbb{R}^n$ is the fact that *bounded sequences may do not have convengent subsequences* (see examples 3.6 and 3.7 below).

EXAMPLE 3.6.– Let us consider $\Omega = (0, 2\pi)$,

$$(u, v) = \int_\Omega u(x)v(x)dx$$

and $S = \{v \in V: v \text{ } is \text{ } continuous \text{ } on \text{ } \Omega \text{ and } (v, v) < \infty\}$. Let us consider the sequence defined by

$$u_n(x) = \cos(nx)$$

Let $v_0 = 1$. We have

$$\int_0^{2\pi} u_n(x)v_0(x)dx = \frac{\sin(nx)}{n}\bigg|_{x=0}^{x=2\pi} = 0 \ .$$

Let $v_k(x) = x^k$ $(k \geq 1)$. We have:

$$\int_0^{2\pi} u_n(x)v_k(x)dx = -\underbrace{\frac{x^k \sin(nx)}{n}\bigg|_{x=0}^{x=2\pi}}_{=0} + (k-1)\int_0^{2\pi} \frac{x^{k-1}\sin(nx)}{n}dx,$$

Thus, $\forall k \geq 1$,

$$\left|\int_0^{2\pi} u_n(x)v_k(x)dx\right| \leq \frac{(k-1)}{n}\int_0^{2\pi} x^{k-1}dx \leq \frac{(k-1)(2\pi)^k}{kn} \to 0 \; for \; n \to +\infty .$$

Thus, $(u_n, v_k) \longrightarrow 0, \forall k \geq 0$, so that

$$(u_n, P) \longrightarrow 0, \forall \; polynomial \; function \; P\colon (0,2\pi) \to \mathbb{R} .$$

Let $v\colon (0, 2\pi) \longrightarrow \mathbb{R}$ be a continuous function. From Stone–Weierstrass approximation theorem (see theorem 3.13), there is a polynomial $P_\varepsilon\colon (0,2\pi) \to \mathbb{R}$ such that $\|v - P_\varepsilon\|_\infty \leq \varepsilon$. Considering that

$$u_n(x)v(x) = u_n(x)\big(v(x) - P_\varepsilon(x)\big) + u_n(x)P_\varepsilon(x),$$

we have

$$\left|\int_0^{2\pi} u_n(x)v(x)dx\right| \leq 2\pi\varepsilon + \left|\int_0^{2\pi} u_n(x)P_\varepsilon(x)dx\right|$$

and

$$0 \leq \lim_{n\to+\infty}|(u_n, v)| = \lim_{n\to+\infty}\left|\int_0^{2\pi} u_n(x)v(x)dx\right| \leq 2\pi\varepsilon .$$

Since $\varepsilon > 0$ is arbitrary, we have $(u_n, v) \longrightarrow 0$. Assume that $u_{n_k} \longrightarrow u \in S$. $\{(u_{n_k}, v)\colon k \in \mathbb{N}^*\}$ is a subsequence of $\{(u_n, v)\colon n \in \mathbb{N}^*\}$, so that $(u_{n_k}, v) \longrightarrow 0$. We also have

$$|(u_{n_k}, v) - (u, v)| \leq |(u_{n_k} - u, v)| \leq \|u_{n_k} - u\|\|v\| \longrightarrow 0,$$

so that $(u_{n_k}, v) \longrightarrow (u, v)$. Thus, $(u, v) = 0, \forall v \in S$. Taking $v = u$, we have $(u, u) = 0$. This equality implies that $\|u\| = 0$. But

$$\|u_n\|^2 = (u_n, u_n) = \int_\Omega \cos^2(x)\, dx = \pi,$$

so that $\|u_{n_k}\| = \sqrt{\pi} \longrightarrow \sqrt{\pi} = \|u\|$ and we have $0 = \sqrt{\pi}$, which is a contradiction. We conclude that the sequence $\{u_n\colon n \in \mathbb{N}^*\}$ has no converging subsequence. ■

3.4.3. *Weak convergence*

DEFINITION 3.11.– Let V be a vector space having scalar product $(\blacksquare, \blacksquare)$. $\{u_n : n \in \mathbb{N}^*\} \subset V$ weakly converges to $u \in V$ if and only if $\forall v \in V: (u_n - u, v) \longrightarrow 0$, when $n \longrightarrow +\infty$. Notation: $u_n \rightharpoonup u$. ■

EXAMPLE 3.7.– Let us consider $\Omega = (0,1)$

$$(u, v) = \int_0^1 u(x)v(x)dx$$

and $u_n : (0,1) \longrightarrow \mathbb{R}$ given by $u_n(x) = \cos\,(2n\pi x)$. Let $v_0 = 1$. We have

$$\int_0^1 u_n(x)v_0 dx = \frac{\sin(2n\pi x)}{2n\pi}\bigg|_{x=0}^{x=1} = 0\,.$$

Let $v_k(x) = x^k$ $(k \geq 1)$. We have:

$$\int_0^1 u_n(x)v_k(x)dx = -\underbrace{\frac{x^k \sin(2n\pi x)}{2n\pi}\bigg|_{x=0}^{x=1}}_{=0} + (k-1)\int_0^1 \frac{x^{k-1}\sin(2n\pi x)}{2n\pi}dx,$$

So that

$$\left|\int_0^1 u_n(x)v_k(x)dx\right| \leq \frac{(k-1)}{2kn\pi} \rightarrow 0\; for\; n \rightarrow\; +\infty\,.$$

Thus,

$$(u_n, P) \longrightarrow 0, \forall\; polynomial\; function\; P: (0,1) \rightarrow \mathbb{R}\,.$$

Let v be continuous on $[0,1]$ and $\varepsilon > 0$. From Stone–Weierstrass approximation theorem 3.13, there is a polynomial $P_\varepsilon : (0,1) \rightarrow \mathbb{R}$ such that $\|v - P_\varepsilon\|_\infty \leq \varepsilon$. Thus,

$$\left|\int_0^1 u_n(x)v(x)dx\right| \leq \varepsilon + \left|\int_0^1 u_n(x)P_\varepsilon(x)dx\right|$$

and we have

$$\lim_{n\to+\infty}|(u_n, v)| = \lim_{n\to+\infty}\left|\int_0^1 u_n(x)v(x)dx\right| \leq \varepsilon .$$

Since $\varepsilon > 0$ is arbitrary, we have $(u_n, v) \longrightarrow 0$, so that $u_n \rightharpoonup 0$ (weakly) on $V = \{v: (0,1) \to \mathbb{R}: v \text{ is continuous on } [0,1] \text{and } \|v\|_2 < \infty\}$, with the scalar product under consideration. Notice that $u_n \nrightarrow 0$ (strongly) since

$$\|u_n\|_2 = \frac{1}{\sqrt{2}}, \forall n > 0 . \qquad \blacksquare$$

We have

THEOREM 3.7.– $u_n \to u$ strongly if and only if $u_n \rightharpoonup u$ weakly and $\|u_n\| \longrightarrow \|u\|$ when $n \longrightarrow +\infty$. $\blacksquare$

PROOF.– Assume that $u_n \to u$ strongly. Then,

$$|(u_n - u, v)| \leq \|u_n - u\|\|v\| \longrightarrow 0; \ |\|u_n\| - \|u\|| \leq \|u_n - u\| \longrightarrow 0.$$

Thus, $u_n \rightharpoonup u$ weakly and $\|u_n\| \longrightarrow \|u\|$when $n \longrightarrow +\infty$. Converse is obtained by using the equality

$$\|u_n - u\|^2 = \|u_n\|^2 + \|u\|^2 - 2(u_n, u) . \qquad \blacksquare$$

THEOREM 3.8.– Let $\dim(V) = n < \infty$. Then $u_n \to u$ strongly if and only if $u_n \rightharpoonup u$ weakly. $\blacksquare$

PROOF.– Assume that $u_n \rightharpoonup u$ weakly. Let $F = \{\varphi_1, \dots, \varphi_n\}$ be an orthonormal basis of V. Then $(u_n, \varphi_i) \longrightarrow (u, \varphi_i), 1 \leq i \leq n$, so that

$$\|u_n\|^2 = \sum_{i=1}^n |((u_n, \varphi_i))|^2 \longrightarrow \sum_{i=1}^n |((u, \varphi_i))|^2 = \|u\|^2 . \qquad \blacksquare$$

3.4.4. *Compactness*

DEFINITION 3.12.– Let $S \subset V$. S is *compact* if and only if any sequence $\{u_n: n \in \mathbb{N}^*\} \subset S$ has a subsequence $\{u_{n_k}: k \in \mathbb{N}^*\}$ such that $u_{n_k} \longrightarrow u \in S$. S is *weakly compact* if and only if any sequence $\{u_n: n \in \mathbb{N}^*\} \subset S$ has a subsequence $\{u_{n_k}: k \in \mathbb{N}^*\}$ such that $u_{n_k} \rightharpoonup u \in S$. ■

We have:

THEOREM 3.9.– Let V be a vector space having norm $\| \blacksquare \|$. If $S \subset V$ is compact then S is closed and bounded (i.e. there exists $M \in \mathbb{R}$ such that $\|u\| \leq M, \forall u \in S$). ■

PROOF.– Assume that S is not closed. Then, there exists a sequence $\{u_n: n \in \mathbb{N}^*\} \subset S$, such that $u_n \longrightarrow u$ (resp. $u_n \rightharpoonup u$) and $u \notin S$. Thus, for any subsequence, $u_{n_k} \longrightarrow u$ (resp. $u_{n_k} \rightharpoonup u$). Since S is compact, $u \in S$. Then, $u \in S$ and $u \notin S$, what is a contradiction.

Assume that S is not bounded. Then, $\forall n > 0 : \exists u_n \in S : \|u_n\| > n$. Since S is compact, $\{u_n: n \in \mathbb{N}^*\} \subset S$ has a subsequence $\{u_{n_k}: k \in \mathbb{N}^*\}$ such that $u_{n_k} \longrightarrow u \in S$ (resp. $u_{n_k} \rightharpoonup u \in S$). Thus, $\{u_{n_k}: k \in \mathbb{N}^*\}$ is bounded and $\exists M \in \mathbb{R}$ such that $\|u_{n_k}\| \leq M, \forall k > 0$. Then $M > n_k, \forall k > 0$. Since $n_k \rightarrow +\infty$, there is $k(M)$ such that $n_{k(M)} > M$, so that $M > n_{k(M)} > M$, what is a contradiction. ■

In finite dimensional situations, we have the converse:

THEOREM 3.10.– Let V be a vector space having norm $\| \blacksquare \|$ and $\dim(V) = n < \infty$. Then $S \subset V$ is compact if and only if S is closed and bounded. ■

PROOF.– From theorem 3.9, we have the direct implication: if S is compact, then S is closed and bounded. For the opposite, assume that S is closed and bounded. Let $F = \{\varphi_1, \dots, \varphi_n\}$ be an orthonormal basis of V and $T: V \rightarrow \mathbb{R}^n$ be the linear map $T(u) = x$, with $x_i = (u, \varphi_i), 1 \leq i \leq n$. T is bijective, since, on the one hand, $\forall x \in \mathbb{R}^n : \exists u = x_1\varphi_1 + \dots + x_n\varphi_n$ such that $T(u) = x$ and, on the other hand, $T(u) = T(v) \Leftrightarrow (u, \varphi_i) = (v, \varphi_i), 1 \leq i \leq n \Leftrightarrow u = v$. Moreover,

$$\|u\|^2 = \sum_{i=1}^{n} \left|\left((u, \varphi_i)\right)\right|^2 = \|T(u)\|^2,$$

so that $T(S) \subset \mathbb{R}^n$is closed and bounded. As a consequence, $T(S) \subset \mathbb{R}^n$is compact (Heine–Borel–Lebesgue theorem. See [HOF 07], for instance).

Let $\{u_n: n \in \mathbb{N}^*\} \subset S$: $\{x_n = T(u_n): n \in \mathbb{N}^*\} \subset T(S) \subset \mathbb{R}^n$ has a subsequence $\{x_{n_k}: k \in \mathbb{N}^*\}$ such that$x_{n_k} \to x \in T(S)$. Then $u_{n_k} = T^{-1}(x_{n_k}) \to T^{-1}(x) = u \in S$. ■

The preceding example, 3.7, shows that, contrarily to finite dimensional situations, a bounded sequence of functions does not form a compact set. In the following we will see that, for a particular class of Hibert spaces, a bounded sequence forms a weakly compact set.

3.5. Hilbert spaces and completeness

One of the main difficulties in the practical manipulation of sequences is the verification of their convergence. In order to apply definition 3.9, we must know the limit u – otherwise, we cannot evaluate $\|u_n - u\|$. An attempt to obtain a way to verify the convergence without knowing the limit is furnished by Cauchy sequences:

DEFINITION 3.13.– Let V be a vector space having norm $\| \blacksquare \|$. $\{u_n: n \in \mathbb{N}^*\} \subset V$ is a *Cauchy sequence* if and only if: $\forall \varepsilon > 0, \exists\, n_0(\varepsilon)$ *such that* $m, n \geq n_0(\varepsilon) \Longrightarrow \|u_m - u_n\| \leq \varepsilon$.

We have:

PROPOSITION 3.2.– Let V be a vector space having norm $\| \blacksquare \|$ and $\{u_n: n \in \mathbb{N}^*\} \subset V$. If $u_n \to u$ strongly then $\{u_n: n \in \mathbb{N}^*\}$ is a Cauchy sequence. ■

PROOF.– Assume that $u_n \to u$ strongly. Then,

$$\forall \varepsilon > 0: \exists n_0(\varepsilon) > 0 \; such \; that \; n \geq n_0(\varepsilon) \Longrightarrow \|u_n - u\| \leq \frac{\varepsilon}{2}.$$

Thus,

$$m,n \geq n_0(\varepsilon) \Longrightarrow \|u_m - u_n\| \leq \|u_m - u\| + \|u_n - u\| \leq \varepsilon \,. \blacksquare$$

Unfortunately, the opposite is not true: example 3.5 shows that a Cauchy sequence may have as limit an element that does not belong to V. Only at particular spaces, the limits of their Cauchy sequences remain in the space:

DEFINITION 3.14.– Let V be a vector space having norm $\| \blacksquare \|$. V is *complete* if and only if all its Cauchy sequences converge to elements of V: $\{u_n : n \in \mathbb{N}^*\} \subset V$ is a Cauchy sequence$\Longrightarrow \exists u \in V$ *such that* $u_n \to u$ strongly. ■

DEFINITION 3.15.– A space V that possesses a scalar product and is complete is called a Hilbert space. ■

3.5.1. *Fixed points*

A fixed point of a map, as it names explicitly mentions, is an invariant point – i.e. a point such that its image by the map is itself:

DEFINITION 3.16.– Let $F: V \to V$ be a map. We say that $u \in V$ is a fixed point of F if and only if $u = F(u)$. ■

A large number of numerical methods is based in the construction of fixed points, such as, for instance, Newton's method for algebraic equations.

The following result is known as Banach's fixed point theorem:

THEOREM 3.11.– Let V be a Hilbert space. Let $F: V \to V$ be continuous on V. F is a *contraction*, *i.e.* if there exists there exists $M < 1$ such that

$$\forall u, v \in V : \ \|F(u) - F(v)\| \leq M\|u - v\| \,.$$

If F is a contraction, then there exists an unique $u \in V$ such that $u = F(u)$. Moreover, the sequence $\{u_n : n \in \mathbb{N}^*\} \subset V, u_{n+1} = F(u_n)$ converges to $u, \forall u_0 \in V$. ■

PROOF.– We have

$$\|u_{n+1} - u_n\| \leq \|F(u_n) - F(u_{n-1})\| \leq M\|u_n - u_{n-1}\|,$$

so that

$$\|u_{n+1} - u_n\| \leq M^n\|u_1 - u_0\|$$

and

$$\|u_{n+p} - u_n\| \leq \sum_{i=n}^{n+p-1}\|u_{i+1} - u_i\| \leq \|u_1 - u_0\| \sum_{i=n}^{n+p-1} M^i .$$

Thus,

$$\|u_{n+p} - u_n\| \leq \|u_1 - u_0\| \frac{M^n - M^{n+p}}{1-M} \leq M^n \frac{\|u_1 - u_0\|}{1-M} .$$

Let $\varepsilon > 0$. There exists $n_0(\varepsilon)$ such that

$$n \geq n_0(\varepsilon) \Longrightarrow M^n \frac{\|u_1 - u_0\|}{1-M} \leq \varepsilon .$$

Thus,

$$m, n \geq n_0(\varepsilon) \Longrightarrow \|u_m - u_n\| \leq \varepsilon$$

and the sequence is a Cauchy sequence. As a consequence, there exists an element $u \in V$ such that $u_n \to u$. We obtain $u = F(u)$ from $u_{n+1} = F(u_n)$. Moreover, let $v = F(v)$. Then

$$\|u - v\| = \|F(u) - F(v)\| \leq M\|u - v\| ,$$

so that $(1 - M)\|u - v\| \leq 0 \Longrightarrow \|u - v\| = 0 \Longrightarrow u = v$. ■

The reader will find in the literature a large number of Fixed Point Theorems, established by Luitzen Egbertus Jan Brouwer [BRO 11],

Juliusz Paweł Schauder [SHA 30], Andrei Nikolaievich Tikhonov [TIK 35] and Shizuo Kakutani [KAK 41]. Other developments may be found in [FAN 52, RIL 66, EAR 70]. A complete panorama of fixed points is given in [SMA 74].

3.6. Open and closed sets

Open and closed sets are a fondamental concept in variational methods. In fact, an alternative way to define a topology consists of giving the families of open and closed sets. Naively, a closed set is a set that contains its boundary. In mathematical terms, we talk about cluster points, i.e. points which cannot be separated from the set, since all their neighborhoods contain points of the set.

3.6.1. *Closure of a set*

DEFINITION 3.17.– Let V be a vector space having scalar product $(\blacksquare, \blacksquare)$. Let $S \subset V$. $u \in V$ is a *cluster point* (or adhering point) of S if and only if

$$\exists\{u_n : n \in \mathbb{N}^*\} \subset S \; such\ that\ u_n \to u.$$

Analogously, $u \in V$ is a *weak cluster point* (or weak adhering point) of S if and only if

$$\exists\{u_n : n \in \mathbb{N}^*\} \subset S \; such\ that\ u_n \rightharpoonup u.$$ ■

DEFINITION 3.18.– Let V be a vector space having scalar product $(\blacksquare, \blacksquare)$. Let $S \subset V$. The *closure* of S is denoted by $\overline{S}$ and defined as:

$$\overline{S} = \{u \in V : u\ is\ a\ cluster\ point\ of\ S\}.$$

The weak closure of S is

$$\overline{S} = \{u \in V : u\ is\ a\ weak\ cluster\ point\ of\ S\}.$$ ■

We have

PROPOSITION 3.3.– Let $S \subset U$. Then $S \subset \overline{S} \subset \overline{U}$. ■

PROOF.– We have $S \subset \overline{S}$: $\forall u \in S : u_n = u, \forall n$ verifies $\{u_n : n \in \mathbb{N}^*\} \subset S$ and $u_n \to u$.

Consider $\{u_n : n \in \mathbb{N}^*\} \subset S$ such that $u_n \to u$. Then $\{u_n : n \in \mathbb{N}^*\} \subset U$, so that $u \in \overline{U}$. ■

3.6.2. *Open and closed sets*

DEFINITION 3.19.– Let $S \subset V$. S is *closed* (respectively, *weakly closed*) if and only if S contains all its cluster points (resp., weak cluster points), i.e. $S = \overline{S}$. S is *open* (respectively, *weakly open*) if and only if $V - S$ is closed (resp. weakly closed). ■

We have:

THEOREM 3.12.– S is open if and only if:

$$\forall\, u \in S : \exists \varepsilon > 0 \; such\; that \; \|v - u\| \leq \varepsilon \Longrightarrow v \in S\,. \qquad ■$$

PROOF.– Assume that S is open. Let us consider $u \in S : \forall \varepsilon > 0, \exists\, v_\varepsilon \in S \; such\; that\; \|v_\varepsilon - u\| \leq \varepsilon \; and\; v_\varepsilon \notin S$. Let $\varepsilon = \frac{1}{n}$and $u_n = v_\varepsilon$. We have $\|u_n - u\| \leq \frac{1}{n} \longrightarrow 0$, so that $u_n \to u$. Moreover, $u_n = v_\varepsilon \notin S$, $\forall S$. Thus, $\{u_n : n \in \mathbb{N}^*\} \subset V - S$. Since S is open, $V - S$ is closed. Thus, $u \in V - S$: then, we have $u \in S$ and $u \in V - S$, which is a contradiction.

However, assume that

$$\forall\, u \in S : \exists \varepsilon > 0 \; such\; that \; \|v - u\| \leq \varepsilon \Longrightarrow v \in S\,,$$

but $V - S$ is not closed. The, $\exists \{u_n : n \in \mathbb{N}^*\} \subset V - S \; such\; that\; u_n \to u \notin V - S$. Thus, on the one hand, $u \in S$ and $\exists \varepsilon > 0 \; such\; that\; \|v - u\| \leq \varepsilon \Longrightarrow v \in S$. On the other hand, $\exists n_0(\varepsilon) : n \geq n_0(\varepsilon) \Longrightarrow \|u_n - u\| \leq \varepsilon$. Thus: $u_n \in S, \forall n \geq n_0(\varepsilon)$ and $\{u_n : n \in \mathbb{N}^*\} \subset V - S$, so that $\forall n \geq n_0(\varepsilon) : u_n \in S \cap V - S = \emptyset$, which is a contradiction. Thus, S is closed. ■

Moreover:

PROPOSITION 3.4.– If S is weakly closed, then S is closed. ■

PROOF.– Assume that S is weakly closed. Consider $\{u_n : n \in \mathbb{N}^*\} \subset S$ such that $u_n \to u$. From theorem 3.7, $u_n \rightharpoonup u$, so that $u \in S$. ■

PROPOSITION 3.5.– $S^\perp$ is weakly closed. ■

PROOF.– Consider $\{u_n : n \in \mathbb{N}^*\} \subset S^\perp$ such that $u_n \rightharpoonup u$,. Then: $\forall v \in S : (u_n, v) = 0, \forall n > 0$.
Thus, $\forall v \in S : (u, v) = \lim_{n \to +\infty}(u_n, v) = 0$, so that $u \in S^\perp$. ■

COROLLARY 3.2.– $S^{\perp\perp} = (S^\perp)^\perp$ is weakly closed and $S \subset \overline{S} \subset S^{\perp\perp}$. ■

PROOF.– $S^{\perp\perp}$ is weakly closed, from proposition 3.5: then, $\overline{S^{\perp\perp}} = S^{\perp\perp}$. Moreover, for $u \in S$,$(u, v) = 0 , \forall\, v \in S^\perp$, so that $u \in S^{\perp\perp}$ and we have $S \subset S^{\perp\perp}$. From proposition 3.3, $S \subset \overline{S} \subset S^{\perp\perp}$. ■

PROPOSITION 3.6.– Let $S \subset V$. Then $\overline{S}$ is closed. If S is a vector subspace, then $\overline{S} \subset V$ is a closed subspace. ■

PROOF.– See [SOU 10]. ■

3.6.3. *Dense subspaces*

DEFINITION 3.20.– Let V be a vector space having scalar product $(\blacksquare, \blacksquare)$. Let $S \subset V$. S is *dense* on V if and only if $\overline{S} = V$. ■

A first example of dense subspace is furnished by the theorem of Stone-Weierstrass:

THEOREM 3.13.– Let $\Omega \subset \mathbb{R}^n$ and $F : \Omega \to \mathbb{R}$ be a continuous function. Then, $\forall \varepsilon > 0, \exists\, P_\varepsilon , P_\varepsilon\ polynomial$, such that $\|F - P_\varepsilon\|_0 \leq \varepsilon$. ■

PROOF.– See [DUR 12]. ■

3.7. Orthogonal projection

Orthogonal projections are a basic tool in variational methods, often invoked and used in practice. *In the sequel, V is a Hilbert space having scalar product*$(\blacksquare, \blacksquare)$.

3.7.1. *Orthogonal projection on a subspace*

DEFINITION 3.21.– Let $S \subset V$. $S \neq \emptyset$. Let $u \in V$. The *orthogonal projection* Pu of u onto S is $Pu \in S$ such that $\|u - Pu\| \leq \|u - v\|, \forall v \in S$, i.e.

$$Pu = \arg\min\{\|u - v\| : v \in S\}. \qquad \blacksquare$$

We have

THEOREM 3.14.– Let S be a closed vector subspace. Then, $\forall u \in V$, there exists an unique $Pu \in S$ such that that $\|u - Pu\| \leq \|u - v\|, \forall v \in S$. ■

PROOF.– Let $d^2 = \inf\{\|u - v\|^2 : v \in S\}$. Let $n > 0$: there exists $u_n \in S$ such that

$$d^2 \leq \|u - u_n\|^2 \leq d^2 + \frac{1}{n}. \qquad [3.1]$$

Thus, for $m, n \geq k$,

$$\|u_m - u_n\|^2 + \|u_m + u_n - 2u\|^2 = 4(\|u - u_n\|^2 + \|u - u_m\|^2),$$

so that

$$\|u_m - u_n\|^2 + \|u_m + u_n - 2u\|^2 \leq 4d^2 + \frac{8}{k}. \qquad [3.2]$$

Since

$$u_m + u_n - 2u = 2\left(\underbrace{\frac{1}{2}u_m + \frac{1}{2}u_n}_{\in S} - u\right), \qquad [3.3]$$

we have $\|u_m + u_n - 2u\|^2 \geq 4d^2$, so that equation [3.2] implies that

$$\|u_m - u_n\|^2 \leq \frac{8}{k}. \qquad [3.4]$$

and the sequence $\{u_n : n \in \mathbb{N}^*\} \subset S$ is Cauchy sequence. Thus, there exists $Pu \in S$ (since S is closed and V is a Hilbert space) such that $u_n \to Pu$. Equation [3.1] shows that $\|u - Pu\| = d \leq \|u - v\|, \forall v \in S$. Moreover, if $Qu \in S$ and $\|u - Qu\| = d$, then, the equality

$$\|Pu - Qu\|^2 + \|Pu + Qu - 2u\|^2 = 4(\|u - Pu\|^2 + \|u - Qu\|^2),$$

shows that $\|Pu - Qu\|^2 \leq 0$, so that $Pu = Qu$ (analogous to equations [3.2], [3.3], [3.4]).

THEOREM 3.15.– Let S be a closed vector subspace and $u \in V$. Then, Pu is the orthogonal projection of u onto S if and only if

$$Pu \in S \;\; and \;\; (u - Pu, v) = 0, \forall v \in S\,. \blacksquare \qquad [3.5]$$

PROOF.– Let $v \in S$ and $f(\lambda) = \|u - Pu - \lambda v\|^2$. We have

$$f(\lambda) = \|u - Pu\|^2 - 2\lambda(u - Pu, v) + \lambda^2\|v\|^2\,.$$

Then

$$f'(\lambda) = -2(u - Pu, v) + 2\,\lambda\|v\|^2\,.$$

If Pu is the orthogonal projection of u onto S, the minimum of f is attained at $\lambda = 0$and $f'(0) = 0$, so that $(u - Pu, v) = 0$. However, if $(u - Pu, v) = 0$, then $f'(\lambda) = 0$ for $\lambda = 0$, so that the minimum is attained at $\lambda = 0$ and we have $\|u - Pu\| = \min\{\|u - v\| : v \in S\}$.

Equation [3.5] is a *variational equation*. It may be interpreted in terms of orthogonality: vector $u - Pu$ is orthogonal to all the elements of the subspace S. A simple geometrical interpretation is given in Figure 3.5: in dimension 2, a vector subspace is a line passing through the origin.

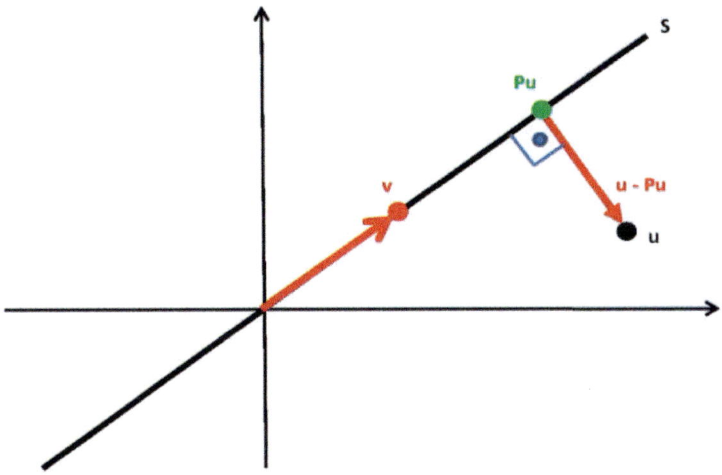

Figure 3.5. *Geometrical interpretation of the orthogonal projection on a vector subspace. For a color version of the figure, see www.iste.co.uk/souzadecursi/variational.zip*

One of the main consequences of the orthogonal projection is [RSZ 07]

THEOREM 3.16.– Let $S \subset V$ be a vector subspace. S is dense on V if and only if $S^{\perp} = \{0\}$. ■

PROOF.– Let S be dense on V and $v \in S^{\perp}$. Then, on the one hand, there exists a sequence $\{v_n : n \in \mathbb{N}^*\} \subset V$ such that $v_n \longrightarrow v$; on the other hand, $(v_n, v) = 0, \forall n \in \mathbb{N}^*$. Thus, from theorem 3.7, $(v, v) = 0 \Longrightarrow v = 0$. Thus, $S^{\perp} = \{0\}$.

Let $S^{\perp} = \{0\}$. Assume that $\overline{S} \neq V$. Then, there exists $u \in V, u \notin \overline{S}$. Since $\overline{S}$ is closed, $V - \overline{S}$ is open: from theorem 3.12, there is $\varepsilon > 0$ such that $\|u - v\| \leq \varepsilon \Longrightarrow v \in V - \overline{S} \Longrightarrow v \notin \overline{S}$. Let Pu be the orthogonal projection of u onto $\overline{S}$: equation [3.5] shows that $u - Pu \in \overline{S}^{\perp}$. Since $S \subset \overline{S}$, we have $u - Pu \in S^{\perp}$. Thus, $u - Pu = 0$, so that $u = Pu \in \overline{S}$. Thus, $u \notin \overline{S}$ and $u \in \overline{S}$, what is a contradiction. ■

3.7.2. Orthogonal projection on a convex subset

DEFINITION 3.22.– Let $S \subset V$. $S \neq \emptyset$. S is convex if and only if

$$u, v \in S, \lambda \in [0,1] \Rightarrow \lambda u + (1-\lambda)v \in S. \qquad \blacksquare$$

THEOREM 3.17.– Let S be a closed non-empty convex. Then, $\forall u \in V$, there exists a unique $Pu \in S$ such that that $\|u - Pu\| \leq \|u - v\|, \forall v \in S$. Moreover, Pu is the orthogonal projection of u onto S if and only if

$$Pu \in S \ \ and \ \ (u - Pu, v - Pu) \leq 0, \forall v \in S. \qquad [3.6]$$

PROOF.– The proof of existence and uniqueness is analogous to these given for theorem 3.14. Since S is convex, equations [3.2], [3.3], [3.4] are verified and the conclusion is obtained by the same way (the sequence is a Cauchy sequence, V is Hilbert, S is closed). The inequality is obtained as follows: let $v \in S$ and $f(\lambda) = \|u - Pu - \lambda(v - Pu)\|^2$. Since

$$\forall \lambda \in (0,1): u + \lambda(v - Pu) = \lambda v + (1-\lambda)Pu \in S,$$

we have $f(0) \leq f(\lambda), \forall \lambda \in (0,1)$, so that $f'(0) = -2(u - Pu, v - Pu) \geq 0$ and we have the inequality. $\blacksquare$

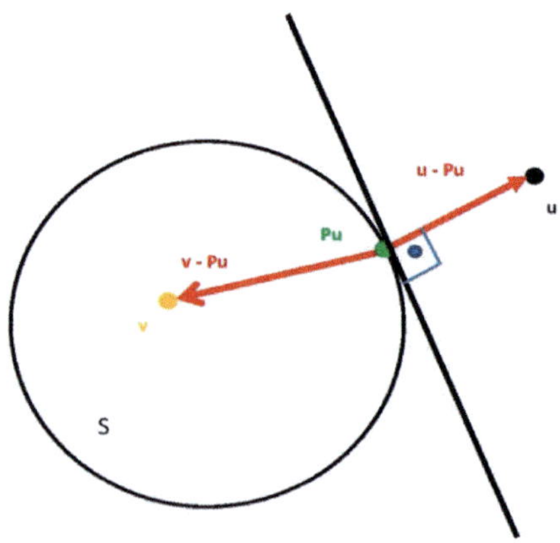

Figure 3.6. *Geometrical interpretation of the orthogonal projection on a convex subset. For a color version of the figure, see www.iste.co.uk/souzadecursi/variational.zip*

Equation [3.6] is a *variational inequality*. It may be interpreted in terms of orthogonality: vector $u - Pu$ is orthogonal to the tangent space of set S. A simple geometrical interpretation in dimension 2 is given in Figure 3.6: we see that the angle between $u - Pu$ and $v - Pu$ is superior to $\pi/2$.

3.7.3. *Orthogonal projection on an affine subspace*

DEFINITION 3.23.– Let $S \subset V$. $S \neq \emptyset$. S is an affine subspace if and only if

$$u, v \in S, \lambda \in \mathbb{R} \Longrightarrow \lambda u + (1 - \lambda)v \in S\,. \qquad \blacksquare$$

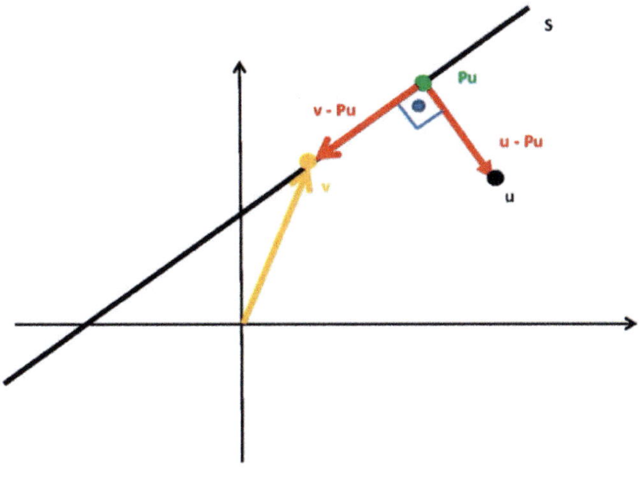

Figure 3.7. *Geometrical interpretation of the orthogonal projection on an affine subspace. For a color version of the figure, see www.iste.co.uk/souzadecursi/variational.zip*

THEOREM 3.18.– Let S be a closed non-empty affine subspace. Then, $\forall u \in V$, there exists an unique $Pu \in S$ such that that $\|u - Pu\| \leq \|u - v\|, \forall v \in S$. Moreover, Pu is the orthogonal projection of u onto S if and only if

$$Pu \in S \;\; and \;\; (u - Pu, v - Pu) = 0, \forall v \in S\,. \blacksquare \qquad [3.7]$$

PROOF.– Notice that S is a closed convex set, so that theorem 3.17 applies. In order to prove the equality, notice that, for $v \in S$, $2Pu - v \in S$, so that we have

$$(u - Pu, Pu - v) = (u - Pu, 2Pu - v - Pu) \leq 0 .$$

Thus, $(u - Pu, v - Pu) \leq 0$ and $(u - Pu, v - Pu) \geq 0$, which establishes the equality.

Equation [3.7] is a *variational equation*. It may be interpreted in terms of orthogonality: vector $u - Pu$ is orthogonal to the differences of elements of S. A simple geometrical interpretation in dimension 2 is given in Figure 3.7: we see that $u - Pu$ is orthogonal to $v - Pu$.

Affine subspaces are generally defined by introducing a translation of a vector subspace:

$$S = S_0 + \{u_0\} = \{v = w + u_0 : w \in S_0\}, S_0 \text{ vector subspace}.$$

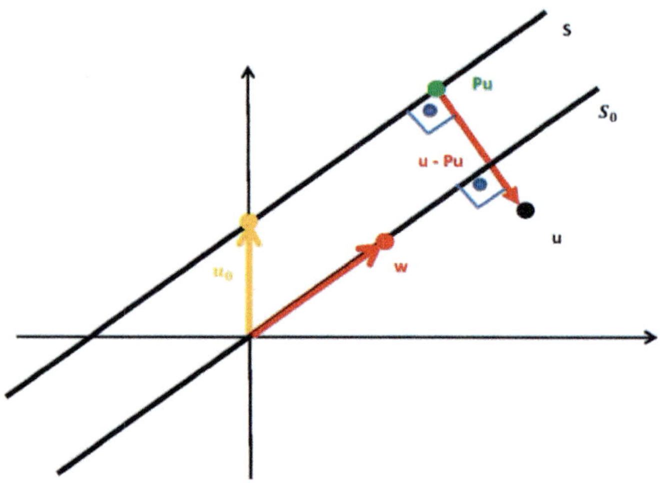

Figure 3.8. *A second interpretation of the orthogonal projection on an affine subspace. For a color version of the figure, see www.iste.co.uk/souzadecursi/variational.zip*

In this case

THEOREM 3.19.– Let $S = S_0 + \{u_0\}$, where S_0 is a closed vector subspace and $u_0 \in V$. Then, $\forall u \in V$, there exists an unique $Pu \in S$ such that that $\|u - Pu\| \leq \|u - v\|, \forall v \in S$. Moreover, Pu is the orthogonal projection of u onto S if and only if

$$Pu \in S \ \ and \ \ (u - Pu, w) = 0, \forall w \in S_0 \, . \blacksquare \qquad [3.8]$$

PROOF.– Notice that S is a closed affine subspace, so that theorem 3.18 applies. For the equality, notice that, for any $w \in S_0$: $w + Pu \in S$. Indeed, $U = Pu - u_0 \in S_0$ and, as a consequence, $w + U \in S_0$, so that $w + Pu = w + U + u_0 \in S$. Thus, equation [3.7] is equivalent to equation [3.8]. ■

Equation [3.8] is a *variational equation.* A simple geometrical interpretation in dimension 2 is given in Figure 3.8: we see that $u - Pu$ is orthogonal to S_0 parallel to S, which is generated by a translation of S_0.

3.7.4. *Matlab® determination of orthogonal projections*

Orthogonal projections may be assimilated to linear equations and determined into a way analogous to the one introduced in section 2.2.4.

Let us consider the situation where S is a subspace: we may consider a basis $\{\varphi_i : i \in \mathbb{N}\} \subset S$ and look for an approximation

$$Pu \approx \sum_{j=1}^{n} u_j \varphi_j \ .$$

Then, the variational equation [3.5] yields that

$$\sum_{j=1}^{n} u_j (\varphi_i , \varphi_j) = (\varphi_i , u) , \ 1 \leq i \leq n \, .$$

Thus, the coefficients $U = (u_1, \dots, u_n)^t$ satisfy a linear system $AU = B$, with $A_{ij} = (\varphi_i , \varphi_j)$ and $B_i = (\varphi_i , u)$.

When using the scalar product `sp0`, defined in section 3.2.5, and the basis defined in section 2.2.6, the coefficients U may be determined by using the method `approx_coeffs`.Notice that matrix A corresponds to methods `varmat_mean`, `varmat_int`, while vector B corresponds to methods `sm_mean` and `sm_int`. The method of evaluation is selected by choosing `'variational_mean'`, `'variational_int'` or `'variational_sp'`.

Otherwise, matrix A and vector B may be determined by using the class introduced in section 3.2.5. We introduce below a new class corresponding to this last situation. Analogously to the approach presented in 3.2.6, we assume that `basis{i}` defines φ_i as a structure having properties `dim`, `dimx`, `component`, where component is a cell array such that each element is a structure having as properties `value` and `grad`. u is assumed to be analogously defined. In this case, an example of class for orthogional projection is given in program 3.6. The class contains two methods: `coeffs`, which evaluates the coefficients of the orthogonal projection and `funct`, which generates a subprogram (anonymous function) evaluating the projection.

```
classdef orthogonal_projection
    methods (Static)
        function c = coeffs(u,basis,sp_space)
            n = length(basis);
            A = zeros(n,n);
            for i = 1: n
                sp = sp_space(basis{i},basis{i});
                A(i,i) = sp;
                for j = i+1: n
                    sp = sp_space(basis{i},basis{j});
                    A(i,j) = sp;
                    A(j,i) = sp;
                end;
            end;
            c = cell(u.dim, 1);
            for k = 1: u.dim
                B = zeros(n,1);
                uu.dim = 1;
                uu.dimx = u.dimx;
                uu.component = cell(1,1);
                uu.component{1} = u.component{k};
                for i = 1: n
```

```
                sp = sp_space(uu,basis{i});
                B(i) = sp;
            end;
            c{k} = A \ B;
        end;
        return;
    end
    function pu = funcproj(c,basis)
        pu.dim = basis{1}.dim;
        pu.dimx = basis{1}.dimx;
        pu.component = cell(pu.dim,1);
        for k = 1: pu.dim
            uu = @(x) 0;
            guu = @(x) zeros(pu.dimx,1);
            cc = c{k};
            for i = 1: length(cc)
                uu = @(x) uu(x) +
cc(i)*basis{i}.component{k}.value(x);
                guu = @(x) guu(x) +
cc(i)*basis{i}.component{k}.grad(x);
            end;
            pu.component{k}.value = uu;
            pu.component{k}.grad = guu;
        end;
    end
  end
end
```

Program 3.6. *A class for orthogonal projection*

EXAMPLE 3.8.– Let us consider the family $\varphi_i(x) = x^{i-1}, 1 \le i \le 4$ and the function $u(x) = x^5$ on $(-1,1)$. Using the scalar product sp0, the orthogonal projection is $Pu = -\frac{5}{21}x + \frac{10}{9}x^3$. The code

```
xlim.lower.x = -1;
xlim.upper.x = 1;
xlim.dim = 1;
u.dim = 1;
u.dimx = 1;
u.component = cell(u.dim,1);
u.component{1}.value = @(x) x^5;
u.component{1}.grad = @(x) 5*x^4;
sp_space = @(u,v)
scalar_product.sp0(u,v,xlim,'subprogram');
```

```
c = orthogonal_projection.coeffs(u,basis,sp_space);
pu = orthogonal_projection.funcproj(c,basis);
```

generates a subprogram pu which evaluates the orthogonal projection. Analogously,

```
U.dim = 1;
U.dimx = 1;
U.component = cell(U.dim,1);
U.component{1}.value =
spam.partition(u.component{1}.value,p);
U.component{1}.grad = cell(U.dim,1);
U.component{1}.grad{1} =
spam.partition(u.component{1}.grad,p);
sp_space = @(u,v)
scalar_product.sp0(u,v,p,'table');
c_tab =
orthogonal_projection.coeffs(U,tb,sp_space);
pu_tab =
orthogonal_projection.funcproj(c_tab,basis);
```

generates a subprogram pu_tab which evaluates the orthogonal projection. The results are shown in Figure 3.9. ■

EXAMPLE 3.9.– Let us consider the family $\varphi_i(x) = x^{i-1}, 1 \le i \le 4$ and the function $u(x) = \sin(x)$ on $(0,2\pi)$. Using the scalar product sp0, the orthogonal projection is $Pu = \frac{5(-21+2\pi^2)}{\pi^3} - \frac{45(-14+\pi^2)}{\pi^4}x + \frac{105(-15+\pi^2)}{2\pi^5}x^2 - \frac{35(-15+\pi^2)}{2\pi^6}x^3$. The code

```
xlim.lower.x = 0;
xlim.upper.x = 2*pi;
xlim.dim = 1;
u.dim = 1;
u.dimx = 1;
u.component = cell(u.dim,1);
u.component{1}.value = @(x) sin(x);
u.component{1}.grad = @(x) cos(x);
```

defines u. pu and pu_tab are generated analogously to example 3.3. The results are shown in Figure 3.10. ■

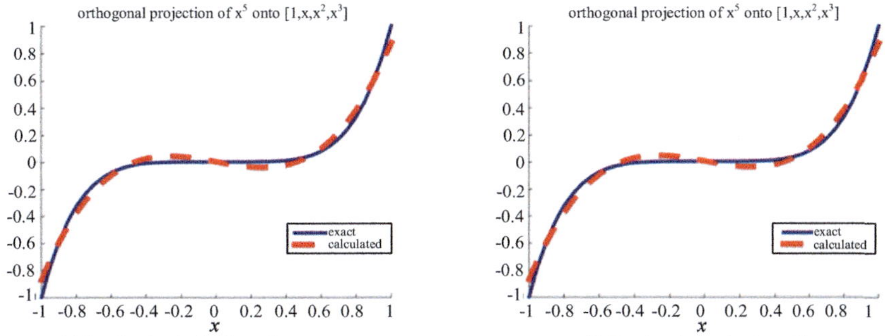

(a) Results obtained using subprograms (b) Results obtained using tables

Figure 3.9. *Orthogonal projection of x^5 onto a polynomial subspace (degree <= 3)*

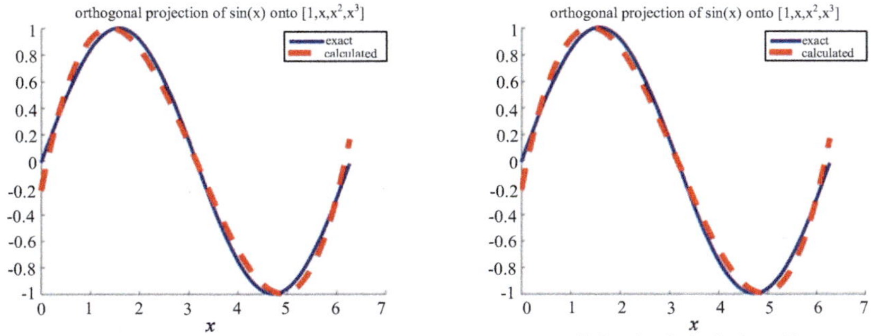

(a) Results obtained using subprograms (b) Results obtained using tables

Figure 3.10. *Orthogonal projection of $sin(x)$ onto a polynomial subspace (degree <= 3). For a color version of the figure, see www.iste.co.uk/souzadecursi/variational.zip*

3.8. Series and separable spaces

A series on V is an infinite sum of elements of V. In the framework of variational methods, the main use of series is the representation of elements of V by infinite sums of a given family $F \subset V$. We are particularly interested in countable families: $F = \{\varphi_n : n \in \mathbb{N}^*\}$, so that an element $u \in V$ may be represented as

$$u = \sum_{n=1}^{+\infty} u_n \varphi_n \ .$$

Such an expansion is usually referred to as a generalized *Fourier series*. Indeed, such an expansion was introduced by Fourier [POI 07], which proposed a solution of heat equation by series of trigonometric functions. His paper was presented by Siméon Denis Poisson and immediately aroused a controversy with Joseph-Louis Lagrange and Pierre-Simon Laplace, due to its mathematical shortcomings. The main question was the sense of the equality and, so, what the functions u admitting such an expansion are. Previously to Fourier, Daniel Bernoulli faced a controversy with Leonhard Euler for analogous reasons: Euler considered that expansions by using trigonometrical function could not represent other functions [BER 53a, BER 53b, EUL 53]. Poisson worked with Jacques Charles François Sturm and Joseph Liouville in order to get results that are more rigorous. However, their work found some difficulty in being accepted by the French Academy of Sciences, due to the incompleteness of the theory – of course, we may blame their referees with a few centuries of additional mathematical work in our possession, but the subject was difficult and about one century has been needed for a more satisfactory solution. A new result arrived with Peter Gustav Lejeune Dirichlet [DIR 29a], but Paul David Gustave du Bois-Reymond produced an example which generates new discussions (a continuous function having a Fourier series that diverges at a point) [DUB 73]. In a further development, these reflections led to the theory of integrals by Riemann and Lebesgue, as introduced in Chapter 1.

3.8.1. *Series*

DEFINITION 3.24.– A series $\sum_{n=1}^{+\infty} v_n$ of elements of V is a sequence $\{S_n : n \in \mathbb{N}^*\}$, such that $S_n = \sum_{i=1}^{n} v_i$, où $\{v_i : i \in \mathbb{N}^*\} \subset V$. If $S_n \longrightarrow v \in V$, we say that the series is convergent and its sum is $v = \sum_{n=1}^{+\infty} v_n$. A series that is not convergent is said to be divergent. ■

S_n is usually referred to as a partial sum. The series

$$R_n = \sum_{i=n}^{+\infty} v_i$$

is the series of the residual of order n. R_n is defined by the sequence $\{R_{n,n+k} = \sum_{i=n}^{n+k} v_i : k \in \mathbb{N}^*\}$.

We have

THEOREM 3.20.– The following assertions are equivalent

i) $\sum_{n=1}^{+\infty} v_n$ converges.

ii) $R_n = \sum_{i=n}^{+\infty} v_i \longrightarrow 0, for\ n \longrightarrow +\infty$

iii) $R_n = \sum_{i=n}^{+\infty} v_i$ converges, $\forall n \in \mathbb{N}^*$. ■

PROOF.– Assume (i). Let $v = \sum_{n=1}^{+\infty} v_n$. Then $S_n = \sum_{i=1}^{n} v_i \longrightarrow v$. Theorem 3.5 shows that all the subsequences of $\{S_n : n \in \mathbb{N}^*\}$ converge to the same limit. Thus, for $n > 1 :$ $R_{n,n+k} = S_{n+k} - S_{n-1} \longrightarrow v - v = 0$, so that we have (ii).

Assume (ii). Then (iii) is immediate.

Assume (iii). We have $S_{n+k} = S_{n-1} + R_{n,n+k} = S_{m-1} + R_{m,m+k}$. Taking the limit for $k \longrightarrow +\infty$, we have$S_{n-1} + R_n = S_{m-1} + R_m$. Thus, $s = S_{n-1} + R_n$ is independent of n and we have $\forall n, k > 1$: $S_{n+k} = S_{n-1} + R_{n,k} \longrightarrow s$, for $k \longrightarrow +\infty$. Thus

$$\forall n > 1: \quad \forall\, \varepsilon > 0: \ \exists k_0(\varepsilon)\ such\ that\ k \geq k_0 \Longrightarrow \|S_{n+k} - s\| \leq \varepsilon\,,$$

what shows that $S_{k+1} \longrightarrow s$, so that $\sum_{n=1}^{+\infty} v_n$ converges. ■

3.8.2. *Separable spaces and Fourier series*

DEFINITION 3.25.– Let $F = \{\varphi_n : n \in \mathbb{N}^*\} \subset V$ and $[F\,]$ be the set of the finite linear combinations of elements of F:

$$[F] = \{v = \textstyle\sum_{i=1}^{n} v_i \varphi_{k_i} : n \in \mathbb{N}^*;\ v_i \in \mathbb{R}\ and\ k_i \in \mathbb{N}^*, 1 \leq i \leq n\}$$

F is a total family if and only if $\overline{[F]} = V$, i.e. $[F]$ is dense on V. We say that F is a Hilbert basis if and only if F is orthonormed and total. ■

DEFINITION 3.26.– V is separable if and only if there exists $F \subset V$ such that F is total. ■

We have:

THEOREM 3.21.– Let $F = \{\varphi_n : n \in \mathbb{N}^*\}$ be a Hilbert basis of V. Let $u, v \in V$. Then

i) $S = [\{\varphi_1, \dots, \varphi_n\}]$ is a closed subspace.

ii) The orthogonal projection $P_n u$ of u onto S is $P_n u = \sum_{i=1}^{n} (u, \varphi_i) \varphi_i$

iii) $u = \sum_{n=0}^{+\infty} (u, \varphi_n) \varphi_n$, i.e. $P_n u \longrightarrow u$ strongly in V.

iv) $\|u\|^2 = \sum_{n=0}^{+\infty} |(u, \varphi_n)|^2$.

v) $(u, v) = \sum_{n=0}^{+\infty} (u, \varphi_n)(v, \varphi_n)$.

vi) $u \in V \Leftrightarrow \sum_{n=0}^{+\infty} |(u, \varphi_n)|^2$converges. ■

PROOF.– See [SOU 10]. ■

$\sum_{n=0}^{+\infty} (u, \varphi_n) \varphi_n$ is the *Fourier series* of u associated with the Hilbert basis F. Equality 3.21. (iv) is known as *Bessel-Parseval equality*.

Separable spaces have the weak Bolzano–Weierstrass property:

THEOREM 3.22.– Let V be a separable Hilbert space. Let $\{u_n : n \in \mathbb{N}^*\} \subset V$ be a bounded sequence ($\exists M \in \mathbb{R} : \forall n \in \mathbb{N}^* : \|u_n\| \leq M$). Then $\{u_n : n \in \mathbb{N}^*\}$ has a subsequence $\{u_{n_k} : k \in \mathbb{N}^*\}$ which is weakly convergent. ■

PROOF.– See [SOU 10]. ■

3.9. Duality

Hilbert spaces are associated with their dual spaces. If we consider Hilbert spaces as spaces of displacements of a mechanical system, dual spaces may be interpreted as the spaces of virtual works associated with these displacements. Riesz's theorem establishes that any virtual work corresponds to a force [RSZ 09].

3.9.1. *Linear functionals*

DEFINITION 3.27.– A linear functional is an application $\ell: V \to \mathbb{R}$ such that

$$\ell(\alpha u + v) = \alpha\ell(u) + \ell(v), \forall u, v \in V, \alpha \in \mathbb{R}. \qquad \blacksquare$$

A linear functional ℓ verifies $\ell(0) = 0$: since $0 = 2 \times 0$, we have $\ell(0) = \ell(2 \times 0)$. Thus, $\ell(0) = 2 \times \ell(0)$ and $\ell(0) = 0$.

DEFINITION 3.28.– The norm of a linear functional ℓ is

$$\|\ell\| = \sup_{v \neq 0} \frac{|\ell(v)|}{\|v\|}. \qquad \blacksquare$$

We have

PROPOSITION 3.7.–

$$\|\ell\| = \sup_{v \neq 0} \frac{\ell(v)}{\|v\|} = \sup_{\|v\|=1} |\ell(v)| = \sup_{\|v\|=1} \ell(v) = \sup_{\|v\| \leq 1} |\ell(v)| = \sup_{\|v\| \leq 1} \ell(v). \qquad \blacksquare$$

and

THEOREM 3.23.– Let $\ell: V \to \mathbb{R}$ be a linear functional. ℓ is continuous if and only if $\|\ell\| < \infty$. ■

PROOF.– Assume that $\|\ell\| < \infty$. Then

$$\forall u, v \in V: \ \|\ell(u) - \ell(v)\| = \|\ell(u - v)\| \leq \|\ell\| \|u - v\|,$$

so that ℓ satisfies Lipschitz's condition and, thus, is continuous (observe that, for $u_n \rightarrow u$, we have $\|\ell(u_n) - \ell(u)\| \leq \|\ell\|\|u_n - u\| \rightarrow 0$, so that $\ell(u_n) \rightarrow \ell(u)$).

Assume that ℓ is continuous. Let us establish that $\|\ell\| < \infty$. Effectively, if $\|\ell\| = \infty$, then:

$$\forall n > 0 : \exists u_n \in V : \|u_n\| = 1 \; and \; \ell(u_n) \geq n^2.$$

Then

$$v_n = \frac{1}{n} u_n \rightarrow 0 \Longrightarrow \ell(v_n) \rightarrow \ell(0) = 0 .$$

But

$$\ell(v_n) = \frac{1}{n} \ell(u_n) \geq \frac{n^2}{n} = n \longrightarrow +\infty .$$

Thus, $0 > +\infty$, what is a contradiction. ■

COROLLARY 3.3.– Let $\ell: V \rightarrow \mathbb{R}$ be a linear functional. ℓ is continuous if and only if there exists $M \in R$ such that $\forall v \in V: |\ell(v)| \leq M\|v\|$. In this case, $\|\ell\| \leq \mathrm{M}$. ■

PROOF.– Assume that ℓ is continuous. Then $\mathrm{M} = \|\ell\|$ verifies

$$\frac{|\ell(v)|}{\|v\|} \leq M , \forall v \neq 0 \Longrightarrow |\ell(v)| \leq M\|v\|, \forall v \in V.$$

Assume that there exists $M \in R$ such that $\forall v \in V: |\ell(v)| \leq M\|v\|$. Then

$$\frac{|\ell(v)|}{\|v\|} \leq M , \forall v \neq 0 \Longrightarrow \|\ell\| \leq M < \infty$$

and ℓ is continuous. ■

EXAMPLE 3.10.– Let us give an example of a linear functional ℓ which is not continuous and has $\|\ell\| = \infty$. Let us consider

$\mathcal{V} = \{v: (-1,1) \to \mathbb{R},\ \ v \text{ is continuous on } \Omega\}$, $(u, v) = \int_{-1}^{1} u(x)v(x)dx.$

Let $\ell: \mathcal{V} \to \mathbb{R}$ be given by $\ell(v) = v(0)$. Let us consider the sequence

$$u_n(x) = \begin{cases} n^\alpha \left(\frac{1}{n} - |x|\right), if\ |x| < \frac{1}{n} \\ 0,\ otherwise. \end{cases}$$

Then, for $0 < \alpha < 1/2$,

$$(u_n, u_n) \leq \frac{n^{2\alpha}}{2n} \leq n^{2\alpha-1} \to 0\,,$$

while

$$\ell(u_n) = n^\alpha \to +\infty\,.$$

Assume that there exists $M \in \mathbb{R}$ such that $\forall v \in V: |\ell(v)| \leq M\|v\|$. Then

$$n^\alpha = |\ell(u_n)| \leq M\|u_n\| \leq M\, n^{\alpha-1/2}$$

So that

$$M \geq \sqrt{n} \longrightarrow +\infty,$$

what is a contradiction, since $M \in \mathbb{R}$. Thus, $\nexists M \in \mathbb{R}$ such that $|\ell(v)| \leq M\|v\|, \forall v \in V$: ℓ is not continuous and $\|\ell\| = \infty$. This functional is kwown as *Dirac's delta* or *Dirac's mass* (see Chapter 6). ■

An important property of continuous linear functionals is the *continuous extension*:

THEOREM 3.24.– Let $S \subset V$ be a dense subspace. Let $\ell: S \to \mathbb{R}$ be such that there exists $M \in \mathbb{R}$ such that $\forall v \in S: |\ell(v)| \leq M\|v\|$. In this case, ℓ has an extension $\ell: V \to \mathbb{R}$ such that $\|\ell\| \leq$ M. ■

PROOF.– Let $v \in V$. Since S is dense on V, there is a sequence $\{v_n : n \in \mathbb{N}^*\} \subset S$ such that $v_n \to v$. Let us consider the sequence of real numbers $\{\ell(v_n) : n \in \mathbb{N}^*\} \subset \mathbb{R}$. This is a Cauchy sequence, since

$$\|v_m - v_n\| \leq \frac{\varepsilon}{M} \Longrightarrow |\ell(v_m) - \ell(v_n)| = |\ell(v_m - v_n)| \leq M\|v_m - v_n\| \leq \varepsilon.$$

Then, $\ell(v_n) \to \lambda \in \mathbb{R}$. Let $\{w_n : n \in \mathbb{N}^*\} \subset S$ such that $w_n \to v$ and $\ell(w_n) \to \eta \in \mathbb{R}$. Then,

$$|\ell(w_n) - \ell(v_n)| = |\ell(w_n - v_n)| \leq M\|w_n - v_n\| \longrightarrow M\|v - v\| = 0,$$

So that $\lambda = \eta$. Thus, we may define

$$\ell(v) = \lim_{n \to \infty} \ell(v_n)\,.$$

ℓ is linear: if $S \ni v_n \to v \in V, S \ni u_n \to u \in V, \alpha \in \mathbb{R}$, then

$$\ell(u_n + \alpha v_n) = \ell(u_n) + \alpha\ell(v_n) \longrightarrow \ell(u) + \alpha\ell(v),$$

so that $\ell(u + \alpha v) = \ell(u) + \alpha\ell(v)$. Finally, by taking the limit for $n \to \infty$,

$$|\ell(v_n)| \leq M\|v_n\| \Longrightarrow |\ell(v)| \leq M\|v\|\,.$$ ■

DEFINITION 3.29.– The *topological dual* space V' associated with V is

$$V' = \{\ell : V \to \mathbb{R},\ \ \ell \textit{ is linear an continuous}\}.$$ ■

3.9.2. *Kernel of a linear functional*

DEFINITION 3.30.– The *Kernel* of a linear functional ℓ is

$$Ker(\ell) = \{v \in V : \ell(v) = 0\}\,.$$ ■

Notice that $Ker(\ell)$ is a subspace, since

$$0 \in Ker(\ell)$$

and, since $\ell(\alpha u + v) = \alpha\ell(u) + \ell(v)$,

$$\alpha \in R;\ u, v \in Ker(\ell) \implies \alpha u + v \in Ker(\ell)\,.$$

We have

PROPOSITION 3.8.– Let $\ell: V \to \mathbb{R}$ be a linear functional. ℓ is continuous if and only if $Ker(\ell)$is closed. ■

PROOF.– Assume that ℓ is continuous. Then

$$u_n \to u \implies \ell(u_n) \to \ell(u),$$

so that

$$\ell(u_n) = 0, \forall n \ and\ u_n \to u \implies \ell(u) = 0$$

and

$$\{u_n\}_{n>0} \subset Ker(\ell)\ and\ u_n \to u \implies u \in Ker(\ell)\,.$$

Thus $Ker(\ell)$ is closed.

Assume that $Ker(\ell)$ is closed. Let us establish that $\|\ell\| < \infty$. This is immediate if $\ell = 0$, since, in this case, $\|\ell\| = 0$. Let us assume $\ell \neq 0$ and $\|\ell\| = \infty$. Then, on the one hand,

$$\forall n > 0 : \exists v_n \in V : \ \|v_n\| = 1\ and\ \ell(v_n) \geq n^2.$$

and, on the other hand, there exists $\bar{u}$ such that $\ell(\bar{u}) = 1$. Let

$$u_n = \bar{u} - \frac{1}{\ell(v_n)} v_n\,.$$

Then

$$\ell(u_n) = \ell(\bar{u}) - \frac{1}{\ell(v_n)}\ell(v_n) = 1 - 1 = 0,$$

so that $\{u_n\}_{n>0} \subset Ker(\ell)$. Moreover,

$$\left\|\frac{1}{\ell(v_n)} v_n\right\| = \frac{1}{\ell(v_n)}\|v_n\| \leq \frac{1}{n^2} \longrightarrow 0\,,$$

so that $u_n \longrightarrow \bar{u}$. Since $Ker(\ell)$ is closed, we have $\bar{u} \in Ker(\ell) \Longrightarrow \ell(\bar{u}) = 0$. Thus, $1 = \ell(\bar{u}) = 0$, what is a contradiction. ■

3.9.3. *Riesz's theorem*

The main result is the *Riesz's theorem* [RSZ 09, FRE 07a]:

THEOREM 3.25.– Let $\ell: V \rightarrow \mathbb{R}$ be a continuous linear functional. Then there exists one and only one $u \in V$ such that:

$$(u, v) = \ell(v), \forall v \in V .$$

Moreover

$$\|\ell\| = \|u\| .$$ ■

PROOF.– The result is immediate when $\ell = 0$, since, in this case, $\|\ell\| = 0$ and $u = 0$. Let us assume $\ell \neq 0$. Then, there exists w such that $\ell(w) = 1$. Let us denote Pw its orthogonal projection onto $Ker(\ell)$. We have $w - Pw \neq 0$, since $\ell(w) = 1 \neq 0 = \ell(Pw)$. Let us consider,

$$u = \frac{w-Pw}{\|w-Pw\|^2} .$$

From the properties of the orthogonal projection, we have $u \perp Ker(\ell)$, so that $(u, Pw) = 0$. Moreover,

$$(u, w) = (u, w - Pw) + (u, Pw) = (u, w - Pw) = \frac{\|w-Pw\|^2}{\|w-Pw\|^2} = 1.$$

Let $v \in V$: we have,

$$v = \ell(v)w + (v - \ell(v)w) .$$

Since $\ell(v - \ell(v)w) = \ell(v) - \ell(v)\ell(w) = \ell(v) - \ell(v) = 0$, we have $v - \ell(v)w \in Ker(\ell)$ and

$$(u, v) = (u, \ell(v)w) = \ell(v)(u, w) = \ell(v) .$$

Moreover,

$$\|\ell\| \geq \frac{\ell(u)}{\|u\|} = \frac{(u,u)}{\|u\|} = \|u\|$$

and

$$\forall v \in V: |\ell(v)| = |(u,v)| \leq \|u\|\|v\| \Longrightarrow \|\ell\| \leq \|u\|$$

so that $\|\ell\| = \|u\|$. ■

Riesz's theorem establishes a connection between linear functionals and scalar products. Since scalar products represent the virtual work of internal efforts and linear functionals represent the virtual work of external forces, Riesz's theorem may be interpreted as a connection between the work of external forces and the variation of internal energy, which corresponds to the work of internal efforts. More precisely, Riesz's theorem connects the virtual work of external forces to the virtual work of internal efforts.

Moreover, Riesz's theorem connects a linear functional and a field: it may also be interpreted as the connection between a virtual work and a force. This connection is formalized by *Riesz canonical isometry*:

$$\Pi: V \to V', \Pi(u) = \ell, \ell(v) = (u,v) .$$

Riesz's theorem shows that Π is a bijection, so that Π^{-1} exists. Moreover, on the one hand, $\|\Pi(u)\| = \|u\|$, while, on the other hand, $\|\Pi^{-1}(\ell)\| = \|\ell\|$.

3.10. Generating a Hilbert basis

There are two basic ways for the generation of a Hilbert basis:

1) If a total family $\{\phi_n\}_{n\in\mathbb{N}}$ is given, we may use the Gram-Schmidt procedure;

2) Otherwise, we may use the eigenfunctions $\phi_n$associated with a pair order differential operator *T*:$\phi_n \neq 0, T(\phi_n) = \lambda_n \phi_n$, with

homogeneous boundary conditions. For instance, on $\Omega = (0,1)$, you may use $T(u) = u''$: $\phi_n \neq 0$, $\phi_n'' = \lambda_n \phi_n$, $\phi_n(0) = \phi_n(1) = 0$.

The use of a differential operator of pair order is justified by the fact that the inverse of T is a linear bounded compact self-adjoint operator. Let us recall that:

1) A linear operator T on V is an application $T: V \to V$ such that $T(\alpha u + v) = \alpha T(u) + T(v), \forall u, v \in V, \alpha \in \mathbb{R}$.

2) A bounded linear operator T on V is a linear operator T on V such that $\exists\, M \in \mathbb{R} : \|T(v)\| \leq M\|v\|, \forall v \in V$.

3) A compact bounded linear operator T on V is a bounded linear operator T on V which transforms weakly convergent sequences into strongly convergent sequences, i.e. $(u_n, v) \to (u, v), \forall\, v \in V \Rightarrow \|T(u_n) - T(u)\| \to 0$.

4) T is self-adjoint if and only if $(T(u), v) = (u, T(v)), \forall\, u, v \in V$.

Compact bounded linear operators may be considered as a generalization of matrices to infinite dimensional spaces. They may be generated by the inversion of differential or partial differential operators. We have

THEOREM 3.26.– Let $T: V \to V$ be a linear, bounded, compact, self-adjoint operator. Then, there exists a hilbertian basis of V formed by eigenvectors of T. ■

3.10.1. *1D situations*

For instance, let us consider $\Omega = (a, b)$ and the space $V = L^2(\Omega)$. The operator defined by:

$$\big(T(u)\big)'' = u, \; T(u)(a) = T(u)(b) = 0$$

is linear, bounded, compact and self-adjoint. Let ϕ be an eigenvector of T, associated with the eigenvalue α:

$$\phi \neq 0 \,, T(\phi) = \alpha\phi \,.$$

We have:

$$\phi \neq 0\,, \alpha\phi'' = \phi, \phi(a) = \phi(b) = 0.$$

Let us observe that $\alpha_n \neq 0$, since,

$$\alpha = 0 \Longrightarrow 0 = \alpha\phi'' = \phi \neq 0\,,$$

which is a contradiction. Thus, denoting $\lambda = 1/\alpha$, we have:

$$\phi \neq 0\,, \phi'' = \lambda\phi, \phi(a) = \phi(b) = 0.$$

In this case, the solution is *a trigonometric family*:

$$\phi_n(x) = A_n sin\left(n\pi\frac{x-a}{b-a}\right).$$

3.10.2. *2D situations*

Analogously, if $\Omega \subset \mathbb{R}^2$ is a regular bounded domain, $H = L^2(\Omega)$,

$$\Delta\big(T(u)\big) = u,\ \ T(u) = 0\ \ on\ \ \partial\Omega$$

is also linear, bounded, compact and self-adjoint. Let ϕ be an eigenvector of T, associated with the eigenvalue α:

$$\phi \neq 0\,, T(\phi) = \alpha\phi\,.$$

We have:

$$\phi \neq 0\,, \alpha\Delta\phi = \phi\ on\ \Omega, \phi = 0\ \ on\ \ \partial\Omega.$$

Here again, $\alpha_n \neq 0$, since,

$$\alpha = 0 \Longrightarrow 0 = \alpha\Delta\phi = \phi \neq 0\,,$$

what is a contradiction? Thus, denoting $\lambda = 1/\alpha$, we have:

$$\phi \neq 0\,, \Delta\phi = \lambda\phi, \phi = 0\ \ on\ \ \partial\Omega.$$

In this case, the solution is not explicitly known for general domains: it is explicitly known only for particular shapes of Ω. Otherwise, it must be numerically determined. For special geometries, the eigenfunctions are often determined by separation of variables.

3.10.2.1. *The case of a rectangle*

If $\Omega = (a_x, b_x) \times (a_y, b_y)$, we may consider $(x, y) = X(x)Y(y)$. Then,

$$\Delta\phi = X''Y + XY''.$$

Thus, we have,

$$XY \neq 0\,, X''Y + XY'' = \lambda XY \;\; on\; \Omega.$$

So,

$$\frac{X''}{X} + \frac{Y''}{Y} = \lambda \;\; on\; \Omega.$$

and:

$$\frac{X''}{X} = \gamma\,, \frac{Y''}{Y} = \beta \quad on\; \Omega(\alpha + \beta = \lambda).$$

Moreover,

$$\phi = 0 \;\; on \;\; \partial\Omega \Leftrightarrow X(a_x) = X(b_x) = Y(a_y) = Y(b_y) = 0\,.$$

Thus, we have:

$$XY \neq 0\,, X''Y + XY'' = \lambda XY, X(a_x) = X(b_x) = Y(a_y) = Y(b_y) = 0\,.$$

So,

$$\frac{X''}{X} = \gamma\,, X(a_x) = X(b_x) = 0\,; \frac{Y''}{Y} = \beta\,, Y(a_y) = Y(b_y) = 0\,.$$

The solution is a *product of trigonometric families:* $X = u_m$ and $Y = u_n$ are solutions of the 1D case,

$$\phi_{mn}(x) = A_{mn} sin\left(m\pi \frac{x-a_x}{b_x-a_x}\right) sin\left(n\pi \frac{y-a_y}{b_y-a_y}\right)$$

3.10.2.2. *The case of a disk*

If $\Omega = \{(x,y) : x^2 + y^2 < a^2\}$, we may consider $(x,y) = R(r)\Theta(\theta), = rcos\theta, y = rsin\,\theta$. Let us denote by μ' the derivative of a function $\mu(\xi)$ with respect to its single real variable $\xi \in \mathbb{R}$ – for instance $R' = dR/dr, \Theta' = d\Theta/d\theta$, etc. Since

$$\Delta\phi = \frac{1}{r}\frac{\partial}{\partial r}\left(r\frac{\partial\phi}{\partial r}\right) + \frac{1}{r^2}\frac{\partial^2\phi}{\partial\theta^2},$$

we have:

$$\frac{1}{r}(rR')'\Theta + \frac{1}{r^2}R\Theta'' = \lambda R\Theta$$

and

$$\frac{(rR')'}{rR} + \frac{\Theta''}{r^2\Theta} = \lambda\,.$$

So,

$$\frac{\Theta''}{\Theta} = \beta\,, \frac{(rR')'}{rR} + \frac{\beta}{r^2} = \lambda\ \ .$$

Thus,

$$r(rR')' + (\beta - \lambda r^2)R = 0\,, R(a) = 0\,;\ \Theta'' = \beta\Theta\,,$$

Let us introduce $\rho = r\sqrt{-\lambda}, y(\rho) = R(\rho/\sqrt{-\lambda})$: we have (recall that $y' = dy/d\rho$, while $R' = dR/dr$)

$$\rho y'(\rho) = rR'(r),\ \rho\big(\rho y'(\rho)\big)' = r(rR'(r))'$$

and

$$\rho\big(\rho y'(\rho)\big)' + (\beta + \rho^2)y = 0\,, y\big(a\sqrt{-\lambda}\big) = 0\,,$$

For $\beta = -m^2$, this is the classical *Bessel's equation* and Θ is a *trigonometric function*. Thus, denoting by J_m the *m-th* Bessel's function and by z_{mn} its *n-th* zero:

$$\phi_{mn}(x) = J_m\left(z_{mn}\frac{r}{a}\right)\left(A_{mn}cos(m\theta) + B_{mn}sin(m\theta)\right) \ , J_m(z_{mn}) = 0 \, .$$

3.10.3. *3D situations*

If $\Omega \subset \mathbb{R}^3$ is a regular bounded domain, $H = L^2(\Omega)$,

$$\Delta\big(T(u)\big) = u, \ T(u) = 0 \ on \ \partial\Omega$$

is again linear, bounded, compact and self-adjoint. Thus, an eigenvector ϕ, associated with the eigenvalue α verifies, with $\lambda = 1/\alpha$,

$$\phi \neq 0 \, , \Delta\phi = \lambda\phi, \phi = 0 \ on \ \partial\Omega.$$

Here, the solution is explicitly known only for particular shapes of Ω and it is often determined by separation of variables.

3.10.3.1. *The case of a rectangular parallelepiped*

If $\Omega = (a_x, b_x) \times (a_y, b_y) \times (a_z, b_z)$, we may consider $(x, y, z) = X(x)Y(y)Z(z)$. Then, we obtain:

$$\phi_{mnk}(x) = A_{mn}sin\left(m\pi\frac{x-a_x}{b_x-a_x}\right)sin\left(n\pi\frac{y-a_y}{b_y-a_y}\right)sin\left(k\pi\frac{z-a_z}{b_z-a_z}\right)$$

3.10.3.2. *The case of a cylinder*

$\Omega = \{(x, y, z) : x^2 + y^2 < a^2, a_z \leq z \leq b_z\}$, we may consider $(x, y, z) = R(r)\Theta(\theta)Z(z), x = rcos\theta, y = rsin\theta$. Since,

$$\Delta\phi = \frac{1}{r}\frac{\partial}{\partial r}\left(r\frac{\partial\phi}{\partial r}\right) + \frac{1}{r^2}\frac{\partial^2\phi}{\partial\theta^2} + \frac{\partial^2\phi}{\partial z^2} \, ,$$

we have:

$$\frac{(rR')'}{rR} + \frac{\Theta''}{r^2\Theta} + \frac{Z''}{Z} = \lambda \, .$$

Thus,

$$\frac{Z''}{Z} = \gamma \, , \frac{\Theta''}{\Theta} = \beta \, , \frac{(rR')'}{rR} + \frac{\beta}{r^2} + \gamma = \lambda \, .$$

So,

$$r(rR')' + (\beta - (\lambda - \gamma)r^2)R = 0 \, , R(a) = 0 \, ;$$

$$\Theta'' = \beta\Theta \; ;$$

$$Z \;\; '' = \gamma Z \, , Zaz = Zb_z = 0 \, ;$$

For $\beta = -m^2 \, , \gamma = -k^2$, we take $\rho = r\sqrt{-\lambda + \gamma}, y(\rho) = R(\rho/\sqrt{-\lambda + \gamma})$ and we obtain, analogously to the 2D-case,

$$\rho\big(\rho y'(\rho)\big)' + (\beta + \rho^2)y = 0 \, , y\big(a\sqrt{-\lambda + \gamma}\big) = 0.$$

Thus, y is again the solution of Bessel's equation, while Z and Θ are trigonometric functions and we have:

$$\phi_{mnk}(x) = J_m\left(z_{mn}\frac{r}{a}\right)\left(A_{mnk}\cos(m\theta) + B_{mnk}\sin(m\theta)\right)\sin\left(k\pi\frac{z - a_z}{b_z - a_z}\right), J_m(z_{mn}) = 0.$$

3.10.3.3. *The case of a ball*

For $\Omega = \{(x, y, z) : x^2 + y^2 + z^2 < a^2\}$, we may consider $\phi(x, y, z) = R(r)\Theta(\theta)\Phi(\varphi), x = rcos\theta sin\varphi, y = rsin\theta sin\varphi, z = rcos\varphi$ associated with the eigenvalue α. Since,

$$\Delta\phi = \frac{1}{r^2}\frac{\partial}{\partial r}\left(r^2\frac{\partial\phi}{\partial r}\right) + \frac{1}{r^2 sin^2\varphi}\frac{\partial^2\phi}{\partial\theta^2} + \frac{1}{r^2 sin\varphi}\frac{\partial}{\partial\varphi}\left(sin\varphi\frac{\partial\phi}{\partial\varphi}\right),$$

we have, for $\lambda = 1/\alpha$,

$$\frac{(r^2R')'}{r^2R} + \frac{\Theta''}{r^2\Theta sin^2\varphi} + \frac{(\Phi' sin\varphi)'}{r^2\Phi sin\varphi} = \lambda$$

and

$$\frac{\Theta''(\theta)}{\Theta(\theta)} = \beta \, ; \; \frac{(\Phi'(\varphi)sin\varphi)'}{\Phi(\varphi)sin\varphi} + \frac{\beta}{sin^2\varphi} = \gamma \; ; \; \frac{(r^2R')'}{r^2R} + \frac{\gamma}{r^2} = \lambda$$

For $\beta = -m^2$, $\gamma = -k(k+1)$, $s = cos\varphi$, $y(s) = \Phi\left(cos^{-1}(s)\right)$, we have (recall that $y' = dy/ds$, while $\Phi' = d\Phi/d\varphi$)

$$\sqrt{1-s^2}y'(s) = -\Phi'\left(cos^{-1}(s)\right) = -\Phi'(\varphi)$$

and

$$(1-s^2)y'(s) = -\Phi'(\varphi)sin\varphi.$$

Thus,

$$\left((1-s^2)y'(s)\right)' = \frac{1}{\sqrt{1-s^2}}(\Phi'(\varphi)sin\varphi)' = \frac{(\Phi\prime(\varphi)sin\varphi)\prime}{sin\varphi}$$

and

$$\left((1-s^2)y'(s)\right)' + \left(\frac{\beta}{1-s^2} - \gamma\right)y(s) = 0.$$

This equation is known as *Legendre's equation*. The solutions are the *associated Legendre functions* $P_k^m(s)$. Moreover, Θ is a trigonometric function and R verifies

$$(r^2R')' + (\gamma - \lambda r^2)R = 0\,, R(a) = 0\,.$$

Taking again $\rho = r\sqrt{-\lambda}, y(\rho) = R(\rho/\sqrt{-\lambda})$, we have:

$$r^2R'(r) = \frac{1}{\sqrt{-\lambda}}\rho^2y'(\rho) \Rightarrow (r^2R')' = (\rho^2y')'$$

and

$$(\rho^2y')' + (\gamma - \rho^2)y = 0\,, y\left(a\sqrt{-\lambda}\right) = 0\,.$$

The solutions of this equation are the *spherical Bessel's functions*. Thus, the eigenfunctions are:

$$\phi_{mnk}(x) = \frac{1}{\sqrt{r}}J_{m+1/2}\left(z_{m+1/2,n}\frac{r}{a}\right)\left(A_{mnk}cos(m\theta) + B_{mnk}sin(m\theta)\right)P_m^k(cos\varphi)\,, J_{m+1/2}\left(z_{m+1/2,n}\right) = 0\,.$$

$z_{m+1/2,n}$ is the n-th zero of $J_{m+1/2}$. The angular term $\left(A_{mnk}cos(m\theta) + B_{mnk}sin(m\theta)\right)P_m^k(cos\varphi)$ is called a *spherical harmonic*.

3.10.4. *Using a sequence of finite families*

An alternative to the generation of a hilbertian basis consists of the use of *a family of finite families*, i.e. $F \subset V$ such that,

$$F = \cup_{n=n_0}^{\infty} F_n \ , \ F_n = \{\phi_{n,1}, \dots, \phi_{n,n}\}, \ F \text{ dense on } H.$$

For instance, families of finite elements may be used: in such a case, F_n corresponds to a particular discretization (i.e. a particular mesh) and F is the family of all the possible discretizations (i.e. of all the possible meshes). Finite volume approximations or wavelets may be interpreted by the same way.

3.11. Exercises

EXERCISE 3.1.– Let V be a vector space on $\mathbb{R}$.

1) Show that $-1.u = -u, \forall u \in V$. (Hint: $1.u = u$)

2) Show that $0.u = 0, \forall u \in V$. (Hint: $0 = 0 + 0$). ■

EXERCISE 3.2.– Let $\Omega \subset \mathbb{R}^n$ and $V = \{v: \Omega \to \mathbb{R}^k\}$.

1) Show that V is a vector space on $\mathbb{R}$.

2) Show that

$$S = \left\{v \in V : \int_\Omega |v(x)|^2 dx < \infty\right\}$$

is a vector subspace of V. ■

EXERCISE 3.3.– Let d be a distance on V and $\eta: \mathbb{R} \to \mathbb{R}$ be a strictly increasing application such that $\eta(0) = 0$ and $\eta(a + b) \leq \eta(a) +$

$\eta(b)$. Show that $d_\eta(u,v) = \eta\big(d(u,v)\big)$ is a distance on V. (Hint: $\eta(d(u,v)) \leq \eta\big(d(u,w) + d(w,v)\big)\big)$. ■

EXERCISE 3.4.– Show that $|\|u\| - \|v\|| \leq \|u - v\|$ ■

EXERCISE 3.5.– Consider

$$(u,v) = \int_0^T u(x)v(x)dx$$

Let $\theta(m,n)$ be the angle between t^m and t^n. Show that

$$\theta(m,n) = cos^{-1}\left(\frac{\sqrt{(2n+1)(2m+1)}}{m+n+1}\right).$$ ■

EXERCISE 3.6.– Consider

$$(u,v) = \int_0^T u(x)v(x)dx$$

Let θ be the angle between $sin(t)$ and $cos(t)$. Show that:

$$\theta = cos^{-1}\left(\frac{sin^2(T)}{\sqrt{sin^2(T)(T - sin(T)cos(T))}}\right).$$ ■

EXERCISE 3.7.– Let $x \in \mathbb{R}^n$ et $A \subset \mathbb{R}^n$, $d(x,A) = \inf\{\| x - a \| : a \in A\}$. Determine $d(0, \mathbb{R} - \mathbb{Q}), d(\sqrt{2}, \mathbb{Q})$, $d(M,D)$, where $M = (x,y,z) \in \mathbb{R}^3$ and D is the line $D = \{\lambda(a,b,c) : \lambda \in \mathbb{R}\}$. ■

EXERCISE 3.8.– Let $A \subset \mathbb{R}^n$, $B \subset \mathbb{R}^n$, $d(A,B) = \inf\{\| x - a \| : a \in A, b \in B\}$. Find $d(A,B)$ when A is the hyperbole $A = \{(x,y) \in \mathbb{R}^2 : x_1x_2 = 1\}$ and B is one of its asymptotes. ■

EXERCISE 3.9.– Let $D = \{x \in \mathbb{R} \,|\, x = p + q\sqrt{2},\ p \in \mathbb{Z}, q \in \mathbb{Z}\}$. Show that:

1) $x \in D, y \in D \Rightarrow x + y \in D$.

2) $x \in D, y \in D \Rightarrow xy \in D$.

3) $x \in D \Rightarrow x^n \in D, \forall n \geq 1$.

4) Let $u = \sqrt{2} - 1$. Show that, for any $(a, b) \in \mathbb{R}^2$, $a < b$, there exists $n \geq 1$ such that$0 < u^n < b - a$. Conclude that there exists $m \in \mathbb{Z}$ such that$a < mu^n < b$.

5) Show that: $\forall x \in \mathbb{R} : \forall \varepsilon > 0, \exists x_\varepsilon \in D, |x - x_\varepsilon| \leq \varepsilon$. Conclude that $d(x, D) = 0$. ■

EXERCISE 3.10.– Let $d(x, y) = |x - y|$, $f: \mathbb{R} \to \mathbb{R}$ satisfy: $\forall a > 0$, $f^{-1}((a, +\infty))$ is finite. Let $D = \{x \in R | f(x) = 0\}$. Show that $\forall x \in R: \forall \varepsilon > 0, \exists x_\varepsilon \in D, |x - x_\varepsilon| \leq \varepsilon$. ■

EXERCISE 3.11.– Let $E = C^0(I; R^n)$; $I = (0,1)$. We denote, for $p \geq 1$ and $u \in E$,

$$\|u\|_p = \left(\int_I |u|^p\right)^{1/p} .$$

The aim is to establish that $\|\bullet\|_p$ is a norm E and that E is not a complete space for this norm.

1) As a first step, establish Hölder's inequality: for $p, q \geq 1$ such that $\frac{1}{p} + \frac{1}{q} = 1$ (conjugate exponents),

$$\forall u, v \in E \ : \quad \int_I |u||v| \leq \|u\|_p \|v\|_q .$$

a) Let $f(t) = \log(t)$ (natural basis). Show that $f'' < 0$ on $\mathbb{R}_+^* = (0, +\infty)$. Conclude that f is concave on $\mathbb{R}_+^*$ and that:

$$\forall \theta \in (0,1), t_1, t_2 \in \mathbb{R}_+^* \ : \ f(\theta t_1 + (1 - \theta) t_2) \geq \theta f(t_1) + (1 - \theta) f(t_2) .$$

b) Show that

$$\forall t_1, t_2 \in \mathbb{R}_+^* \ : \ t_1^{1/p} t_2^{1/q} \leq \frac{1}{p} t_1 + \frac{1}{q} t_2 .$$

c) Conclude that

$$\forall a, b \in \mathbb{R}_+^* \ : \ ab \leq \frac{1}{p} a^p + \frac{1}{q} b^q .$$

d) Extend the result to$\forall a, b \in \mathbf{R}_+ = [0, +\infty)$.

e) Take $= |u|/\|u\|_p$, $b = |v|/\|v\|_q$ and conclude.

2) As a second step, establish Minkowski's inequality:

$$\forall u, v \in E \ : \quad \|u+v\|_p \leq \|u\|_p + \|v\|_p \ .$$

a) Set $r = (|u| + |v|)^p$, $s = (|u| + |v|)$, $t = (|u| + |v|)^{p-1}$. Verify that $r = st = |u|t + |v|t$.

b) Verify that

$$\textstyle\int_I |u|t \leq \|u\|_p \|t\|_q \quad \text{and} \quad \int_I |v|t \leq \|v\|_p \|t\|_q.$$

c) Conclude that

$$\textstyle\int_I r \leq \left(\|u\|_p + \|v\|_p\right)\|t\|_q.$$

d) Verify that $(p-1)q = p$. Conclude that $\|t\|_q = \left(\|s\|_p\right)^{p-1}$.

e) Show Minkowski's inequality.

3) Establish th $\|\bullet\|_p$ atis a norm on E.

4) Consider the sequence $\{u_n : n \geq 0\}$ given by

$$u_n(x) = \begin{cases} 1, \text{if } 0 < x \leq 1/2 \ ; \\ 1 - n(x - 1/2), \text{if } \frac{1}{2} < x \leq \frac{1}{2} + \frac{1}{n}; \\ 0, \text{if } x \geq \frac{1}{2} + \frac{1}{n} \end{cases} .$$

a) Let

$$\overline{u}(x) = \begin{cases} 1, \text{si } 0 < x < 1/2 \ ; \\ 0, \qquad \text{si} x \geq \frac{1}{2} \end{cases} .$$

Show that $\|u_n - \overline{u}\|_p \leq \left(\frac{1}{n}\right)^{1/p}$.

b) Conclude that $\{u_n : n \geq 0\}$ is a Cauchy sequence of elements of $(E, \|\bullet\|_p)$.

c) Let $u \in E$ verify $u_n \underset{n \to +\infty}{\to} u$ on $(E, \|\bullet\|_p)$. Show that $\|u - \overline{u}\|_p = 0$. Conclude that $u \notin E$ and that on $(E, \|\bullet\|_p)$ is not complete. ■

5) Let $b \in \mathbb{R}$, $b \geq 1$ and $a \in \mathbb{R}$, $1 \leq a < b$.

a) Let $p = \frac{b}{b-a}$, $q = \frac{b}{a}$. Verify that $p \geq 1$, $q \geq 1$, $\frac{1}{p} + \frac{1}{q} = 1$ and use Holder's inequality to show that

$$(\|u\|_a)^a \leq \|\mathbf{1}\|_p (\|u\|_b)^a$$

b) *Since I is bounded*, show that there exists a constant $C_b(a, I)$ such that

$$\|u\|_a \leq C_b(a, I)\ \|u\|_b\ .$$

6) Let

$$\|u\|_\infty = \sup\{|u(x)| : x \in I\}\ .$$

a) Show that $\|\bullet\|_\infty$ is a norm on E.

b) Verify that

$$\forall u, v \in E\ :\quad \int_I |u||v| \leq \|u\|_\infty \|v\|_1\ .$$

This inequality justifies the generalization where the conjugate exponent of $p = \infty$ is $q = 1$ and the convention $\frac{1}{\infty} = 0$.

c) *Since I is bounded*, show that there exists a constant $C_\infty(a, I)$ such that, for $a \in \mathbb{R}$, $a \geq 1$,

$$\|u\|_a \leq C_\infty(a, I)\ \|u\|_\infty\ .$$

7) Let $p \geq 1$, $q \geq 1$ be conjugate exponents, eventually taking the value ∞, and u regular enough such that $u(0) = 0$. By convention $\frac{1}{\infty} = 0$.

a) Show that

$$\forall\ x, y \in I\quad :\quad |u(x) - u(y)| \leq \|u'\|_p |x - y|^{1/q}\ .$$

b) Conclude that

$$\forall \; x \in I \quad : \quad |u(x)| \leq \|u'\|_p |x|^{1/q} \qquad .$$

c) *Since I is bounded*, show that there exists a constant $M\ (a, I)$ such that

$$\|u\|_\infty \leq M(a, I) \quad \|u'\|_a \ .$$

Conclude that there exists a constant $M\ (a, p, I) \in \mathbb{R}$ such that:

$$\|u\|_a \leq M(a, p, I) \ \ \|u'\|_p \quad .$$

d) Let u be a regular function such that $u(0) = u'(0) = 0$. Verify that:

$$\|u'\|_a \leq M(a, p, I) \ \ \|u''\|_p \ .$$

Show that there exists a constant $N(a, p, I) \in \mathbb{R}$ such that:

$$\|u\|_a \leq N(a, p, I) \|u''\|_p \quad .$$

■

EXERCISE 3.12.– Let $I = (0, T)$, $f: I \times \mathbb{R}^n \to \mathbb{R}^n$, $y_0 \in \mathbb{R}^n$, $E = [L^1(I)]^n$. We consider the ordinary differential equation (ODE)

$$y' = f(t, y) \ \ (t \in I) \quad ; \quad y(0) = y_0 \ .$$

Assume that f satisfies a Lipschitz condition:

$$\forall \; t \in I; \; \xi_1, \xi_2 \in \mathbb{R}^n \quad : \quad |f(t, \xi_1) - f(t, \xi_2)| \leq K |\xi_1 - \xi_2| \ .$$

1) Show that:

$$\forall \; t \in I \; : \; y(t) = y_0 + \int_0^t f(s, y(s)) ds \ .$$

2) Let $F: E \to E$ be defined as follows :

$$F(y)(t) = y_0 + \int_0^t f(s, y(s)) ds \ .$$

a) Let $u, v \in E$. Verify that:

$$\forall \; s \in I : \; |f(s, v(s)) - f(s, u(s))| \; \leq K |v(s) - u(s)|.$$

b) Conclude that

$\forall\ t \in I\ :\ |F(v(t)) - F(u(t))|\ \le K\|v - u\|_1.$

c) Show that:

$\|F(v) - F(u)\|_1\ \le KT\|v - u\|_1.$

d) Let $\{u_n : n \ge 0\} \subset E$verify $u_n \to u$ on $(E, \|\blacksquare\|_1)$. Show that:

$\|F(u_n) - F(u)\|_1\ \le KT\|u_n - u\|_1.$

e) Conclude that F is continuous.

3) Let $T < \frac{1}{K}$.

a) Show that F is a contraction.

b) Conclude that there exists one and only one $y \in E$ such that $y = F(y)$.

c) Conclude that the ODE has a unique solution $y \in E$.

EXERCISE 3.13.– Let $\Omega = (0,1)$ and $(u, v) = \int_\Omega uvdx$, $\mathrm{C} = \{\mathrm{v}: \Omega \to \mathbb{R}: \mathrm{v} \ge 0 \text{ on } \Omega\}$.

1) Show that C is convex.

2) Let $u: \Omega \to \mathbb{R}$, $u^+(x) = max\ (u(x),0)$, $u^-(x) = max\ (-u(x),0)$. Verify that $u^+ \in C$, $u^- \in C$, $u^+u^- = 0$, $u = u^+ - u^-$.

3) Show that $(u - u^+, v - u^+) \le 0, \forall v \in C$.

4) Conclude that u^+ is the orthogonal projection of u onto C.

EXERCISE 3.14.– Let $\Omega = (0,1)$ and $(u, v) = \int_\Omega (uv + u'v')dx$. Let $S = \{v: \Omega \to \mathbb{R}: v(0) = 1\}$.

1) Show that S is an affine subspace.

2) Verify that $\{1\} + S_0$, $S_0 = \{v: \Omega \to \mathbb{R}: v(0) = 0\}$.

1) Let $U: \Omega \to \mathbb{R}$ satisfy $U(0) = 1, U'(1) = 0, U - U'' = 0 \; on \; \Omega$. Verify that:

$$\forall w \in S_0:\; U'(1)w(1) + \int_0^1 (U - U'')w dx = 0.$$

2) Conclude that $\forall w \in S_0: (U, w) = 0$.

3) Conclude that $\forall v \in S: (U, v - U) = 0$.

4) Show that U is the orthogonal projection of 0 onto S.

EXERCISE 3.15.– Let $\Omega = (0,1)$ and $(u, v) = \lambda u(0)v(0) + \int_\Omega u'v'dx$ $(\lambda > 0)$. Let $S = \{v: \Omega \to \mathbb{R}: v(1) = 1\}$.

1) Show that S is an affine subspace.

2) Verify that= $\{1\} + S_0$, $S_0 = \{v: \Omega \to \mathbb{R}: v(1) = 0\}$.

3) Let $U: \Omega \to \mathbb{R}$ satisfy $U(1) = 1, \lambda U(0) - U'(0) = 0, U'' = 0 \; on \; \Omega$. Verify that:

$$\forall w \in S_0:\; \big(\lambda U(0) - U'(0)\big)w(1) - \int_0^1 U''w dx = 0.$$

4) Conclude that $\forall w \in S_0: (U, w) = 0$.

5) Conclude that $\forall v \in S: (U, v - U) = 0$.

6) Show that U is the orthogonal projection of 0 onto S.

EXERCISE 3.16.– Let $\Omega = (0,1) \times (0,1)$ and $(u, v) = \lambda \int_{\partial\Omega} uv + \int_\Omega \nabla u \nabla v dx$ $(\lambda > 0)$. Let $S = \{v: \Omega \to \mathbb{R}: v = u_0 \; on \; \partial\Omega\}$.

1) Show that S is an affine subspace.

2) Verify that= $\{u_0\} + S_0$, $S_0 = \{v: \Omega \to \mathbb{R}: v = 0 \; on \; \partial\Omega\}$.

3) Let $U: \Omega \to \mathbb{R}$ satisfy $U = u_0 \; on \; \partial\Omega, \Delta U = 0 \; on \; \Omega$. Verify that:

$$\forall w \in S_0:\; \int_0^1 w\Delta U dx = 0.$$

4) Conclude that $\forall w \in S_0: (U, w) = 0$.

5) Conclude that $\forall v \in S\colon (U, v - U) = 0$.

6) Show that U is the orthogonal projection of 0 onto S.

EXERCISE 3.17.– Let $\Omega = (0,1) \times (0,1)$ and $(u, v) = \int_\Omega (uv + \nabla u \nabla v) dx$. Let $S = \{v\colon \Omega \to \mathbb{R}\colon v = u_0 \ on\ \partial\Omega\}$.

1) Show that S is an affine subspace.

2) Verify that= $\{u_0\} + S_0$, $S_0 = \{v\colon \Omega \to \mathbb{R}\colon v = 0\ on\ \partial\Omega\}$.

3) Let $U\colon \Omega \to \mathbb{R}$ satisfy $U = u_0\ on\ \partial\Omega, U - \Delta U = 0\ on\ \Omega$. Verify that:

$$\forall w \in S_0\colon \int_0^1 w(U - \Delta U) dx = 0.$$

4) Conclude that $\forall w \in S_0\colon (U, w) = 0$.

5) Conclude that $\forall v \in S\colon (U, v - U) = 0$.

6) Show that U is the orthogonal projection of 0 onto S

4

Functional Spaces for Engineers

Functional spaces are Hilbert spaces formed by functions. For engineers, these spaces represent spaces of quantities having a spatial or temporal distribution, i.e. *fields*. For instance, the field of temperatures on a region of the space, the field of velocities of a continuous medium, the history of the velocities of a particle is functions. As previously observed, these spaces have some particularities connected to the fact that they are infinite dimensional.

From the beginning of the 20th Century, the works of Richard Courant, David Hilbert, Kurt Friedrichs and other mathematicians of Gottingen have pointed the necessity of a redefinition of derivatives and of the functional spaces involving derivatives.

Otto Nykodim introduced a class named "BL" (class of Bepo-Levi), involving $(u, v) = \int \nabla u . \nabla v \, dx$ [NIK 33]. He constructed a theory about this class and brought it to the framework of Hilbert spaces two years later [NIK 35], but the complete theory arrived later with the works of Serguei Sobolev, a Russian mathematician that introduced functional spaces fitting to variational methods.

Sobolev was closely connected to Jacques Hadamard, who faced the mathematical difficulties arising in fluid mechanics, when studying the Navier–Stokes equations [HAD 03a, HAD 03b]. Far from Hadamard, the Russian Nikolai Gunther, a former PhD student of Andrey Markov, was interested in potential theory and, particularly, in extensions of Kirchhoff formula and derivatives of irregular functions by a new smoothing method. Gunther suggested, in collaboration with Vladimir Smirnov, a promising PhD student – Serguei Sobolev. Hadamard was a frequent visitor of Sobolev and was in contact with him since 1930. At these times, Sobolev was working at the Seismological Institute of the Academy of Sciences in Leningrad (now Saint Petersburg), under the direction of Smirnov, and studied the propagation of waves in inhomogeneous media [SOB 30, SOB 31]. During this work, he formulated a method of generalized solutions [SOB 34a, SOB 34b, SOB 35a, SOB 35b], which was rapidly diffused in France by Jacques Hadamard and Jean Leray. This last mathematician then applied the nascent – and imperfect – theory to fluid mechanics and obtained a fundamental result, by using a new notion of derivative [LER 34].

World War II interrupted the advances of variational methods, notably in the works of Laurent Schwartz and Israel Gelfand, who popularized the works of Sobolev among mathematicians and developed the extension of their applications considerably. The end of the war was to produce their more interesting works. For instance, the fundamental work of Schwartz appeared in 1945 [SCH 45] and the complete theory was published in 1950–1951 [SCH 50, SCH 51]. Gelfand published his major works in 1958–1962 [GEL 64a, GEL 68b, GEL 63c, GEL 64b, GEL 66]. The works of Schwarz and Gelfand diffused the new theory worldwide and led to fantastic developments in terms of solution of differential equations – we will study this aspect in the next chapter.

These works did not signal the end in the development of functional spaces connected to variational methods. Namely, the works of Jean-François Colombeau [COL 84] led to a response to the objections of Schwarz in [SCH 54] and the theory of hyper functions generalized his works [SAT 59a, SAT 60]. These new developments still need to find engineering applications and will thus not be studied in this book.

4.1. The $L^2(\Omega)$ space

The construction of complete function spaces is complex: as shown in the preceding examples, there is a contradiction between the assumptions of regularity used in order to verify that $(u, u) = 0 \Longrightarrow$

$u = 0$ and the assumption that Cauchy sequences have a limit in the space.

Such a difficulty has been illustrated, for instance, in example 3.5, with $\mathcal{V} = \{v: \Omega \subset \mathbb{R}^n \to \mathbb{R}: v \text{ is continuous on } \Omega \text{ and } (v, v) < \infty\}$ and

$$(u, v) = \int_{\Omega} u(x)v(x)dx$$

In this example, we have shown a sequence $\{u_n: n \in \mathbb{N}^*\}$ such that $u_n \to u$ strongly, but $u \notin \mathcal{V}$. In order to overcome this difficulty, the method of *completion* has been introduced. Completion consists of generating a new vector space, referred to as $L^2(\Omega)$ and formed by sets of Cauchy sequences on V (see [SOU 10]). $L^2(\Omega)$ may be interpreted as follows: let V be the set of continuous square summable functions:

$$V = \left\{v \in \mathcal{V}: \int_{\Omega} |v(x)|^2 dx < \infty\right\}.$$

and

$$[v] = \left\{w \in V: \int_{\Omega} |w(x) - v(x)|^2 dx = 0\right\}.$$

Then

$$L^2(\Omega) = \{[v]: v \in V\}; \ ([u], [v]) = \int_{\Omega} u(x)v(x)dx.$$

is a Hilbert space (see [SOU 10]). Indeed, in this case,

$$([u], [u]) \geq 0 \ \text{ and } \ ([u], [u]) = 0 \Leftrightarrow \int_{\Omega} |u(x)|^2 dx \Leftrightarrow [u] = [0].$$

In general, in order to alleviate the notation, the square brackets are not written and $[v]$ is denoted by v. When using this simple notation, we must keep in mind that v (i.e. $[v]$) contains infinitely many elements, so that $u(x)$ has no meaning. For instance, let us consider,

$$u_\alpha(x) = 0, if\ x \neq \frac{1}{2};\ u_\alpha\left(\frac{1}{2}\right) = \alpha\,.$$

Then, $\forall \alpha \in \mathbb{R}, u_\alpha \in [0]$. So, $[u_\alpha] = [0]$ and, $u_\alpha(x) = \alpha$, so that the class of the null function contains an infinite amount of elements and the value of $[0](\frac{1}{2})$ has no meaning. When denoting $[0]$ by 0, we must keep in mind that the values of 0 in a particular point vary. By these reasons, elements of $L^2(\Omega)$ are sometimes referred to as *generalized functions* and completion implies a loss of information: *point values are lost* when passing to the complete space $L^2(\Omega)$. In order to help the reader to keep in mind the difference between u and $[u]$, we will use the notation:

$$(u, v)_0 = ([u], [v]).$$

Thus, index zero recalls that the elements on the scalar products are not standard functions, but classes, for which point values are not defined.

It must be noted that the construction of $L^2(\Omega)$ may be carried using other spaces $\mathcal{V}$, such as, for instance,

$$\mathcal{V} = C^\infty(\Omega) = \{v: \Omega \to \mathbb{R}: v \textit{ is infinitely derivable on } \Omega \text{ and } (v, v) < \infty\},$$

$$\mathcal{V} = C_0^\infty(\Omega) = \{v \in C^\infty(\Omega): v = 0 \textit{ on } \partial\Omega\},$$

or

$$\mathcal{V} = \mathcal{D}(\Omega) = \{v \in C^\infty(\Omega): \exists A \subset \Omega, \text{A } \textit{closed bounded}, v(x) = 0, \textit{if } x \notin A\}$$

$\mathcal{D}(\Omega)$ is often referred to as the set of test functions. A square summable function may be approximated by elements of these spaces, namely by using *mollifiers* introduced by Sobolev [SOB 38] and Friedrichs [FRI 44, FRI 53]. Moreover, these spaces are dense in $L^2(\Omega)$.

In the following, this approach is generalized by Sobolev spaces, and a way to retrieve the notion of point values is given.

4.2. Weak derivatives

As observed above, passing from standard functions to generalized functions involves the loss of the notion of point values. This implies that the standard notion of derivative does not apply. Indeed, let $\Omega = (a,b), v \in L^2(\Omega)$: the quotient used to define the derivative at point $x \in \Omega$ reads as $\big(v(x+h) - v(x)\big)/h$ and has no meaning, since the values of $v(x)$ and $v(x+h)$ are undefined. Distribution theory proposed by Schwarz introduces a new concept, based on the integration by parts. Indeed, assume that $v: \Omega \to \mathbb{R}$ is regular enough. Then, we have:

$$\int_a^b v'(x)\varphi(x)dx = v(b)\varphi(b) - v(a)\varphi(a) - \int_a^b v(x)\varphi'(x)dx$$

and, for a regular function φ such that $\varphi(a) = \varphi(b) = 0$,

$$\int_a^b v'(x)\varphi(x)dx = -\int_a^b v(x)\varphi'(x)dx\,,$$

so that v' is characterized by the linear functional

$$\ell(\varphi) = -\int_a^b v(x)\varphi'(x)dx\,.$$

This remark suggests the following definition:

DEFINITION 4.1.– Let $\Omega \subset \mathbb{R}^n$ be a regular domain and $u \in L^2(\Omega)$. The *weak derivative* $\frac{\partial u}{\partial x_i}$ is the linear functional $\ell: \mathcal{D}(\Omega) \to \mathbb{R}$ defined by:

$$\ell_i(\varphi) = -\left(u, \frac{\partial \varphi}{\partial x_i}\right)_0. \qquad \blacksquare$$

Weak derivatives are also referred to as *distributional derivatives* or *generalized derivatives*. We have:

THEOREM 4.1.– Let $\ell_i: \mathcal{D}(\Omega) \to \mathbb{R}$ be the weak derivative $\partial u/\partial x_i$ and verify:

$$\exists M \in \mathbb{R} : \ |\ell_i(\varphi)| \leq M\|\varphi\|_0\,, \forall \varphi \in \mathcal{D}(\Omega)\,.$$

Then, there exists a unique $w_i \in L^2(\Omega)$ such that,

$$\ell_i(v) = (w_i, v)_0, \forall v \in L^2(\Omega).$$

In this case, we say that $w_i = \partial u/\partial x_i$. ■

PROOF–. Theorem 3.23 shows that ℓ_i extends to a linear continuous functional $\ell_i : L^2(\Omega) \to \mathbb{R}$. Therefore, the result is a consequence of Riesz (theorem 3.24). ■

COROLLARY 4.1.– Let $\psi \in C^1(\Omega)$ be such that both ψ and $\nabla\psi$ are square summable. Let $u = [\psi] \in L^2(\Omega)$. Then $\partial u/\partial x_i = \partial\psi/\partial x_i$, for $1 \le i \le n$, i.e. the weak and classical derivatives coincide. ■

PROOF.– We observe that:

$$\ell_i(\varphi) = -\left(u, \frac{\partial\varphi}{\partial x_i}\right)_0 = -\int_\Omega u\, div(w)dx \;, w_j = \delta_{ij}\varphi.$$

Thus, Green's formula:

$$\int_\Omega u\, div(w)dx = -\int_\Omega w\,\nabla u dx = -\int_\Omega \varphi \frac{\partial\psi}{\partial x_i} dx$$

yields:

$$\ell_1(\varphi) = \left(\frac{\partial\psi}{\partial x_i}, \varphi\right)_0,$$

so that, taking $M = \|\partial\psi/\partial x_i\|_0$,

$$|\ell_1(\varphi)| = \left|\left(\frac{\partial\psi}{\partial x_i}, \varphi\right)_0\right| \le M\|\varphi\|_0 \,, \forall\varphi \in \mathcal{D}(\Omega)$$

and the result follows from theorem 4.1. ■

REMARK 4.1.– Standard derivation rules apply to weak derivatives, under the assumption that each element is defined. For instance, if $u, v, \frac{\partial u}{\partial x_i}, \frac{\partial v}{\partial x_i} \in L^2(\Omega)$, then,

$$\frac{\partial}{\partial x_i}(uv) = u\frac{\partial v}{\partial x_i} + v\frac{\partial u}{\partial x_i}$$

However, in derivatives of composite functions, change of variables have the same properties as classical derivatives, provided each element is defined. ■

EXAMPLE 4.1.– Let $\Omega = (-1, 1)$, $u(x) = |x|$. Let us show that $u'(x) = sign(x)$. Indeed,

$$\ell_1(\varphi) = -(u, \varphi')_0 = -\int_{-1}^{1} u(x)\varphi'(x)dx$$

and we have (recall that $\varphi(-1) = \varphi(1) = 0$).

$$\ell_1(\varphi) = \int_{-1}^{0} x\varphi'(x)dx - \int_{0}^{1} x\varphi'(x)dx = -\int_{-1}^{0} \varphi(x)dx + \int_{0}^{1} \varphi(x)dx$$

Thus,

$$\ell_1(\varphi) = -(u, \varphi')_0 = (sign(x), \varphi)_0.$$ ■

EXAMPLE 4.2.– Let $\Omega = (-1, 1)$, $u(x) = sign(x)$. Let us show that $u' = 2\delta_0$, with $\delta_0(\varphi) = \varphi(0)$ [Dirac's delta – see Chapter 6]. Indeed,

$$\ell_1(\varphi) = -(u, \varphi')_0 = -\int_{-1}^{1} u(x)\varphi'(x)dx$$

and we have (recall that $\varphi(-1) = \varphi(1) = 0$).

$$\ell_1(\varphi) = \int_{-1}^{0} \varphi'(x)dx - \int_{0}^{1} \varphi'(x)dx = 2\varphi(0)\,.$$

In this case, there is no $w \in L^2(\Omega)$ such that $\ell(v) = (w, v)_0, \forall v \in L^2(\Omega)$.

■

4.2.1. *Second-order weak derivatives*

We have, for $\Omega = (a, b)$, $\varphi \in \mathcal{D}(\Omega)$ (thus, $\varphi(a) = \varphi(b) = \varphi'(a) = \varphi'(b) = 0$)

$$\int_{a}^{b} v''(x)\varphi(x)dx = -\int_{a}^{b} v'(x)\varphi'(x)dx = \int_{a}^{b} v(x)\varphi''(x)dx\,.$$

Then, we may define:

DEFINITION 4.2.– Let $\Omega \subset \mathbb{R}^n$ be a regular domain and $u \in L^2(\Omega)$. The *weak derivative* $\partial^2 u/\partial x_i \partial x_j$ is the linear functional $\ell_{ij}: \mathcal{D}(\Omega) \to \mathbb{R}$ defined by:

$$\ell_{ij}(\varphi) = \left(u, \frac{\partial^2 \varphi}{\partial x_i \partial x_j}\right)_0 .$$ ■

Analogous to theorem 4.1, we have:

THEOREM 4.2.– Let $\ell_{ij}: \mathcal{D}(\Omega) \to \mathbb{R}$ be the weak derivative $\partial^2 u/\partial x_i \partial x_j$ and verify:

$$\exists M \in \mathbb{R} : \left|\ell_{ij}(\varphi)\right| \leq M \|\varphi\|_0 , \forall \varphi \in \mathcal{D}(\Omega) .$$

Then, there exists a unique $w_{ij} \in L^2(\Omega)$ such that:

$$\ell_{ij}(v) = \left(w_{ij}, v\right)_0, \forall v \in L^2(\Omega).$$ ■

In this case, we say that $w_{ij} = \partial^2 u/\partial x_i \partial x_j$.

The proof is analogous to these given in theorem 4.1.

EXAMPLE 4.3.– Let $\Omega = (-1, 1)$, $u(x) = |x|$. Let us show that $u'' = 2\delta_0$, with $\delta_0(\varphi) = \varphi(0)$ (Dirac's delta, see Chapter 6). Indeed,

$$\ell_{11}(\varphi) = -\int_{-1}^{0} x\varphi''(x)dx + \int_0^1 x\varphi''(x)dx$$

and we have (recall that $\varphi'(-1) = \varphi'(1) = 0$):

$$\ell_{11}(\varphi) = \int_{-1}^{0} \varphi'(x)dx - \int_0^1 \varphi'(x)dx = 2\varphi(0).$$ ■

4.2.2. *Gradient, divergence, Laplacian*

Let $\Omega \subset \mathbb{R}^n$ be a regular domain. Let $u \in L^2(\Omega)$. The gradient of u, denoted by ∇u, is defined as $\ell_\nabla: [\mathcal{D}(\Omega)]^n \to \mathbb{R}$, given by:

$$\ell_{\nabla}(\boldsymbol{\varphi}) = -\int_{\Omega} u\, div(\boldsymbol{\varphi})dx = -\sum_{i=1}^{n}\int_{\Omega} u\frac{\partial\varphi_i}{\partial x_i}dx\ ,\forall\boldsymbol{\varphi} = (\varphi_1,\ldots,\varphi_n) \in [\mathcal{D}(\Omega)]^n.$$

The Laplacian of u, denoted by Δu, is defined as $\ell_{\Delta}: \mathcal{D}(\Omega) \to \mathbb{R}$, given by:

$$\ell_{\Delta}(\varphi) = \int_{\Omega} u\,\Delta\varphi\, dx = \sum_{i=1}^{n}\int_{\Omega} u\frac{\partial^2\varphi}{\partial x_i^2}dx\,,\forall\varphi \in \mathcal{D}(\Omega).$$

Let $\boldsymbol{u} = (u_1,\ldots,u_n) \in [L^2(\Omega)]^n$. The divergence of u, denoted by div(u), is defined as $\ell_{\text{div}}: \mathcal{D}(\Omega) \to \mathbb{R}$, given by:

$$\ell_{\text{div}}(\varphi) = -\int_{\Omega} \boldsymbol{u}.\nabla\varphi\, dx = -\sum_{i=1}^{n}\int_{\Omega} u_i\frac{\partial\varphi}{\partial x_i}dx\ ,\forall\varphi \in \mathcal{D}(\Omega).$$

REMARK 4.2.– Classical results such as Green's formulas or Stokes' formula remain valid when using weak derivatives, under the assumption of each term has a meaning. For instance,

$$\int_{\Omega} (v\Delta u - u\Delta v)dx = \int_{\partial\Omega}\left(v\frac{\partial \mathrm{u}}{\partial \mathrm{n}} - u\frac{\partial \mathrm{v}}{\partial \mathrm{n}}\right)dS,$$

under the condition that each term (including products) is defined. ■

EXAMPLE 4.4.– Let $\Omega = \{x \in \mathbb{R}^2: x_1^2 + x_2^2 < 1\}$, $u(x) = \ln r$, $r = \sqrt{x_1^2 + x_2^2}$. Let us show that $\Delta u = 2\pi\delta_0$. Let $\Omega_\varepsilon = \left\{x \in \mathbb{R}^2: \sqrt{x_1^2 + x_2^2} < \varepsilon\right\}$, $\varphi \in \mathcal{D}(\Omega)$. Notice that $u \in C^{\infty}(\Omega_\varepsilon)$, so that the weak and classical derivatives coincide on Ω_ε. Then, Green's formula shows that:

$$\int_{\Omega-\Omega_\varepsilon} u\Delta\varphi\, dx = \int_{\Omega-\Omega_\varepsilon} \varphi\Delta u\, dx + \int_{\partial(\Omega-\Omega_\varepsilon)}\left(u\frac{\partial\varphi}{\partial n} - \varphi\frac{\partial u}{\partial n}\right)dS$$

Since $\Delta u = 0$ on $\Omega - \Omega_\varepsilon$, $\varphi = \frac{\partial\varphi}{\partial n} = 0$ on $\partial\Omega$,

$$\int_{\Omega-\Omega_\varepsilon} u\Delta\varphi\, dx = -\int_{\partial\Omega_\varepsilon}\left(u\frac{\partial\varphi}{\partial n} - \varphi\frac{\partial u}{\partial n}\right)dS.$$

Using polar coordinates (r, θ) and taking into account that $\frac{\partial \varphi}{\partial n} = \frac{\partial \varphi}{\partial r}$, $\frac{\partial u}{\partial n} = \frac{\partial u}{\partial r} = \frac{1}{r}$, $dS = r d\theta$ and that $r = \varepsilon$on $\partial \Omega_\varepsilon$, we have:

$$\int_{\Omega - \Omega_\varepsilon} u \Delta \varphi \, dx = - \int_{\partial \Omega_\varepsilon} \left(\varepsilon \ln \varepsilon \frac{\partial \varphi}{\partial n} - \varphi \right) d\theta.$$

Since $\varepsilon \ln \varepsilon \to 0$ and

$$\left| \int_{\partial \Omega_\varepsilon} (\varphi(x) - \varphi(0)) d\theta \right| \leq 2\pi\varepsilon \max\{|\nabla \varphi(x)| : x \in \Omega\} \to 0,$$

we have:

$$\int_{\Omega - \Omega_\varepsilon} u \Delta \varphi \, dx \to 2\pi \varphi(0).$$

Thus,

$$\int_\Omega u \Delta \varphi \, dx = \lim_{\varepsilon \to 0+} \int_{\Omega - \Omega_\varepsilon} u \Delta \varphi \, dx = 2\pi \varphi(0). \qquad \blacksquare$$

EXAMPLE 4.5.– Let $\Omega = \{x \in \mathbb{R}^3 : x_1^2 + x_2^2 + x_3^2 < 1\}$, $u(x) = \frac{1}{r}$, $r = \sqrt{x_1^2 + x_2^2 + x_3^2}$. Analogously to the preceding situation, we have $\Delta u = -4\pi \delta_0$. In this case, $\Omega_\varepsilon = \left\{ x \in \mathbb{R}^2 : \sqrt{x_1^2 + x_2^2 + x_3^2} < \varepsilon \right\}$, $\varphi \in \mathcal{D}(\Omega)$ and, using spherical coordinates (r, θ, ψ), we have $\frac{\partial \varphi}{\partial n} = \frac{\partial \varphi}{\partial r}$, $\frac{\partial u}{\partial n} = \frac{\partial u}{\partial r} = -\frac{1}{r^2}$, $dS = r^2 sin\psi d\theta d\psi$ and that $r = \varepsilon$ on $\partial \Omega_\varepsilon$, so that

$$\int_{\Omega - \Omega_\varepsilon} u \Delta \varphi \, dx = - \int_{\partial \Omega_\varepsilon} \left(\varepsilon \frac{\partial \varphi}{\partial n} + \varphi \right) sin\psi d\theta d\psi.$$

Thus,

$$\int_{\Omega - \Omega_\varepsilon} u \Delta \varphi \, dx \to -4\pi \varphi(0).$$

and

$$\int_\Omega u \Delta \varphi \, dx = \lim_{\varepsilon \to 0+} \int_{\Omega - \Omega_\varepsilon} u \Delta \varphi \, dx = -4\pi \varphi(0). \qquad \blacksquare$$

4.2.3. *Higher-order weak derivatives*

We have, for $\Omega = (a, b)$, $\varphi \in \mathcal{D}(\Omega)$,

$$\int_a^b v^{(n)}(x)\varphi(x)dx = (-1)^n \int_a^b v(x)\varphi^{(n)}(x)dx.$$

Thus, we may define:

DEFINITION 4.3.– Let $\Omega \subset \mathbb{R}^n$ be a regular domain and $v \in L^2(\Omega)$. Let $\alpha = (\alpha_1, \dots, \alpha_n)$ and $|\alpha| = \alpha_1 + \dots + \alpha_n$. The *weak derivative* $\frac{\partial^{|\alpha|}u}{\partial x_1^{\alpha_1} \dots \partial x_n^{\alpha_n}}$ is the linear functional $\ell_\alpha: \mathcal{D}(\Omega) \rightarrow \mathbb{R}$ defined by:

$$\ell_\alpha(\varphi) = (-1)^{|\alpha|}\left(u, \frac{\partial^{|\alpha|}\varphi}{\partial x_1^{\alpha_1} \dots \partial x_n^{\alpha_n}}\right)_0. \qquad \blacksquare$$

We have:

THEOREM 4.3.– Let $\ell_\alpha: \mathcal{D}(\Omega) \rightarrow \mathbb{R}$ be the weak derivative $\frac{\partial^{|\alpha|}u}{\partial x_1^{\alpha_1} \dots \partial x_n^{\alpha_n}}$ and verify:

$$\exists M \in \mathbb{R} : |\ell(\varphi)| \leq M\|\varphi\|_0, \forall \varphi \in \mathcal{D}(\Omega).$$

Then, there exists a unique $w_\alpha \in L^2(\Omega)$ such that:

$$\ell_\alpha(v) = (w_\alpha, v)_0, \forall v \in L^2(\Omega).$$

In this case, we say that $w_\alpha = \frac{\partial^{|\alpha|}u}{\partial x_1^{\alpha_1} \dots \partial x_n^{\alpha_n}}$. $\blacksquare$

The proof is analogous to those given in theorem 4.1.

4.2.4. *Matlab® determination of weak derivatives*

Approximations of classical derivatives may be obtained by using the method `derive` defined in the class `basis` introduced in section 2.2.6. The evaluation of weak derivatives requests the use of scalar products, analogous to those defined in section 3.2.5. We present below a Matlab implementation using this last class. Recall that $du = u'$verifies:

$$\int_a^b du(x)\varphi(x)dx =$$
$$u(b)\varphi(b) - u(a)\varphi(a) - \int_a^b u(x)\varphi'(x)dx.$$

Let us consider a basis $\{\varphi_i : i \in \mathbb{N}\} \subset S$ and

$$du \approx \sum_{j=1}^n du_j\varphi_j.$$

We have:

$$\sum_{j=1}^n du_j(\varphi_j,\varphi_i)_0 = u(b)\varphi_i(b) - u(a)\varphi_i(a) - (u,\varphi_i)_0.$$

Thus, vector $DU = (du_1, \dots, du_n)$ is the solution of a linear system $A.DU = B$, where $A_{ij} = (\varphi_j,\varphi_i)_0$, $B_i = u(b)\varphi_i(b) - u(a)\varphi_i(a) - (u,\varphi_i)_0$.

Let us assume that the family $F = \{\varphi_i : 1 \le i \le n\}$ is defined by a cell array `basis` such that `basis{i}` defines φ_i according to the standard defined in section 3.2.6: `basis{i}` is a structure having properties `dim`, `dimx`, `component`, where component is a cell array such that each element is a structure having as properties `value` and `grad`. u is assumed to be analogously defined. In this case, the weak derivative is evaluated as follows (Program 4.1):

```
classdef weak_derivative
    methods (Static)
        function c =
coeffs(u,basis,sp_space,a,b,ua,ub,met)
                n = length(basis);
                A = zeros(n,n);
                for i = 1: n
                    sp = sp_space(basis{i},basis{i});
                    A(i,i) = sp;
                    for j = i+1: n
                        sp = sp_space(basis{i},basis{j});
                        A(i,j) = sp;
                        A(j,i) = sp;
                    end;
                end;
                c = cell(u.dim, 1);
                for k = 1: u.dim
                    B = zeros(n,1);
                    uu.dim = 1;
                    uu.dimx = u.dimx;
                    uu.component = cell(1,1);
                    uu.component{1} = u.component{k};
                    switch met
                        case 'subprogram'
```

```
                        for i = 1: n
                            vv.dim = 1;
                            vv.dimx = 1;
                            vv.component = cell(1,1);
                            aux =
basis{i}.component{k}.grad;
                vv.component{1}.value = @(x)
select_index(1,aux,x);
                            sp = sp_space(uu,vv);
                            aa =
ua*basis{i}.component{k}.value(a);
                            bb =
ub*basis{i}.component{k}.value(b);
                            B(i) = bb - aa - sp;
                        end;
                    case 'table'
                        for i = 1: n
                            vv.dim = 1;
                            vv.dimx = 1;
                            vv.component = cell(1,1);
                            aux =
basis{i}.component{k}.grad{1};
                            vv.component{1}.value = aux;
                            sp = sp_space(uu,vv);
                            aa =
ua*basis{i}.component{k}.value(1);
                        bb =
ub*basis{i}.component{k}.value(end);
                            B(i) = bb - aa - sp;
                        end;
                    otherwise
                        c{k} = [];
                        return;
                end;
                c{k} = A \ B;
            end;
            return;
        end
        function pu = funcwd(c,basis)
            pu.dim = basis{1}.dim;
            pu.dimx = basis{1}.dimx;
            pu.component = cell(pu.dim,1);
            for k = 1: pu.dim
                uu = @(x) 0;
                guu = @(x) zeros(pu.dimx,1);
                cc = c{k};
                for i = 1: length(cc)
            uu = @(x) uu(x) +
cc(i)*basis{i}.component{k}.value(x);
            guu = @(x) guu(x) +
cc(i)*basis{i}.component{k}.grad(x);
                end;
                pu.component{k}.value = uu;
                pu.component{k}.grad = guu;
            end;
        end
    end
end
```

Program 4.1. *A class for weak derivatives of functions of one variable*

EXAMPLE 4.6.– Let us consider the family $\varphi_i(x) = x^{i-1}$ and $u(x) = x^4$. The weak derivative is $u'(x) = 4x^3$. We consider $a = -1, b = 1$. We generate the function by using the commands:

```
xlim.lower.x = a;
xlim.upper.x = b;
xlim.dim = 1;
u.dim = 1;
u.dimx = 1;
u.component = cell(u.dim,1);
u.component{1}.value = @(x) x^4;
ua = u.component{1}.value(a);
ub = u.component{1}.value(b);
```

Then, the commands

```
sp_space = @(u,v) scalar_product.sp0(u,v,xlim,'subprogram');
c = weak_derivative.coeffs(u,basis,sp_space,a,b,ua,ub,'subprogram');
du = weak_derivative.funcwd(c,basis);
```

generate the weak derivative by using the method "subprogram". We may also generate a table containing the values of u as follows:

```
x = a:0.01:b;
p.x = x;
p.dim = 1;
U.dim = 1;
U.dimx = 1;
U.component = cell(U.dim,1);
U.component{1}.value = spam.partition(u.component{1}.value,p);
```

The commands

```
sp_space = @(u,v) scalar_product.sp0(u,v,p,'table');
ctab = weak_derivative.coeffs(U,tb,sp_space,a,b,ua,ub,'table');
dutab = weak_derivative.funcwd(ctab,basis);
```

generate the weak derivative by using the method "table". The results are shown in Figure 4.1. ■

EXAMPLE 4.7.– Let us consider the family $\varphi_i(x) = \sin(ix)$ and $u(x) = |x|$. The weak derivative is $u'(x) = sign(x)$. We consider $a = -1, b = 1$. The results are exhibited in Figure 4.2. ■

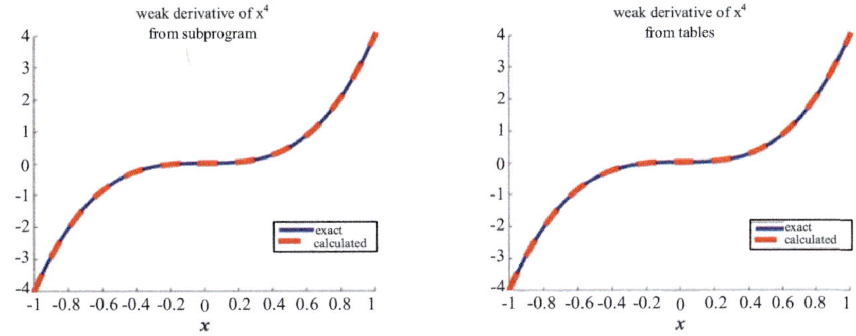

Figure 4.1. *Evaluation of the weak derivative of x^4 using subprograms and tables*

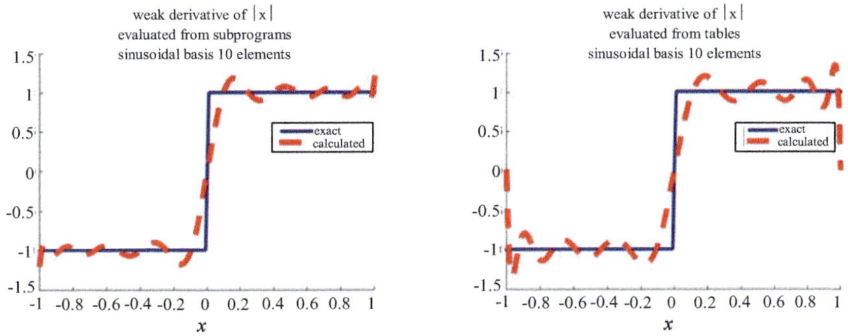

Figure 4.2. *Evaluation of the weak derivative of $|x|$ using subprograms and tables*

4.3. Sobolev spaces

As previously indicated, Serguei Lvovitch Sobolev was the Russian mathematician that introduced functional spaces fitting to variational methods, giving a complete mathematical framework. In his works, Sobolev formulated a theory of variational solutions and introduced Hilbert spaces involving derivatives [SOB 64], such as, for instance,

$$H^1(\Omega) = \{v \in L^2(\Omega): \nabla v \in [L^2(\Omega)]^n\},$$

$$(u,v)_1 = (u,v)_0 + (\nabla u, \nabla v)_0 = \int_\Omega (uv + \nabla u.\nabla v)dx$$

and

$$H_0^1(\Omega) = \{v \in L^2(\Omega): |\nabla v| \in L^2(\Omega), v = 0 \ on \ \partial\Omega\},$$

$$(u,v)_{1,0} = (\nabla u, \nabla v)_0 = \int_\Omega (\nabla u.\nabla v)dx.$$

We may also consider $\Gamma \subset \partial\Omega$ such that $meas(\Gamma) \neq 0$, i.e. Γ has a strictly positive length if its dimension is 1 and a strictly positive surface if its dimension is 2:

$$H_{0,\Gamma}^1(\Omega) = \{v \in L^2(\Omega): |\nabla v| \in L^2(\Omega), v = 0 \ on \ \Gamma\},$$

$$(u,v)_{1,0} = (\nabla u, \nabla v)_0 = \int_\Omega (\nabla u.\nabla v)dx.$$

Sobolev spaces are generated by completion of particular spaces [SOB 38]. For instance, by considering

$$\mathcal{V} = \{v: \Omega \to \mathbb{R}: v \ and \ \nabla v \ are \ continuous \ on \ \Omega \text{ and } (v,v)_1 < \infty\}$$

with the scalar product

$$(u,v) = \int_\Omega (uv + \nabla u.\nabla v)dx,$$

the completion method generates $H^1(\Omega)$. Analogously, the completion of

$$\mathcal{V} = \{v: \Omega \to \mathbb{R}: \nabla v \ is \ continuous \ on \ \Omega, v = 0 \text{ on } \Gamma \text{ and } (\nabla v, \nabla v) < \infty\}$$

generates $H_{0,\Gamma}^1(\Omega)$. Taking $\Gamma = \partial\Omega$ generates $H_0^1(\Omega)$.

The generic Sobolev space is denoted by:

$$H^m(\Omega) = \left\{v \in L^2(\Omega): \frac{\partial^{|\alpha|}u}{\partial x_1^{\alpha_1} \ldots \partial x_n^{\alpha_n}} \in L^2(\Omega), \forall\, \alpha \text{ such that } |\alpha| \leq m\right\}$$

and has a scalar product:

$$(u,v)_m = \sum_{|\alpha|\leq m}\left(\frac{\partial^{|\alpha|}u}{\partial x_1^{\alpha_1}\dots\partial x_n^{\alpha_n}},\frac{\partial^{|\alpha|}v}{\partial x_1^{\alpha_1}\dots\partial x_n^{\alpha_n}}\right)_0.$$

It is obtained by completion of a convenient space of continuous functions. We have $H^0(\Omega) = L^2(\Omega)$.

Sobolev spaces are often invoked when solving partial differential equations. *Note that they are formed by generalized functions.*

One of the main properties of elements of Sobolev spaces are inequalities connecting an element and its weak derivatives. These inequalities are often referred to as *Sobolev imbedding inequalities*. For instance, we have Morrey's inequality (see [EVA 10]).

THEOREM 4.4.– There exists $C \in \mathbb{R}$ such that

$$\forall u \in H^1(\mathbb{R}) : \ |u(x) - u(y)| \leq C\sqrt{|x-y|}\|u\|_{1,0} = C\sqrt{|x-y|}\|\nabla u\|_0 . \qquad \blacksquare$$

and Poincaré's inequality (see [EVA 10]).

THEOREM 4.5.– Let $\Omega \subset \mathbb{R}^n$ be open bounded, not empty. There exists $C \in \mathbb{R}$ such that:

$$\forall u \in H_0^1(\Omega) : \ \|u\|_0 \leq C\ \|u\|_{1,0} = C\ \|\nabla u\|_0 . \qquad \blacksquare$$

Sobolev spaces are separable (see, for instance, [LEB 96]).

4.3.1. *Point values*

The point value may have a meaning for elements from Sobolev spaces other than $H^0(\Omega)$. For instance, let us consider $\Omega = (a,b)$ and $u \in H^1(\Omega)$. We have:

$$\int_a^b u(x)\varphi''(x)dx = u(b)\varphi(b) - u(a)\varphi(a) - \int_a^b u'(x)\varphi'(x)dx,$$

Let $\lambda \in \mathbb{R}$ and

$$\varphi_\lambda'' = \varphi_\lambda \ on(a,b), \ \ \varphi_\lambda(a) = 0\ , \varphi'_\lambda(b) = \lambda.$$

We have:

$$\int_a^b u(x)\varphi_\lambda(x)dx = \int_a^b u(x)\varphi_\lambda''(x)dx = \lambda u(b) - \int_a^b u'(x)\varphi_\lambda'(x)dx,$$

so that,

$$(u,\varphi_\lambda)_1 = \lambda u(b).$$

This equality may be used to determine $u(b)$. Let us define:

$$T_b: \mathbb{R} \to H^1(\Omega)\ \ ,\ T_b(\lambda) = \varphi_\lambda$$

and $\ell_b: \mathbb{R} \to \mathbb{R}$ given by:

$$\ell_b(\lambda) = \left(u, T_b(\lambda)\right)_1.$$

T_b is linear, so that ℓ_b is also linear. We have:

$$|\ell_b(\lambda)| \leq \|u\|_1 \|T_b(\lambda)\|_1$$

Moreover, there exists $C \in \mathbb{R}$ such that,

$$\forall \lambda \in \mathbb{R} : \ \|T_b(\lambda)\|_1 \leq C\lambda$$

Then, $\ell_b: \mathbb{R} \to \mathbb{R}$ is linear and continuous. From the Riesz theorem, there exists $\beta \in \mathbb{R}$ such that $\ell_b(\lambda) = \beta\lambda$. We define $u(b) = \beta$.

The map $H^1(\Omega) \ni u \longrightarrow \beta \in \mathbb{R}$ is the *trace* of u at b., denoted by γu.

It generalizes to higher dimension by considering $\Omega \subset \mathbb{R}^n, u \in H^1(\Omega), \Gamma \subset \partial\Omega$ – in this case, we use the Green's formula:

$$\int_\Omega \nabla\varphi . \nabla u \, dx = \int_{\partial\Omega} u \frac{\partial\varphi}{\partial n}\, dS - \int_\Omega u\Delta\varphi dx$$

We may consider, for instance, $f \in L^2(\Gamma)$ and

$$\varphi_f - \Delta\varphi_f = 0 \ on \ \Omega, \frac{\partial\varphi_f}{\partial n} = 0 \ on \ \partial\Omega - \Gamma, \frac{\partial\varphi_f}{\partial n} = f \ on \ \Gamma.$$

In this case,

$$\left(u, \varphi_f\right)_1 = \int_\Gamma uf \, dS$$

and this equality may be used in order to define u on Γ, into an analogous way. Here,

$$T_\Gamma: L^2(\Gamma) \to H^1_{0,\partial\Omega-\Gamma}(\Omega)\,,\ T_\Gamma(f) = \varphi_f\,, \ell_\Gamma(f) = \left(u, \varphi_f\right)_1.$$

Again, T_Γ is linear and

$$|\ell_b(\lambda)| \leq \|u\|_1 \|T_\Gamma(f)\|_1,$$

so that the existence of $C \in \mathbb{R}$ such that:

$$\forall f \in L^2(\Gamma): \ \|T_\Gamma(f)\|_1 \leq C\|f\|_0$$

yields that $\ell_\Gamma: L^2(\Gamma) \to \mathbb{R}$ is linear and continuous. In this case, from the Riesz theorem, there exists $g \in L^2(\Gamma)$ such that,

$$\ell_\Gamma(f) = \int_\Gamma gf \, dS$$

We define $\gamma u = g$ on Γ.

4.4. Variational equations involving elements of a functional space

Variational equations were introduced in Chapter 3 when considering orthogonal projections (equations [3.5], [3.7], [3.8]) and in Chapter 2.

The standard form of a variational equation is:

$$u \in H \ and \ a(u, v) = \ell(v), \forall v \in H\,; \qquad [4.1]$$

where $H \subset V$ is a linear subspace, $\ell: V \to \mathbb{R}$ is linear continuous and $a: V \times V \to \mathbb{R}$ is such that $v \to a(u, v)$ is linear. The standard formulation includes situations where the variational equation is a *nonlinear* one. For orthogonal projections, only linear or affine situations have to be considered:

1) In the case of the orthogonal projection Px of $x \in V$ onto a vector subspace H, $u = Px \in H, a(u, v)$ is the scalar product $(u, v), \ell(v) = (x, v)$ and $H = S$.

2) In the case of the orthogonal projection Px of $x \in V$ onto an affine subspace $S = S_0 + \{x_0\}$ (S_0 vector subspace, $x_0 \in V$), $u = Px - u_0$, $a(u, v)$ is the scalar product (u, v), $\ell(v) = (x - x_0, v)$ and $H = S_0$.

Approximations of the solution of the variational equation [4.1] are generated by considering a total family on S, denoted by $F = \{\varphi_i\}_{i \in \mathbb{N}^*}$ and looking for an approximated solution having the form:

$$u \approx Pu = \sum_{j=1}^{k} u_j \varphi_j ,$$

where the coefficients $\boldsymbol{U} = (u_1 , \dots , u_k)^t \in \mathbb{R}^k$ have to be determined. More generally, we may consider a sequence of finite families $F_k = \{\psi_{k,i}: 1 \le i \le N_k\}$ such that $F = \cup_{k \in \mathbb{N}} F_k$ is a total family on S – this is the situation when finite element approximations are used. In both these situations, S is approximated by:

$$S \approx S_k = [\{\varphi_1 , \dots , \varphi_k\}]$$

and the variational equation becomes:

$$u \approx Pu \in S_k \; and \; a(Pu, v) = \ell(v), \forall v \in S_k.$$

Since both $v \to a(u, v)$ and $v \to \ell(v)$ are linear, we have

$$a(Pu, \varphi_i) = \ell(\varphi_i) , \; 1 \le i \le k.$$

Let us introduce $\boldsymbol{G}: \mathbb{R}^k \longrightarrow \mathbb{R}^k$ given by:

$$G_i(\boldsymbol{U}) = a\left(\sum_{j=1}^{k} u_j \varphi_j , \varphi_i\right) - \ell(\varphi_i).$$

Then, $\boldsymbol{U}$ is the solution of the algebraical system:

$$\boldsymbol{U} \in \mathbb{R}^k and \;\; \boldsymbol{G}(\boldsymbol{U}) = \boldsymbol{0},$$

i.e.

$$G_i(u_1, \dots, u_k) = 0, \;\; 1 \le i \le k.$$

If

$$a(u, v) = a_1(u, v) + a_0(v)$$

with $a_1$bilinear and a_0 linear, we have:

$$a\left(\sum_{j=1}^{k} u_j\varphi_j, \varphi_i\right) = \sum_{j=1}^{k} u_j a_1\left(\varphi_j, \varphi_i\right) + a_0(\varphi_i)$$

and

$$\boldsymbol{G}(\boldsymbol{U}) = \boldsymbol{AU} - \boldsymbol{B},$$

where $\boldsymbol{A}$ is the $n \times n$ matrix given by:

$$A_{ij} = a_1\left(\varphi_j, \varphi_i\right), \;\; 1 \le i, j \le k;$$

while B is the $k \times 1$ matrix given by:

$$B_i = \ell(\varphi_i) - a_0(\varphi_i).$$

In this case, U is the solution of the linear system

$$AU = B.$$

4.5. Reducing multiple indexes to a single one

The main applications of variational equations in engineering concern Sobolev spaces such as, for instance, $V = H^0(\Omega) = L^2(\Omega), V = H^1(\Omega), V = {}_0^1(\Omega)$.

For $\Omega \subset \mathbb{R}^d$, with $d > 1$, these spaces become Cartesian products of spaces $V = \mathbb{V}^d$, such as, for instance, $\mathbb{V} = H^0(\Omega) = L^2(\Omega)$, $\mathbb{V} = H^1(\Omega)$ or $\mathbb{V} = H_0^1(\Omega)$.

In these situations, the simplest way to generate a total multidimensional family for a space V is to consider products of total families on $\mathbb{V}$. For instance, let $F_m = \{\psi_j^m : 1 \leq j \leq k_m\}$ be such that $\cup_{m\in\mathbb{N}} F_m$ is a total family on $\mathbb{V}$. Then we may consider:

$$\phi_{i_1 i_2 \dots i_d}(x_1, \dots, x_d) = \psi_{i_1}^1(x_1)\psi_{i_2}^2(x_2) \dots \psi_{i_d}^d(x_d), 1 \leq i_j \leq k_m^j \text{ and } 1 \leq j \leq d.$$

Then, $F_m = \{\phi_{i_1 i_2 \dots i_n} : 0 \leq i_j \leq k_m^j \ \ and\ 1 \leq j \leq d\}$. In order to apply the methods previously exposed, it becomes necessary to transform the multi-index $\boldsymbol{i} = (i_1, i_2, \dots, i_d)$ into a single index. Such a transformation is standard in the framework of finite elements. For instance, we may use the transformations:

$$ind_1(i) = i,$$

$$ind_2(i_1, i_2) = (i_2 - 1)ind_1(k_m^1) + i_1,$$

$$ind_3(i_1, i_2, i_3) = (i_3 - 1)ind_2(k_m^1, k_m^2) + ind_2(i_2, i_3)$$

For a two-dimensional (2D) region $\Omega \subset \mathbb{R}^2$, these transformations lead to:

$$\varphi_j(x_1, x_2) = \phi_{i_1 i_2}(x_1, x_2);\ j = (i_2 - 1)k_m^1 + i_1.$$

The number of unknowns to be determined is $= k_m^1 k_m^2$. Into an analogous way, the coefficient u_j correspond to a multidimensionally indexed $u_{i_1 i_2}$ such that

$$Pu = \sum_{j=1}^{k} u_j \varphi_j = \sum_{i_1=1}^{k_m^1} \sum_{i_2=1}^{k_m^2} u_{i_1 i_2}\, \phi_{i_1 i_2} = \sum_{i_1=1}^{k_m^1} \sum_{i_2=1}^{k_m^2} u_{i_1 i_2}\, \psi_{i_1}^1 \psi_{i_2}^2.$$

For a three-dimensional (3D) region $\Omega \subset \mathbb{R}^3$, the transformation leads to:

$$\varphi_j(x_1, x_2, x_3) = \phi_{i_1 i_2 i_3}(x_1, x_2, x_3); j = (i_3 - 1)k_m^1 k_m^2 + (i_2 - 1)k_m^1 + i_1.$$

The number of unknowns to be determined is $k = k_m^1 k_m^2 k_m^3$. In an analogous way, the coefficient u_j corresponds to a multidimensionally indexed $u_{i_1 i_2 i_3}$ such that:

$$Pu = \sum_{j=1}^{k} u_j \varphi_j = \sum_{i_1=1}^{k_m^1} \sum_{i_2=1}^{k_m^2} \sum_{i_3=1}^{k_m^3} u_{i_1 i_2 i_3} \phi_{i_1 i_2 i_3} = \sum_{i_1=1}^{k_m^1} \sum_{i_2=1}^{k_m^2} \sum_{i_3=1}^{k_m^3} u_{i_1 i_2 i_3} \psi_{i_1}^1 \psi_{i_2}^2 \psi_{i_3}^3 .$$

4.6. Existence and uniqueness of the solution of a variational equation

The reader will find in the literature a large number of results concerning the existence and uniqueness of solutions of variational equations, among all the result known as the *theorem of Lax-Milgram* [LAX 54], from Peter David Lax and Arthur Norton Milgram.

We give below one of these results, containing the essential assumptions and which applies to nonlinear situations. Let us consider the variational equation:

$$u \in H \text{ and } a(u, v) = \ell(v), \forall v \in H$$

We have:

THEOREM 4.6.– Assume that:

i) H is a Hilbert space or a closed vector subspace of a Hilbert space;

ii) $v \to a(u, v)$is linear, $\forall v \in H$;

iii) $\ell: H \to \mathbb{R}$is linear continuous;

iv) $\exists M \in \mathbb{R}$ such that:

$$\forall u, v, w \in H: |a(u,w) - a(v,w)| \leq M\|w\|\|u - v\|;$$

v) $\exists \alpha \in \mathbb{R}, \alpha > 0$ such that:

$$\forall u, v \in H: a(v, v - u) - a(u, v - u) \geq \alpha\|v - u\|^2.$$

Then there exists a unique $u \in V$ such that:

$$u \in H,\ a(u,v) = \ell(v),\ \forall u, v \in H.$$ ■

PROOF.– From (vi) and Riesz's theorem, we have:

$$a(u,v) = (A(u), v);\ \ell(v) = (f_\ell, v);\ f_\ell, A(u) \in H.$$

and the variational equation reads as $A(u) = f_\ell$ or, equivalently,

$$u \in H \ and\ (A(u) - f_\ell, v) = 0, \forall v \in H.$$

This equation may be reformulated as:

$$u \in H \ and\ (\ u - \theta(A(u) - f_\ell) - u, v) = 0, \forall v \in H,$$

i.e.

$$u = F(u),\ F(u) = P_H(\theta(A(u) - f) - u).$$

where P_H is the orthogonal projection onto H. We have:

$$F(u) - F(v) = P_H(\theta(A(u) - f_\ell) - u) - P_H(\theta(A(v) - f_\ell) - v).$$

Thus,

$$\|F(u) - F(v)\|^2 = \|P_H(\theta(A(u) - f_\ell) - u) - P_H(\theta(A(v) - f_\ell) - v)\|^2$$

so that

$$\|F(u) - F(v)\|^2 = \left\|\theta\big(A(u) - A(v)\big) - (u - v)\right\|^2$$

and

$$\|F(u) - F(v)\|^2 \leq \theta^2\|A(u) - A(v)\|^2 + \|u - v\|^2 - 2\theta(A(u) - A(v), u - v).$$

Thus,

$$\|F(u) - F(v)\|^2 \leq \theta^2\|A(u) - A(v)\|^2 + \|u - v\|^2 - 2\theta\alpha\|u - v\|^2$$

Moreover,

$$\|A(u) - A(v)\|^2 = a\big(u, A(u) - A(v)\big) - a\big(v, A(u) - A(v)\big)$$

and we have:

$$\|A(u) - A(v)\|^2 \leq M\|u - v\|\|A(u) - A(v)\|,$$

so that,

$$\|A(u) - A(v)\| \leq M\|u - v\|$$

and

$$\|F(u) - F(v)\|^2 \leq (1 - 2\theta\alpha + \theta^2 M^2)\|u - v\|^2.$$

Taking

$$0 < \theta < \frac{2\alpha}{M} \Rightarrow L = \sqrt{1 - 2\theta\alpha + \theta^2 M^2} < 1,$$

We have,

$$\|F(u) - F(v)\| \leq L\|u - v\|,$$

so that F is a contraction and the result yields from Banach's fixed point theorem (theorem 3.11). ■

4.7. Linear variational equations in separable spaces

Let us consider the situation where $a(\blacksquare, \blacksquare)$ is bilinear, i.e.

$u \to a(u,v)$ *is linear*, $\forall v \in V$ and $v \to a(u,v)$ *is linear*, $\forall u \in V$.

In this case, conditions (iv)–(v) in theorem 4.6 are equivalent to:

$$\forall u, v \in H: |a(u,v)| \leq M\|u\|\|v\|, a(u,u) \geq \alpha\|u\|^2.$$

This situation corresponds to the classical result [LAX 54].

When the space H is separable, we may consider a Hilbert basis $F = \{\varphi_n : n \in \mathbb{N}^*\} \subset H$ and consider the sequence $\{u_n : n \in \mathbb{N}^*\}$ defined as:

$$u_n \in H_n = [\{\varphi_k : 1 \leq k \leq n\}] \textit{ and } a(u_n, v) = \ell(v_n), \forall v \in H_n.$$

We observe that u_n is determined by solving a linear system:

$$AU = B,$$

where

$$A_{ij} = a(\varphi_i, \varphi_j)\,, B_i = \ell(\varphi_i)\,, U = (u_1, \ldots, u_n)^t.$$

Then, u_n is uniquely determined and is given by:

$$u_n = \sum_{i=1}^{n} u_i \varphi_i\,.$$

Moreover,

$$\alpha\|u_n\|^2 \leq a(u_n, u_n) = \ell(u_n) \leq \|\ell\|\|u_n\|,$$

so that,

$$\|u_n\| \leq \frac{\|\ell\|}{\alpha}$$

and $\{u_n : n \in \mathbb{N}^*\}$ is bounded. As a consequence, there is a subsequence $\{u_{n_k} : k \in \mathbb{N}^*\}$ such that $u_{n_k} \rightharpoonup \hat{u} \in H$ (weakly). Thus, Riezs's theorem shows that:

$$\ell(u_{n_k} - \hat{u}) = \ell(u_{n_k}) - \ell(\hat{u}) \to 0.$$

Indeed, there exists $w_\ell \in H$ such that $\ell(v) = (w_\ell, v), \forall v \in H$ and, as a consequence, $(w_\ell, u_{n_k} - \hat{u}) \to 0$. Moreover, we have:

$$\alpha \|u_{n_k} - \hat{u}\|^2 = a(u_{n_k} - \hat{u}, u_{n_k} - \hat{u}) = \ell(u_{n_k} - \hat{u}) \to 0,$$

so that $u_{n_k} \to \hat{u} \in H$ (strongly). Thus, $\forall v \in H$:

$$|a(u_{n_k} - \hat{u}, v)| \leq M \|u_{n_k} - \hat{u}\| \|v\| \to 0.$$

Let us denote by $P_n v$ the orthogonal projection of $v \in H$ on to H_n. Then, $\forall v \in H$:

$$a\left(u_{n_k}, P_n(v)\right) = \ell(P_n(v)) \Longrightarrow a(\hat{u}, P_n(v)) = \ell(P_n(v)).$$

From theorem 3.21 (iii), we obtain:

$$\hat{u} \in H \ \ and \ \ \forall v \in H : a(\hat{u}, v) = \ell(v).$$

The uniqueness of u shows that $\hat{u} = u$. Thus, theorem 3.5 shows that $u_n \to u$.

4.8. Parametric variational equations

In some situations, we are interested in the solution of variational equations depending upon a parameter. For instance, let us consider the situation where a and/or ℓ depends on a parameter θ:

$$u_\theta \in H \ and \ a(u_\theta, v; \theta) = \ell(v; \theta), \forall v \in H$$

In this situation, the solution $u = u_\theta$ depends upon θ and we may consider:

$$u_\theta(x) = \sum_{j=1}^{+\infty} u_j(\theta) \varphi_j(x) \approx P u_\theta(x) = \sum_{j=1}^{n} u_j(\theta) \varphi_j(x),$$

i.e. the coefficients of the expansion depend upon θ. In this case, the approach presented in section 4.4 leads to parametric equalities which may be solved by the methods introduced in section 2.2 (other methods of approximation may be found in [SOU 15]). Indeed, using the same notation of section 4.4, we have

$$G_i(U,\theta) = a\left(\sum_{j=1}^{n} u_j(\theta)\varphi_j, \varphi_i; \theta\right) - \ell(\varphi_i, \theta),$$

so that $U_\theta = U(\theta) = \left(u_1(\theta), \dots, u_n(\theta)\right)^t$ is the solution of the algebraical system:

$$U_\theta \in \mathbb{R}^k \text{and } G(U;\theta) = 0,$$

which may be solved as in section 2.2: assuming that the functional space where $u_j = u_j(\theta)$ is chosen is a separable Hilbert space, we may consider a convenient Hilbert basis $\{\psi_n : n \in \mathbb{N}^*\}$ and the expansion:

$$u_j(\theta) = \sum_{i=1}^{+\infty} u_{i,j}\psi_i(\theta) \approx \sum_{i=1}^{k} u_{i,j}\psi_i(\theta).$$

The coefficients may be determined as in section 2.2. This approach may be reinterpreted as follows: we consider the expansion as:

$$u_\theta(x) \approx \sum_{i=1}^{+\infty} \sum_{j=1}^{+\infty} u_{i,j}\psi_i(\theta)\varphi_j(x).$$

Finite dimensional approximations are obtained by truncation. For instance,

$$u_\theta(x) \approx \sum_{i=1}^{k} \sum_{j=1}^{n} u_{i,j}\psi_i(\theta)\varphi_j(x).$$

Let us collect all the unknowns $\left(u_{i,j} : 1 \le i \le k, 1 \le j \le n\right)$ in a vector U of kn elements: we may use the map $ind_2(i,j) = (i-1)n + j$, so that, for $m = ind_2(i,j)$, $U_m = u_{i,j}$, $\phi_m = \psi_i\varphi_j$. Let $\theta \in (\alpha, \beta)$. Then, we look for a vector U such that:

$$G_q(U) = \int_\alpha^\beta \left[a\left(\sum_{p=1}^{kn} U_p\phi_p, \phi_q; \theta\right) - \ell(\phi_q, \theta)\right] d\theta = 0, 1 \le q \le kn.$$

Then, U is the solution of the algebraical system,

$$U \in \mathbb{R}^{kn} \text{and } G(U) = 0,$$

4.9. A Matlab® class for variational equations.

The numerical solution of variational equations is performed analogously to the determination of orthogonal projections or weak derivatives. Let us assume that a subprogram *avareq* evaluates the value of $a(u,v)$ and a subprogram *ell* evaluates $\ell(v)$. Both $v \to a(u,v)$ and $v \to \ell(v)$ are linear.

In the general situation, $u \to a(u,v)$ is nonlinear and we must solve a system of nonlinear algebraic equations $G(U) = 0$. Assume that a use a subprogram *eqsolver* solves these equations – for instance, we may look for the minimum of $\|G(U)\|$or call the Matlab® program *fsolve*, if the Optimization Toolbox is available. If $u \to a(u,v)$ is linear, $G(U) = AU - B$ and the equations correspond to a linear system, so that we may use linear solvers, such as the anti-slash operator \ of Matlab®.

Under these assumptions, we may use the class below:

```
classdef variational_equation
    methods (Static, Access = private)
        function v = nonlineqs(avareq,ell, basis, c)
            n = length(basis);
            eqs = zeros(n,1);
            cc{1} = c;
            pu =
variational_equation.funcveq(cc,basis);
            for i = 1: n
                sp1 = avareq(pu,basis{i});
                sp2 = ell(basis{i});
                eqs(i) = sp1 - sp2;
            end;
            v = eqs;
        end
    end
    methods (Static)
        function c = lincoeffs(avareq,ell, basis)
            n = length(basis);
            A = zeros(n,n);
            B = zeros(n,1);
            for i = 1: n
                sp = ell(basis{i});
                B(i) = sp;
                sp = avareq(basis{i},basis{i});
                A(i,i) = sp;
                for j = i+1: n
                    sp = avareq(basis{i},basis{j});
                    A(i,j) = sp;
                    A(j,i) = sp;
                end;
            end;
```

```
                c = cell(1, 1);
                c{1} = A \ B;
            end
            function c = nonlincoeffs(avareq,ell, basis,
eqsolver)
   eqs = @(c)
variational_equation.nonlineqs(avareq,ell, basis, c);
                cm = eqsolver(eqs);
                c{1} = cm;
            end
            function pu = funcveq(c,basis)
                pu.dim = 1;
                pu.dimx = 1;
                pu.component = cell(1,1);
                uu = @(x) 0;
                guu = @(x) 0;
                for i = 1: length(c{1})
             uu = @(x) uu(x) +
c{1}(i)*basis{i}.component{1}.value(x);
             guu = @(x) guu(x) +
c{1}(i)*basis{i}.component{1}.grad(x);
                end;
                pu.component{1}.value = uu;
                pu.component{1}.grad = guu;
            end;
        end
end
```

Program 4.2. *A class for one-dimensional variational equations*

EXAMPLE 4.8.– Let us consider $a = 0$, $b = 1$ and the variational equation

$$\forall v\colon\ u(b)v(b) - \int_a^b u(x)v'(x)dx - \int_a^b u^2(x)v(x)dx = \int_a^b v(x)dx$$

The variational equation is nonlinear and its solution is $u(x) = \tan(x)$. We have:

$$a(u,v) = u(b)v(b) - \int_a^b u(x)v'(x)dx - \int_a^b u^2(x)v(x)dx\ ;\ \ell(v) = \int_a^b v(x)dx$$

a and e may be evaluated as follows:

```
sp = @(u,v) scalar_product.sp0(u,v,xlim,'subprogram');
avareq = @(u,v) a1(u, v,sp,b);
ell = @(v) ell1(v,sp);
```

where `a1` and `ell1` are defined in Program 4.3 below.

```
function val = a1(u, v,sp,b)
aux1 =
u.component{1}.value(b)*
v.component{1}.value(b);
vv.dim = 1;
vv.dimx = u.dimx;
vv.component = cell(1,1);
vv.component{1}.value =
u.component{1}.grad;
aux2 = sp(u,vv);
uu.dim = 1;
uu.dimx = u.dimx;
uu.component = cell(1,1);
uu.component{1}.value =
@(x)1 +
u.component{1}.value(x)^2;
aux3 = sp(uu,v);
val = aux1 - aux2 - aux3;
end
```

```
function val = ell1(v,sp)
uu.dim = 1;
uu.dimx = 1;
uu.component = cell(1,1);
uu.component{1}.value =
@(x)1;
val = sp(uu,v);
end
```

Program 4.3. *Definition of* a *and* ℓ *in example 4.8*

Results obtained using the family $\varphi_i(x) = x^{i-1}$ and the family $\varphi_1(x) = 1$, $\varphi_i(x) = \sin((i-1)x), i > 1$ are shown in Figure 4.3. ■

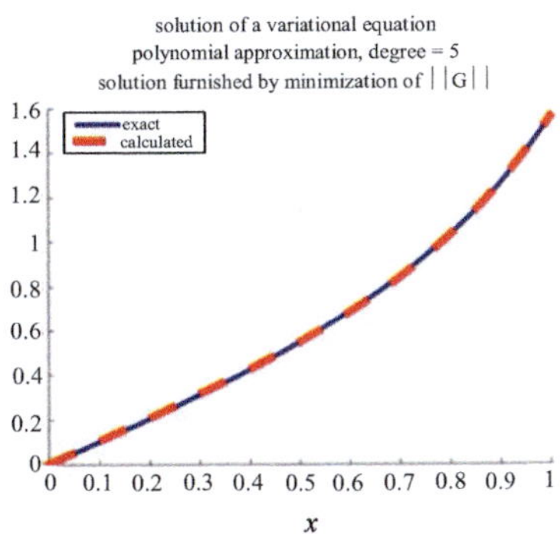

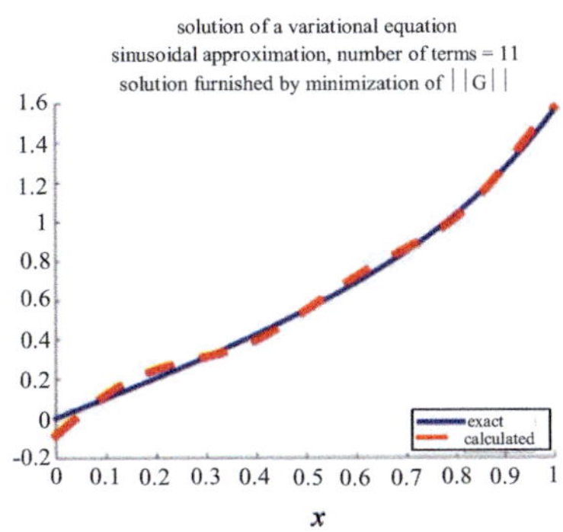

Figure 4.3. *Solution of the nonlinear variational equation in example 4.8*

EXAMPLE 4.9.– Let us consider $a = 0$, $b = 1$ and the variational equation:

$$\forall v: u(0) = 0, \quad \int_a^b u'(x)v'(x)dx = v\left(\frac{1}{2}\right) + \int_a^b v(x)dx$$

Here,

$$a(u, v) = \int_a^b u'(x)v'(x)dx \,; \ell(v) = v\left(\frac{1}{2}\right) + \int_a^b v(x)dx$$

In this case, the problem is linear. Since u(0) = 0, we eliminate the function constant equal to one from the basis and we consider the families $\varphi_i(x) = x^i$ and $\varphi_i(x) = \sin(ix)$. The results are shown in Figure 4.4. ■

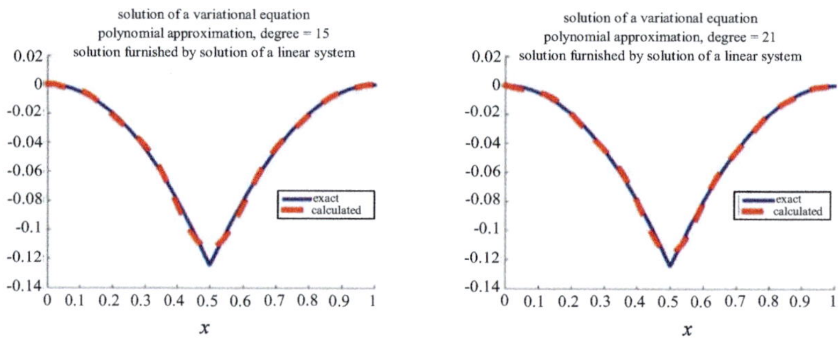

Figure 4.4. *Solution of the linear variational equation in example 4.9*

4.10. Exercises

EXERCISE 4.1.– Let $\Omega = (0,1)$ and $\ell(v) = \int_\Omega v dx$.

1) Show that ℓ is linear.

2) Let $V = L^2(\Omega)$:

a) Show that there exists $M \in \mathbb{R}$ such that: $\forall v \in V: |\ell(v)| \leq M\|v\|_0$.

b) Conclude that there exists one and only one $u \in V$ such that $(u, v)_0 = \ell(v), \forall v \in V$.

c) Show that $u(x) = 1$ on Ω.

3) Let $V = H^1(\Omega)$:

a) Show that there exists $M \in \mathbb{R}$ such that: $\forall v \in V$: $|\ell(v)| \leq M\|v\|_{1,1}$.

b) Conclude that there exists one and only one $u \in V$ such that $(u, v)_{1,1} = \ell(v), \forall v \in V$.

c) Show that $u - u'' = 1$ on Ω, $u'(0) = u'(1) = 0$.

4) Let $V = H^1(\Omega)$, but $(u, v)_V = \int_\Omega u'v' + \lambda u(0)v(0) (\lambda > 0)$.

a) Let $v \in V$.

Verify that $v(x) = v(0) + \int_0^x v', \forall x \in \Omega$:

– show that $|v(x)| \leq |v(0)| + \sqrt{x}\sqrt{\int_0^x (v')^2}$;

– conclude that $|v(x)| \leq \sqrt{2}\sqrt{|v(0)|^2 + \int_0^x (v')^2}, \forall x \in \Omega$;

– show that $|v(x)| \leq max\left\{\sqrt{2}, \sqrt{\frac{2}{\lambda}}\right\} \|v\|_V, \forall x \in \Omega$.

b) Conclude that there exists one and only one $u \in V$ such that $(u, v)_V = \ell(v), \forall v \in V$.

c) Show that $-u'' = 1$ on Ω, $\lambda u(0) - u'(0) = u'(1) = 0$.

5) Let $V = \{v \in H^1(\Omega): v(0) = 0\}$, with the scalar product $(\blacksquare, \blacksquare)_{1,0}$.

a) Le t$v \in V$:

– verify that $v(x) = \int_0^x v', \forall x \in \Omega$;

– conclude that $|v(x)| \leq \sqrt{x}\|v\|_{1,0}, \forall x \in \Omega$;

– show that there exists $M \in \mathbb{R}$ such that: $|v(x)| \leq M\|v\|_{1,0}, \forall x \in \Omega$.

b) Conclude that there exists one and only one $u \in V$ such that $(u,v)_{1,0} = \ell(v), \forall v \in V$.

c) Show that $-u'' = 1$ sur Ω, $u(0) = u'(1) = 0$. ■

EXERCISE 4.2.– Let $\Omega = (-1,1)$ and $\ell(v) = v(0)$.

1) Show that $\boldsymbol{\ell}$ is linear.

2) Let $V = L^2(\Omega)$ and $v_n(x) = 0, if\ |x| \geq \frac{1}{n}$, $v_n(x) = 2n, if\ |x| < \frac{1}{n}$. Show that $\|v_n\|_0 = 1, \forall n \in \mathbb{N}$. Conclude $\ell: V \to \mathbb{R}$ verifies $\|\ell\| = +\infty$.

3) Let $V = H^1(\Omega)$:

a) Verify that:

$$v(0) = v(x) - \int_0^x v'(s)ds\,.$$

b) Show that:

$$|v(0)| \leq |\, v(x)| + \sqrt{|x|}\sqrt{\int_0^x [v'(s)]^2 ds}\,.$$

c) Conclude that:

$$|v(0)| \leq \frac{1}{2}\left(\sqrt{2}\sqrt{\int_0^1 [v(s)]^2 ds} + 2\sqrt{\int_0^1 [v'(s)]^2 ds}\right)$$

and show that $|v(0)| \leq \sqrt{2}\|v\|_1$.

d) Conclude that $\boldsymbol{\ell}: \boldsymbol{V} \to \mathbb{R}$ is continuous.

4) Let $V = H^1(\Omega)$, but consider $(u,v)_V = \lambda u(-1)v(-1) + \int_\Omega u'v'dx\ (\lambda > 0)$.

a) Verify that $|v(-1)| \leq \|v\|_V/\sqrt{\lambda}$.

b) Use the equality $v(0) = v(-1) + \int_{-1}^0 v'dx$ to show that $|v(0)| \leq \left(1 + \frac{1}{\sqrt{\lambda}}\right)\|v\|_V$.

c) Conclude that there exists one and only one u ∈ V such that $(u,v)_V = \ell(v), \forall v \in V$. ■

EXERCISE 4.3.– Let $\Omega = (0,1), a \in \Omega$ and $\ell(v) = Pv(a) + \int_\Omega fvdx$. Let $V = \{v \in H^1(\Omega): v(0) = v'(0) = 0\}$ and $(u,v)_V = \int_\Omega u''v''\text{dx}$.

1) Verify that $(\blacksquare, \blacksquare)_V$ is a scalar product on V.

2) Show that: $\forall x \in \Omega, v \in V : |v'(x)| \leq \left|\int_0^x v''dx\right| \leq \sqrt{x}\sqrt{\int_0^x [v'']^2 dx} \leq \|v\|_V$.

3) Show that $\forall v \in V : |v(x)| \leq \|v\|_V$.

4) Show that $\forall v \in V : \|v\|_0 \leq \|v\|_V$.

5) Conclude that there exists $M \in \mathbb{R}$ such that $\forall v \in V : |\ell(v)| \leq \|v\|_V$.

6) Show that there exists one and only one $u \in V$ such that $(u,v)_V = \ell(v), \forall v \in V$. ■

EXERCISE 4.4.– Let us consider the variational equation ($\boldsymbol{f} \in \mathbb{R}$)

$$\boldsymbol{u(0) = 0 \; and \; \int_0^1 u'(x)v'(x)dx = fv\left(\frac{1}{2}\right) + \int_0^1 v(x)dx} .$$

1) Let us consider the trigonometrical basis $F = \{\varphi_n : n \in \mathbb{N}\}$, with $\varphi_0 = 1$ and, for $k > 0$:

$$\varphi_{2k-1} = \sqrt{2}\sin(2k\pi x), \varphi_{2k} = \sqrt{2}\cos(2k\pi x).$$

Let $u = \sum_{n=1}^{\infty} u_n \varphi_n$. Determine the coefficients u_n as functions of f. Draw a graphic of the solution for $f = -1$ and compare it to:

$$u^1(x) = \begin{cases} -x^2/2, & x < 1/2 \\ (1-x)^2/2, & x > 1/2 \end{cases}$$

2) Let us consider the polynomial basis $F = \{\varphi_n : n \in \mathbb{N}\}$, with $\varphi_n = x^n$. Let $u = \sum_{n=1}^{\infty} u_n \varphi_n$. Determine the coefficients u_n as functions of f. Draw a graphic of the solution for $f = -1$ and compare it to u^1.

3) Consider the family $F = \{\varphi_n : n \in \mathbb{N}\}$ of the P1 shape functions

$$\varphi_n(x) = \begin{cases} 1 - |x - nh|/h, & |x - nh| \leq h \\ 0, & |x - nh| > h \end{cases}, \quad h = \frac{1}{N}$$

Determine the coefficients u_n as functions of f. Draw a graphic of the solution for $f = -1$ and compare it to u^1.

5

Variational Methods for Differential Equations

Differential equations are a major tool for the analysis of real systems, used by scientists, engineers, economists and many people interested in quantitative or qualitative analysis. Nowadays, different areas of knowledge use differential equations in order to generate mathematical models.

The history of differential equations is subject to controversy, since a seminal dispute opposed Isaac Newton and Gottfried Leibniz concerning the origin of differential and integral calculus. In 1675, Leibniz wrote in an unpublished letter the equality

$$\int x dx = \frac{x^2}{2},$$

which may be considered as the resolution of a differential equation. In fact, the problem of the determination of the tangent to a given curve leads to differential equations and has been studied since ancient times, even if the main tools of differential calculus were introduced by Newton and Leibniz. Until these times, general concepts and analysis concerning differential equations were not available, limiting developments concerning differential equations. Since Archimedes, researchers worked on the determination of the tangent to curves. For instance, Gilles de Roberval developed an ingenious method, based on the motion of a particle along the curve. Roberval observed –

analogously to Aristotle, but using more developed mathematical tools – that the direction of the velocity of such a particle is in tangent to the curve [DER 34, DER 66, DER 93]. Based on other research, Pierre de Fermat studied the determination of tangents as a way to find maxima and minima. This research is led to a method for the determination of the tangents by geometrically solving some differential equations [DEF 38, DEF 79].

Isaac Newton introduced classification of differential equations and looked for general methods of solution, namely by using series [NEW 36]. Leibniz used the formula for the derivative of a product in order to produce a method for the solution of linear first order differential equations with constant coefficients and introduced methods connected to changes of variables and complete differentials. Many of these works were either stated in letters to other researchers such as, for instance, Newton, l'Hospital and the Bernoulli family or published in *Acta Eroditorum* – the first German scientific journal, which also published works by Newton, Bernoulli, Huygens, Laplace and other eminent researchers. New developments arise with the formalism of integrating factors [EUL 32, CLA 42] and the variation of the constants [LAG 09, LAG 10]. Jean Le Rond D'Alembert established conditions took an impulsion on Lagrange's works and established many results on differential equations. He also started the study of systems of linear differential equations and worked on vibrating strings, which led D'Alembert to partial differential equations [DAL 47, DAL 50]. At this moment, boundary value problems began to be studied and led to differential equations of hyper geometric type, which ultimately resulted in the modern numerical methods. For instance, these works led to Bessel's equation, Laguerre, Legendre, and Hermite polynomials.

The variational point of view for differential equations appears with the concept of *virtual works* and *virtual motions*, which led to *analytical mechanics*, where the equilibrium and the motion of mechanical systems are described by considering *variations of energy rather than forces* (as in vectorial mechanics). When the point of view of analytical mechanics is adopted, we do not characterize the dynamics of a mechanical system by writing Newton's law, but we examine the variations of energy associated to all the possible variations of a configuration of the system: the real configuration corresponds to a balance where the variation of the internal energy equilibrates the work of the external forces in all its possible variations. The idea of virtual motions is ancient; such a formulation may be dated from Jean Bernoulli (in a letter to Pierre de Varignon [VAR 25]) and was exploited by Pierre-Louis Moreau de Maupertuis [DEM 48], Jean Le Rond D'Alembert [DAL 43], until the advances introduced by the young Italian Giuseppe Ludovico de la Grange Tournier [LAG 61a, LAG 61b, LAG 64, LAG 80], later Joseph Louis Comte de Lagrange when he publishes his masterpiece [LAG 88], where the complete formalism for virtual motions is presented.

John William Strutt (Lord Rayleigh) proposed a variational formulation for Boundary Value Problems [RAY 94, RAY 96], improved by Walter Ritz [RIT 08, RIT 09]. Ritz proposed approximations by choosing finite linear combinations of elements from a convenient family in such a way that a quadratic functional could be minimized. A major success for Ritz was the numerical determination of the Chladni figures [CHL 87], a problem that had seen several failures or only partial solutions by Wheatstone, Kirchhof, and Rayleigh. Ritz died at the age of 31, from tuberculosis. Perhaps because of his illness, his works went relatively unnoticed in Western Europe. Despite his physical presence in Gottingen, mathematicians orbiting Hilbert have overlooked the importance of his work. Elsewhere, in the east, the ideas of Ritz were immediately taken onboard and began to be used. Step an Prokofievitch Timoshenko [TIM 10] and especially Ivan Grigoryevich Bubnov [BUB 13, BUB 14] applied the approach proposed by Ritz and got spectacular results. One of the originalities of the revisions of Ritz by Bubnov was the requirement of orthogonality in the family of functions. In the research that followed, Boris Grigoryevich Galerkin published his main work [GAL 15], where he presented a new method which he called "Ritz's method", but nowadays is referred as "Galerkin's method". In 1909, Galerkin travelled across Western Europe in search of knowledge and scientific contacts. It was then that he was probably put in contact with modern developments and was able to recognize the importance of the work produced by Ritz. The fact is that Galerkin successfully applied the approach to problems considered difficult at that time, and abundantly cited Ritz, which contributed to popularize his name. One of the main contributions of Galerkin is the extension to non-orthogonal functions and the introduction of mass and stiffness matrices. Galerkin's method is still used today. For time-dependent problems, a modification of the basic method is usually referred as the Faedo–Galerkin method [FAE 49], or Petrov-Galerkin [PET 40]. The results presented by Galerkin were impressive and have prompted many engineers and researchers to delve into variational methods. A large number of researchers have successfully applied Galerkin's approach to difficult variational problems and great strides have been taken thanks to it. A new step was given by Richard Courant – a former PhD student of David Hilbert – which devised the use of piecewise linear functions based on triangulations of the domain under consideration [COU 43]. The paper by Courant is the starting point for the Finite Element Method, even if the real development arrives with the generalization of computers. This arrives with the action of John Hadji Argyris [ARG 54, ARG 55a, ARG 55b], Ray William Clough [TUR 56] and Olgierd Cecil Zienkiewicz [ZIE 64a, ZIE 64b], which popularized variational methods – namely finite elements in industry, with a large development that led to the actual situation where software based on variational methods is available in all the industries around the world. It is interesting to notice that Galerkin's method is periodically reinvented and adapted to actual problems – for instance, in the framework of wavelets, proper orthogonal or generalized decompositions, boundary element methods and so on – but that their basis remain the same.

5.1. A simple situation: the oscillator with one degree of freedom

Let us illustrate the variational approach by considering the simple system formed by a mass m and a spring having rigidity k and natural length ℓ, under the action of a force f.

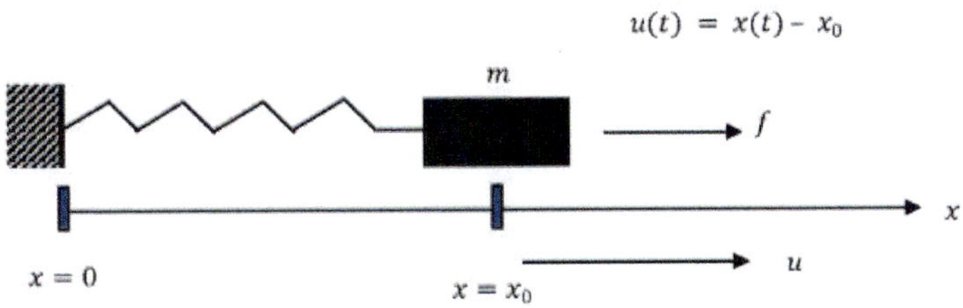

Figure 5.1. *A simple harmonic oscillator*

The position of the mass at time t is x(t) and its initial position is $x(0) = x_0$. The displacement of the mass is $u(t) = x(t) - x_0$.

5.1.1. *Newton's equation of motion*

Isaac Newton enunciated [NEW 87]:

> *The change of motion is proportional to the motive force impressed; and is made in the direction of the right fine in which that force is impressed.*

Newton's book is mainly geometrical as he did not have simple ways for the conceptualization of second derivatives at his disposition. We do not know if Newton abstracted acceleration as the second derivative of displacement with respect to time – this step implies the use of symbolic calculus and was performed by Jacob Hermann [HER 16] and, later, by Euler [EUL 50b]. Today, we write Newton's second law as force = change in motion = change in [mass x velocity], i.e.

$$\frac{d}{dt}\left(m\frac{dx}{dt}\right) = \frac{d}{dt}\left(m\frac{du}{dt}\right) = F ,$$

where F is the force applied to the system. Here, the force applied by the spring on the mass is $-k(x - \ell) = -k(u + x_0 - \ell)$ and we have $F = -k(u + x_0 - \ell) + f$. By assuming that $x_0 = \ell$, this expression reduces to $F = f - ku$. When the mass is constant, we have

$$\frac{d}{dt}\left(m\frac{dx}{dt}\right) = m\frac{d^2x}{dt^2} = m\frac{d^2u}{dt^2} ,$$

which leads to

$$m\frac{d^2u}{dt^2} = F \ ,$$

i.e. force = mass x change in velocity = mass x acceleration.

5.1.2. *Boundary value problem and initial condition problem*

In the framework of Ordinary Differential Equations, it is usual to distinguish two kinds of differential equations: boundary value problems (BVP) and initial condition problems (ICP).

For a BVP, the values of the unknown are given on the extremities (boundaries) of the region of interest. For instance, if the analysis is performed on the time interval $(0, T)$, we give the values of x or u at times $t = 0$ and $t = T$. For instance,

$$u(0) = u_0; \ u(T) = u_T.$$

For ICP, we give the values at the left value of the interval (i.e. at $t = 0$):

$$u(0) = u_0; \ \frac{du}{dt}(0) = \dot{u}_0.$$

As shown in the following, variational methods do not distinguish these two situations: both are treated by the same approach.

5.1.3. *Generating a variational formulation*

Variational formulations – also referred as *variational equations* – are obtained by performing three steps:

1) Multiplying the equation by a *virtual motion* y and integrating the result on the domain of interest (here, the interval $(0, T)$).

2) Using integration by parts (or Green's Theorem, in the multidimensional case) in order to reduce the degree of derivation

of x. Ideally, x and y must have the same degree in the final formulation, in order to get a symmetric expression.

3) Choosing the space V of the *virtual motions*. Ideally, the unknown x to be determined and the virtual motion y are both members of the same space – what may request some transformation on x – and the integrals correspond to scalar products on V. This step is not uniquely determined: different choices for V are possible and lead to different formulations. *In this step, we must keep in mind that the differential equation is valid on the open domain* $\boldsymbol{\Omega}$ *but not on its boundary* $\boldsymbol{\partial\Omega}$*: values on the boundary are to be considered as unknowns, except if they are given by a boundary condition.*

For instance, let us consider the BVP described above.

$$m\frac{d^2u}{dt^2} = f - ku \ on \ (0,T); \ \ u(0) = u_0 \ ; \ \ u(T) = u_T.$$

In this case, Step 1 corresponds to write the equation as

$$y\frac{d}{dt}\left(m\frac{du}{dt}\right) = y(f - ku), \ \ \forall y$$

and

$$\int_0^T y\frac{d}{dt}\left(m\frac{du}{dt}\right)dt = \int_0^T yf dt - \int_0^T kuy dt\,, \ \forall y.$$

Step 2 corresponds to the integration by parts

$$my(T)\frac{du}{dt}(T) - my(0)\frac{du}{dt}(0) - \int_0^T m\frac{du}{dt}\frac{dy}{dt}dt = \int_0^T yf dt - \int_0^T kuy dt, \ \ \forall y.$$

Step 3 is not uniquely determined. For instance, we may set

$$m\frac{du}{dt}(T) = P_T \ ; \ m\frac{du}{dt}(0) = P_0 \ ; V = \ \{y \colon (0,T\,) \to \mathbb{R}\}$$

and we obtain a first variational formulation that reads as: *find the unknowns* P_T, P_0 *and* $u \in V$ *such that*

$$P_T y(T) - P_0 y(0) + \int_0^T \left(kuy - m\frac{du}{dt}\frac{dy}{dt}\right)dt = \int_0^T yf dt, \ \ \forall y \in V;$$

$$u(0) = u_0 \,; \;\; u(T) = \, u_T.$$

As an alternative, we may set:

$$w = u - \bar{u} \;; \bar{u} = \, u_0 + (u_T - u_0)\frac{t}{T}$$

and

$$V = \, \{y: (0, T\,) \to \mathbb{R} : y(0) = y(T) = 0\}$$

Then, a second variational formulation is given by: *find* w *such that,*

$$w \in V \;\; and \;\; \int_0^T \left(kwy - m\frac{dw}{dt}\frac{dy}{dt}\right) dt = \int_0^T y(f - k\bar{u})dt, \forall y \in V,$$

then determine $u = w + \bar{u}$.

Now, let us consider the ICP

$$m\frac{d^2u}{dt^2} = f - ku \; on \;\; (0, T); \;\; u(0) = u_0 \,; \;\; \dot{u}(0) = \, \dot{u}_0.$$

In this case, Step 1 corresponds to

$$y\frac{d}{dt}\left(m\frac{du}{dt}\right) = y(f - ku), \;\; \forall y.$$

and

$$\int_0^T y\frac{d}{dt}\left(m\frac{du}{dt}\right) dt = \int_0^T yFdt = \int_0^T yfdt - \int_0^T kuydt, \forall y.$$

Step 2 corresponds yet to the integration by parts, but taking into account the initial conditions, we have $m\frac{du}{dt}(0) = m\dot{u}_0$:

$$my(T)\frac{du}{dt}(T) - m\dot{u}_0 y(0) - \int_0^T m\frac{du}{dt}\frac{dy}{dt}dt = \int_0^T yfdt - \int_0^T kuydt \,, \;\; \forall y.$$

Again, Step 3 is not uniquely determined. For instance, we may set

$$m\frac{du}{dt}(T) = P_T \ ; V = \{y\colon (0,T) \to \mathbb{R}\}$$

and we obtain a first variational formulation: *find the unknowns* P_T *and* $u \in V$ *such that*

$$P_T y(T) - m\dot{u}_0 y(0) - \int_0^T m\frac{du}{dt}\frac{dy}{dt}dt = \int_0^T yf dt - \int_0^T kuydt,$$
$$\forall y \in V;\ u(0) = u_0.$$

Alternatively, we may set

$$w = u - \bar{u} \ ; \bar{u} = u_0$$

and

$$V = \{y\colon (0,T) \to \mathbb{R} : y(0) = 0\}.$$

Then,

$$w \in V \ \ and\ P_T y(T) - \int_0^T m\frac{dw}{dt}\frac{dy}{dt}dt = \int_0^T y(f - k\bar{u})dt -$$
$$\int_0^T kwydt,\ \forall y \in V;$$

Alternatively, we may set

$$w = u - \bar{u}\ ; \bar{u} = u_0$$

and

$$V = \{y\colon (0,T) \to \mathbb{R} : y(0) = 0\}$$

Then,

$$w \in V \ and\ P_T y(T) - \int_0^T m\frac{dw}{dt}\frac{dy}{dt}dt = \int_0^T y(f - k\,\bar{u})dt - \int_0^T kwydt, \forall y \in V; \frac{dw}{dt}(0) = \dot{u}_0.$$

5.1.4. *Generating an approximation of a variational equation*

Analogously to algebraic equations, the solution is approximated as

$$u(t) \approx \sum_{p=1}^{k} u_p \varphi_p(t)\,, x_p \in \mathbb{R}\ for\ p = 1, \ldots, k,$$

with

$$\mathcal{B} = \{\varphi_i\}_{1 \le i \le k} \subset V = \{v\colon (0,T) \longrightarrow \mathbb{R}\}.$$

The coefficients $u_p \in \mathbb{R}$, for $p = 1, \ldots, k$ are unknowns to be determined. The unknowns may be ranged in a vector *X* including the coefficients and additional unknowns. *X* is solution of a linear system obtained by taking $y(t) = \varphi_j(t)$ in the variational formulation and considering the boundary conditions that were not taken into account in the variational formulation.

5.1.5. *Application to the first variational formulation of the BVP*

For instance, let us consider the first variational formulation of the BVP: we consider

$$\mathcal{B} = \{\varphi_i\}_{1 \le i \le k} \subset V = \{v\colon (0,T) \longrightarrow \mathbb{R}\}.$$

Then, the unknowns are $X = (u_1, \ldots, u_k, P_0, P_T)$ and we have, for $1 \le i \le k$,

$$P_T\varphi_i(T) - P_0\varphi_i(0) + \sum_{j=1}^{k} u_j \int_0^T \left(k\varphi_j\varphi_i - m\frac{d\varphi_j}{dt}\frac{d\varphi_i}{dt}\right) dt = \int_0^T \varphi_i f dt;\ u(0) = u_0;\ u(T) = u_T$$

This set of equations corresponds to a linear system $A.X = B$, where, for $1 \le i \le k$:

$$A_{ij} = \int_0^T \left(k\varphi_j\varphi_i - m\frac{d\varphi_j}{dt}\frac{d\varphi_i}{dt}\right) dt\ for\ 1 \le j \le nb;$$

$$A_{i,k+1} = -\varphi_i(0);\ A_{i,k+2} = \varphi_i(T);\ B_i = \int_0^T \varphi_i f dt.$$

For $i = k+1$ and $i = k + 2$, we have:

$A_{k+1,j} = \varphi_j(0)$ *for* $1 \leq j \leq k$; $A_{k+1,k+1} = A_{k+1,k+2} = 0$; $B_{k+1} = u_0$.

$A_{k+2,j} = \varphi_j(T)$ *for* $1 \leq j \leq k$; $A_{k+2,k+1} = A_{k+2,k+2} = 0$; $B_{k+1} = u_T$.

5.1.6. *Application to the second variational formulation of the BVP*

Now, let us consider the alternative variational formulation of the BVP. In this case, we consider

$$\mathcal{B} = \{\varphi_i\}_{1\leq i\leq k} \subset V = \{v: (0,T) \longrightarrow \mathbb{R}, v(0) = v(T) = 0\}$$

and

$$w(t) \approx \sum_{p=1}^{k} w_p \varphi_p(t).$$

Then, the unknowns are $X = (w_1, \dots, w_k)$ and, recalling that

$$w \in V \quad and \quad \int_0^T \left(kwy - m\frac{dw}{dt}\frac{dy}{dt}\right) dt = \int_0^T y(f - k\bar{u})dt\,, \quad \forall y \in V,$$

we have

$$\sum_{j=1}^{k} w_j \int_0^T \left(k\varphi_j\varphi_i - m\frac{d\varphi_j}{dt}\frac{d\varphi_i}{dt}\right) dt = \int_0^T \varphi_i(f - k\bar{u})dt\,,\ 1 \leq i \leq k$$

This set of equations corresponds to a linear system $A.X = B$, where, for $1 \leq i \leq k$:

$A_{ij} = \int_0^T \left(k\varphi_j\varphi_i - m\frac{d\varphi_j}{dt}\frac{d\varphi_i}{dt}\right) dt$ for $1 \leq j \leq k$; $B_i = \int_0^T \varphi_i(f - k\bar{u})dt$.

5.1.7. Application to the first variational formulation of the ICP

In this case, the unknowns to be determined are $X = (u_1, \dots, u_{nb}, P_T)$ and we have, for $1 \le i \le k$,

$$P_T\varphi_i(T) + \sum_{j=1}^{nb} u_j \int_0^T \left(k\varphi_j\varphi_i - m\frac{d\varphi_j}{dt}\frac{d\varphi_i}{dt}\right) dt = m\dot{u}_0\varphi_i(0) + \int_0^T \varphi_i f dt\,,$$

$$u(0) = u_0.$$

Thus, $A.X = B$, where, for $1 \le i \le k$:

$$A_{ij} = \int_0^T \left(k\varphi_j\varphi_i - m\frac{d\varphi_j}{dt}\frac{d\varphi_i}{dt}\right) dt \; for\; 1 \le j \le k;\; A_{i,k+1} = \varphi_i(T);$$

$$B_i = m\dot{u}_0\varphi_i(0) + \int_0^T \varphi_i f dt.$$

For $i = k + 1$:

$$A_{k+1,j} = \varphi_j(0)\; for\; 1 \le j \le k;\; A_{k+1,k+1} = 0;\; B_{k+1} = u_0.$$

5.1.8. Application to the second variational formulation of the ICP

In the second ICP formulation,

$$\mathcal{B} = \{\varphi_i\}_{1 \le i \le k} \subset V = \{v\colon (0,T) \longrightarrow \mathbb{R}, v(0) = 0\}$$

The unknowns to be determined are $X = (u_1, \dots, u_k, P_T)$ and we have, for $1 \le i \le k$,

$$P_T\varphi_i(T) + \sum_{j=1}^{nb} u_j \int_0^T \left(k\varphi_j\varphi_i - m\frac{d\varphi_j}{dt}\frac{d\varphi_i}{dt}\right) dt = m\dot{u}_0\varphi_i(0) + \int_0^T \varphi_i (f - k\bar{u}) dt.$$

Thus, $A.X = B$, where, for $1 \leq i \leq k$:

$$A_{ij} = \int_0^T \left(k\varphi_j\varphi_i - m\frac{d\varphi_j}{dt}\frac{d\varphi_i}{dt}\right) dt \ for\ 1 \leq j \leq k;$$

$$A_{i,nb+1} = \varphi_i(T); B_i = \int_0^T \varphi_i(f - k\bar{u})dt$$

$$A_{k+1,j} = \dot{\varphi}_j(0)\ for\ 1 \leq j \leq k; A_{k+1,k+1} = 0; B_{k+1} = \dot{u}_0.$$

5.2. Connection between differential equations and variational equations

Let us introduce a general method for the transformation of variational into differential equations and conversely. The transformations are performed by using *virtual motions* (or *test functions*) concomitantly to *integration by parts* applied *to virtual works*.

5.2.1. *From a variational equation to a differential equation*

For an open domain Ω, the set $\mathcal{D}(\Omega)$

$$\mathcal{D}(\Omega) = \{v \in C^\infty(\Omega) : \exists A \subset \Omega, \text{A}\ closed\ bounded,\ v(x) = 0,\ if\ x \notin A\}$$

plays a key role in the transformation. $\mathcal{D}(\Omega)$ is a set of virtual motions (or test functions) having the following useful properties:

1) Any $\varphi \in \mathcal{D}(\Omega)$ has infinitely many derivatives

2) For any $\varphi \in \mathcal{D}(\Omega)$: on the boundary of Ω, φ and all its derivatives are null.

The main result for the transformation of variational equations into differential equations is the following one:

THEOREM 5.1.– Let $V \subset [L^2(\Omega)]^n$ be a subspace such that $[\mathcal{D}(\Omega)]^n \subset V$. Then

$$A \in [L^2(\Omega)]^n\ and \int_\Omega A.v dx = 0, \forall v \in V \Rightarrow A = 0\ on\ \Omega. \quad \blacksquare$$

It admits the following extension

THEOREM 5.2.– Let $V \subset [L^2(\Omega)]^n$ be a subspace such that $[\mathcal{D}(\Omega)]^n \subset V$. Assume that $\Gamma \subset \partial\Omega$, $A \in [L^2(\Omega)]^n$, $B \in [L^2(\Gamma)]^n$, $C_i \in \mathbb{R}^n$, $x_i \in \bar{\Omega} - \bar{\Gamma}$ $(i = 1, \dots, k)$ and

$$\int_\Omega A.vdx + \int_\Gamma B.vdx + \sum_{i=1}^{k} C_i v(x_i) = 0, \forall v \in S .$$

Let $\mathcal{P} = \cup_{i=1}^{k}\{x_i\}$. Then $A = 0 \; on \; \Omega - \mathcal{P}, B = 0 \; on \; \partial\Omega - \Gamma - \mathcal{P}$, $C_i = 0 \; on \; \mathcal{P}, (i = 1, \dots, k)$. ■

In order to apply these theorems, we must eliminate all the derivatives of v from the variational equation. This is usually obtained by using either integration by parts:

$$\int_a^b f'gdx = f(b)g(b) - f(a)g(a) - \int_a^b g'fdx$$

or Green's formula

$$\int_\Omega div(q)\, dx = \int_{\partial\Omega} q.n\, dS.$$

EXAMPLE 5.1.– Let us consider $\Omega = (0,1), \Gamma = \{0\}$, $V = H^1_{0,\Gamma}(\Omega) = \{v \in H^1(\Omega): v(0) = 0\}$ and the variational equation

$$u \in V \; ; \int_0^1 u'v'dx = Fv(\alpha) + \int_0^1 fvdx , \forall v \in V.$$

The first step is the elimination of the derivatives on v: we have

$$\int_0^\alpha u'v'dx = u'(\alpha -)v(\alpha) - u'(0)v(0) - \int_0^\alpha u''vdx;$$

$$\int_\alpha^1 u'v'dx = u'(1)v(1) - u'(\alpha +)v(\alpha) - \int_\alpha^\alpha u''vdx.$$

Thus,

$$u \in V \; ; u'(1)v(1) - [u'(\alpha)]v(\alpha) - \int_0^1 u''vdx = Fv(\alpha) + \int_0^1 fvdx , \forall v \in V$$

and we have

$$u \in V \; ; \; u'(1)v(1) - ([u'(\alpha)] + P)v(\alpha) - \int_0^1 (u'' + f)v dx = 0 \;, \forall v \in V.$$

Since $\mathcal{D}(\Omega) \subset V$, the theorem shows that

$$u \in V \; ; \; u'(1) = 0 \;, [u'(\alpha)] + P = 0, u'' + f = 0.$$

Thus,

$$u(0) = 0 \; ; u'(1) = 0 \; ; [u'(\alpha)] + P = 0; u'' + f = 0 \; on \; (0,1) - \{\alpha\}.\blacksquare$$

EXAMPLE 5.2.– Let us consider $V = H^1_{0,\Gamma}(\Omega)$ and the variational equation

$$u \in V \; ; \int_\Omega \nabla u . \nabla v dx = \int_\Omega f v dx + \int_{\partial\Omega - \Gamma} g v dS \,, \forall v \in V.$$

By using that

$$div(v \nabla u) = \nabla u . \nabla v + v \Delta u$$

and Green's formula, we have

$$\int_\Omega \nabla u . \nabla v dx = - \int_\Omega v \Delta u dx + \int_{\partial\Omega} v \frac{\partial u}{\partial n} dS.$$

Since $v = 0 \; on \; \Gamma$, we have

$$\int_\Omega \nabla u . \nabla v dx = - \int_\Omega v \Delta u dx + \int_{\partial\Omega - \Gamma} v \frac{\partial u}{\partial n} dS$$

and

$$u \in V; - \int_\Omega v \Delta u dx + \int_{\partial\Omega - \Gamma} v \frac{\partial u}{\partial n} dS = \int_\Omega f v dx + \int_{\partial\Omega - \Gamma} g v dS \,, \forall v \in V \; ;$$

what yields

$$u \in V; \; \int_{\partial\Omega - \Gamma} v \left(\frac{\partial u}{\partial n} - g \right) dS - \int_\Omega (f + \Delta u) v dx = 0, \forall v \in V.$$

Since $\mathcal{D}(\Omega) \subset V$, the theorem shows that

$$u \in V;\ \frac{\partial u}{\partial n} - g = 0\ on\ \partial\Omega - \Gamma; f + \Delta u = 0\ on\ \Omega.$$

Thus,

$$u = 0\ on\ \Gamma;\ \frac{\partial u}{\partial n} - g = 0\ on\ \partial\Omega - \Gamma; f + \Delta u = 0\ on\ \Omega. \qquad \blacksquare$$

5.2.2. *From a differential equation to a variational equation*

The converse transformation consists in the transformation of the differential equation into a Principle of Virtual Works. This is achieved by the multiplication by a virtual motion v and the subsequent use of integration by parts or Green's formula. We are generally interested in obtaining a formulation where the virtual motion and the solution are members of the same space V. Initially, the space V is formed by all the functions defined on the domain of interest for which the quantities manipulated have a sense, but conditions satisfied by the solution and that are not used in order to generate the variational formulation have to be introduced in the definition of V.

EXAMPLE 5.3.– Let us consider the differential equation:

$$u(0) = 0; u'(1) = 0; u'' + f = 0\ on\ (0,1).$$

We start by the creation of a table containing:

1) the available equations;

2) the boundary or initial conditions;

3) the tools to be used: virtual works;

4) the steps to be performed:

i) integration by parts or use of Green's formula;

ii) formal definition of the space.

5) The result of the steps performed.

For instance:

Conditions and tools	Not Used	Used
$u(0) = 0$	×	
$u'(1) = 0$	×	
$u'' + f = 0$ *on* $(0,1)$	×	
Virtual work	×	
Integration by parts or Green's Formula	×	
Formal definition of the space	×	
Variational formulation:		

The aim is to use all the conditions. At the end of the transformation, any unused condition will be introduced in the definition of V.

As previously observed, the set of all the virtual motions is *initially*: $V^{(0)} = \{v: (0,1) \to \mathbb{R}\}$. We start the transformation by using the differential equation itself:

Conditions and tools	Not used	Used
$u(0) = 0$	×	
$u'(1) = 0$	×	
$u'' + f = 0$ *on* $(0,1)$		✓
Virtual work	×	
Integration by parts or Green's Formula	×	
Formal definition of the space	×	
Formulation: $u \in V^{(0)},\ u'' + f = 0$ *on* $(0,1)$		

In order to produce a virtual work, two steps are necessary: first, the equation is multiplied by an arbitrary virtual motion v:

$$u''v + fv = 0 \ \ on\ (0,1).$$

Second, the resulting equality is integrated on $\Omega = (0,1)$:

$$\int_\Omega u''vdx + \int_\Omega fvdx = 0.$$

Conditions and tools	Not used	Used
$u(0) = 0$	×	
$u'(1) = 0$	×	
$u'' + f = 0$ *on* $(0,1)$		✓
Virtual work		✓
Integration by parts or Green's Formula	×	
Formal definition of the space	×	
Formulation: $u \in V^{(0)}, \int_\Omega u''vdx + \int_\Omega fvdx = 0, \forall v \in V^{(0)}.$		

Now, we use integration by parts:

$$\int_0^1 u''vdx = u'(1)v(1) - u'(0)v(0) - \int_0^1 u'v'dx.$$

Thus, we have

$$u'(1)v(1) - u'(0)v(0) - \int_0^1 u'v'dx + \int_0^1 fvdx = 0.$$

and

Conditions and tools	Not used	Used
$u(0) = 0$	×	
$u'(1) = 0$	×	
$u'' + f = 0$ *on* $(0,1)$		✓
Virtual work		✓
Integration by parts or Green's Formula		✓
Formal definition of the space	×	
Formulation: $u \in V^{(0)}, -u'(1)v(1) + u'(0)v(0) + \int_0^1 u'v'dx = \int_0^1 fvdx, \forall v \in V^{(0)}.$		

The condition $u(0) = 0$ cannot be used, since there is no $u(0)$ in the variational formulation. However, the expression contains $u'(1)$ and the condition $u'(1) = 0$ may be used:

Conditions and tools	Not used	Used
$u(0) = 0$	×	
$u'(1) = 0$		✓
$u'' + f = 0 \; on \; (0,1)$		✓
Virtual work		✓
Integration by parts or Green's Formula		✓
Formal definition of the space	×	
Formulation: $u \in V^{(0)}, \;\; u'(0)v(0) + \int_0^1 u'v'dx = \int_0^1 fvdx \,, \forall v \in V^{(0)}.$		

The condition $u(0) = 0$ has not been used. Thus, it is introduced in the definition of the space:

$$V^{(1)} = \{v \in V^{(0)}: v(0) = 0\}.$$

Now, we have

$$u \in V^{(1)}; \; \int_0^1 u'v'dx = \int_0^1 fvdx \,, \forall v \in V^{(1)}.$$

Conditions and tools	Not used	Used
$u(0) = 0$		✓
$u'(1) = 0$		✓
$u'' + f = 0 \; on \; (0,1)$		✓
Virtual work		✓
Integration by parts or Green's Formula		✓
Formal definition of the space	×	
Formulation: $u \in V^{(1)}, \; \int_0^1 u'v'dx = \int_0^1 fvdx \,, \forall v \in V^{(1)}.$		

The last step is the formal definition of the space: regarding to the expression, the first derivatives u', v' appear in a scalar product. Thus, a convenient choice is a space where the derivatives are square summable:

$$V = \{v: (0,1) \to \mathbb{R} : v' \in L^2(\Omega), v(0) = 0\} = H^1_{0,\Gamma}(\Omega), \Omega = (0,1), \Gamma = \{0\}.$$

Conditions and tools	Not used	Used
$u(0) = 0$		✓
$u'(1) = 0$		✓
$u'' + f = 0$ *on* $(0,1)$		✓
Virtual work		✓
Integration by parts or Green's Formula		✓
Formal definition of the space		✓
Formulation: $u \in V = H^1_{0,\Gamma}(\Omega)$, $\int_0^1 u'v'dx = \int_0^1 fvdx$, $\forall v \in V$.		

This is the final formulation: all the conditions were used. ■

EXAMPLE 5.4.– Let us consider the differential equation:

$$u = 0 \ on \ \Gamma \ ; \ \frac{\partial u}{\partial n} = g \ on \ \partial\Omega - \Gamma; f + \Delta u = 0 \ on \ \Omega.$$

We have

Conditions and tools	Not used	Used
$u = 0$ *on* Γ	×	
$\frac{\partial u}{\partial n} = g$ *on* $\partial\Omega - \Gamma$	×	
$f + \Delta u = 0$ *on* Ω	×	
Virtual work	×	
Integration by parts or Green's Formula	×	
Formal definition of the space	×	
Formulation:		

We set $V^{(0)} = \{v: \Omega \to \mathbb{R}\}$.

Conditions and tools	Not used	Used
$u = 0$ *on* Γ	×	
$\frac{\partial u}{\partial n} = g$ *on* $\partial\Omega - \Gamma$	×	
$f + \Delta u = 0$ *on* Ω		✓
Virtual work	×	
Integration by parts or Green's Formula	×	
Formal definition of the space	×	
Formulation: $u \in V^{(0)},\ f + \Delta u = 0$ *on* Ω		

The equation is multiplied by an arbitrary virtual motion v:

$$fv + v\Delta u = 0 \text{ on } \Omega$$

The resulting equality is integrated on Ω:

$$\int_\Omega v\Delta u dx + \int_\Omega fv dx = 0.$$

Then, we use Green's formula:

$$\int_\Omega v\Delta u dx = \int_\Omega div(v\nabla u)dx - \int_\Omega \nabla u.\nabla v dx = \int_{\partial\Omega} v\frac{\partial u}{\partial n}dS - \int_\Omega \nabla u.\nabla v dx.$$

Thus, we have

Conditions and tools	Not used	Used
$u = 0$ *on* Γ	×	
$\frac{\partial u}{\partial n} = g$ *on* $\partial\Omega - \Gamma$	×	
$f + \Delta u = 0$ *on* Ω		✓
Virtual work		✓
Integration by parts or Green's Formula	×	
Formal definition of the space	×	
Formulation: $u \in V^{(0)};\int_\Omega fv dx + \int_{\partial\Omega} v\frac{\partial u}{\partial n}dS = \int_\Omega \nabla u.\nabla v dx\,, \forall v \in V^{(0)}$		

Since $\partial u/\partial n$ appears in the formulation, the corresponding condition may be used: we have

$$\int_{\partial\Omega} v\frac{\partial u}{\partial n}dS = \int_{\partial\Omega-\Gamma} v\frac{\partial u}{\partial n}dS + \int_{\Gamma} v\frac{\partial u}{\partial n}dS = \int_{\partial\Omega-\Gamma} gvdS + \int_{\Gamma} v\frac{\partial u}{\partial n}dS$$

and

Conditions and tools	Not used	Used
$u = 0\ on\ \Gamma$	×	
$\frac{\partial u}{\partial n} = g\ on\ \partial\Omega - \Gamma$		✓
$f + \Delta u = 0\ on\ \Omega$		✓
Virtual work		✓
Integration by parts or Green's Formula		✓
Formal definition of the space	×	
Formulation: $u \in V^{(0)}, \int_{\Omega} fvdx + \int_{\partial\Omega-\Gamma} gvdS + \int_{\Gamma} v\frac{\partial u}{\partial n}dS = \int_{\Omega} \nabla u.\nabla vdx\, , \forall v \in V^{(0)}$		

The condition $u = 0\ on\ \Gamma$ has not been used. Thus, it is introduced in the definition of the space:

$$V^{(1)} = \{v \in V^{(0)}: u = 0\ on\ \Gamma\ \}.$$

Now, we have

$$u \in V^{(1)};\ \int_{\Omega} \nabla u.\nabla vdx = \int_{\partial\Omega-\Gamma} gvdS\, , \forall v \in V^{(1)},$$

which corresponds to the following table:

Conditions and tools	Not used	Used
$u = 0\ on\ \Gamma$		✓
$\frac{\partial u}{\partial n} = g\ on\ \partial\Omega - \Gamma$		✓
$f + \Delta u = 0\ on\ \Omega$		✓
Virtual work		✓
Integration by parts or Green's Formula		✓
Formal definition of the space	×	
Formulation: $u \in V^{(1)};\ \int_{\Omega} \nabla u.\nabla vdx = \int_{\partial\Omega-\Gamma} gvdS\, , \forall v \in V^{(1)},$		

The formal definition of the space is obtained by considering the last formulation: gradients $\nabla u, \nabla v$ appear in a scalar product. Thus, analogously to the preceding situation, a convenient choice is a space where the derivatives are square summable:

$$V = \{v: (0,1) \to \mathbb{R} : \nabla u \in L^2(\Omega), v = 0 \; on \; \Gamma\} = H^1_{0,\Gamma}(\Omega).$$

Conditions and tools	Not used	Used
$u = 0 \; on \; \Gamma$		✓
$\frac{\partial u}{\partial n} = g \; on \; \partial\Omega - \Gamma$		✓
$f + \Delta u = 0 \; on \; \Omega$		✓
Virtual work		✓
Integration by parts or Green's Formula		✓
Formal definition of the space		✓
Formulation: $u \in V = H^1_{0,\Gamma}(\Omega)$, $\int_0^1 u'v'dx = \int_0^1 fvdx, \forall v \in V$.		

This is the final formulation: all the conditions were used. ■

5.3. Variational approximation of differential equations

As shown above, differential equations may be transformed in variational equations. Therefore, we may use the approach presented in sections 4.7 and 4.8. The main particularity of differential equations consists in the use of boundary or initial conditions, which:

1) either lead to supplementary equations which must be included in the final algebraic system;

2) or are included in the definition of the functional space, so that they must be taken into account when choosing a basis.

These choices were illustrated in section 5.1.4. Below, we present more in-depth examples.

EXAMPLE 5.5.– Let us consider $\Omega = (0,1)$ and the differential equation

$$(au'')'' + f = 0 \; on \; \Omega, \left[\left(a(\bar{x})u''(\bar{x})\right)'\right] + P = 0,$$

$$u(0) = u(1) = u''(0) = u''(1) = 0, \left(a(0)u''(0)\right)' = R_0, \left(a(1)u''(1)\right)' = R_1$$

Where $a(x) \geq \alpha > 0, \forall\, x \in \Omega$, $P \in \mathbb{R}$, $f: \Omega \longrightarrow \mathbb{R}$, $f \in L^2(\Omega), 0 < \bar{x} < 1$. A variational formulation is obtained by using

$$V = \{v \in L^2(\Omega): v', v'' \in L^2(\Omega)\},\ V_0 = \{v \in V: v(0) = v(1) = 0\}.$$

We have

$$R_0, R_1 \in \mathbb{R};\ u \in V_0\ and$$

$$\int_0^1 (au''v'') + Pv(\bar{x}) - R_0 v(0) + R_1 v(1) + \int_0^1 (fv) = 0\,,\ \forall\, v \in V\,.$$

The unknowns to be determined are R_0, R_1, u. We consider a family $\{\varphi_j: 1 \leq j \leq N\}$ and the approximation $u \approx \sum_{j=1}^{k} u_j \varphi_j(x)$. Taking $X = (u_0, u_1, \dots, u_N, R_0, R_1)^t$, we have $AX = B$, with

$$A_{1,j} = \varphi_{j-1}(0), 1 \leq j \leq N+1; A_{1,N+2} = A_{1,N+3} = 0;$$

$$A_{2,j} = \varphi_{j-1}(0), 1 \leq j \leq N+1; A_{2,N+2} = A_{2,N+3} = 0;$$

$$A_{i,j} = \int_0^1 \left(\alpha \varphi''_{i-3} \varphi''_{j-1}\right), 1 \leq j \leq N+1; A_{i,N+2} = -\varphi_{i-3}(0);$$

$$A_{i,N+3} = \varphi_{i-3}(1);\ (3 \leq i \leq N+3)$$

$$b_1 = b_2 = 0; b_i = -P \quad {}_{i-3}(\bar{x}) - \int_0^1 (f \varphi_{i-3})\,, 3 \leq i \leq N+3.$$

For continuous families, all the elements of A and B may be evaluated by using the Matlab® classes previously defined for integration and scalar products. For instance, we consider the trigonometrical family

$$\varphi_0(x) = 1, \varphi_{2k-1}(x) = sin(kx),\ \varphi_{2k}(t) = cos(kx).$$

We also consider a total family formed by Finite Elements (beam's flexion elements). Finite elements present the advantage that the integrals may be evaluated on each element and, then, assembled in order to furnish the global result – this procedure is called *assembling*. This book is not intended to present finite element methods and the reader interested in this powerful variational method may refer to the large amount a literature in this field. We recall only the general principle: a partition of the domain Ω in elements is introduced. The basis function φ_i has as support a finite number of elements, connected to a node. Then we may evaluate an elementary matrix on each element, which contains the integrals of the basis functions on the element, which furnishes an elementary matrix. The elementary matrices are assembled in order to form a global matrix.

For a beam under flexion, an element is an interval $(x_i, x_{i+1}) \subset \Omega$: the interval is partitioned by using nodes $0 < x_1 < \cdots < x_{n+1} = 1$. For instance, $x_i = (i-1)h, \quad h = \frac{1}{n}$. For $1 \leq i \leq n,$ the approximation reads as ($\xi = \frac{x - x_i}{h}$)

$$u(x) \approx$$
$$u_i N_1(\xi) + \theta_i N_2(\xi) + u_{i+1} N_3(\xi) + \theta_{i+1} N_4(\xi) \ \ on\ (x_i, x_{i+1})$$

where

$$N_1(\xi) = 1 - 3\xi^2 + 2\xi^3, N_2(\xi) = h\xi(1-\xi)^2,$$
$$N_3(\xi) = \xi^2(3 - 2\xi), N_4(\xi) = h\xi^2(\xi - 1) \ .$$

The elementary matrices read as

$$K_e = \frac{\bar{a}}{h^3}\begin{pmatrix} 12 & 6h & -12 & 6h \\ 6h & 4h^2 & -6h & 2h^2 \\ -12 & -6h & 12 & -6h \\ 6h & 2h^2 & -6h & 4h^2 \end{pmatrix}, f_e = \frac{h\bar{f}}{2}\begin{pmatrix} 1 \\ h/6 \\ 1 \\ -h/6 \end{pmatrix},$$

where $\bar{a}$ and $\bar{f}$ are the mean values of a and f on the element (x_i, x_{i+1}). K_e is the elementary stiffness. An example of assembling for a beam under flexion is the following code:

```
function [A,B] = fem_matrices(n,h,x, P, xbar,k)
A = zeros(2*n+4, 2*n+4);
B = zeros(2*n+4,1);
A(1,1)  = 1; %  u(0) = 0
```

```
A(2, 2*n+1) = 1; %  u(1) = 0
k_el = [12     6*h      -12     6*h;
6*h    4*h^2    -6*h    2*h^2;
       -12      -6*h     12     -6*h;
6*h    2*h^2    -6*h    4*h^2]/h^3;
for  i = 1: n
    am = 0.5*(a(x(i),xbar) + a(x(i+1),xbar));
    fm = 0.5*(f(x(i)) + f(x(i+1)));
    b_el = fm*h*[1; h/6; 1; -h/6]/2;
    B(2*i+1:2*i+4,1)=B(2*i+1:2*i+4,1) + b_el;
  A(2*i+1:2*i+4,2*i-1:2*i+2) = A(2*i+1:2*i+4,2*i-1:2*i+2)
+ am*k_el;
end;
A(3,2*n+3)=-1;  %  -R0*phi(0,0)
A(2*n+3,2*n+4)=1; %  R1*phi(n,1)
B(2*k+1) = B(2*k+1) + P;
B = -B;
```

In this code, the integrals over each element are synthesized in matrix k_el, which contains the integrals corresponding to pairs of indexes $(i-1,i)$, (i,i), $(i+1,i)$). Analogously, b_el contains the integrals corresponding to pairs of indexes (i,i), $(i+1,i)$). The reader will find in the literature about finite elements examples of efficient implementations for different families of finite elements. The code assumes that a subprogram a furnishes the values of $a(x)$ and that the values of $\bar{x}$ and P are given by variables are furnished. x is a vector containing the nodes, i.e. the partition.

Let us consider the situation where $f(x) = c$ is constant, $a(x) = 1$ is constant, $P = 1$. In this case, the solution is

$$u(x) = \begin{cases} a_4x^4 + a_3x^3 + a_1x, & si\ x < \bar{x} \\ b_4(x-1)^4 + b_3(x-1)^3 + b_1(x-1), & si\ x > \bar{x} \end{cases}$$

where (notice that $a_2 = b_2 = 0$)

$$a_4 = -\frac{c}{24},\ a_3 = \frac{2+c-2\bar{x}}{12},\quad a_1 = \frac{12\bar{x}^2-4\bar{x}^3-8*\bar{x}-c}{24},$$

$$b_4 = -\frac{c}{24},\ b_3 = -\frac{c+2\bar{x}}{12},\ b_1 = \frac{c+4\bar{x}-4\bar{x}^3}{24}.$$

Results are shown in Figure 5.2. For $c = 0$, both the families furnish $R_0 = 0.5, R_1 = -0.5$. For $c = 1$, both the families furnish $R_0 = 1, R_1 = -1$.

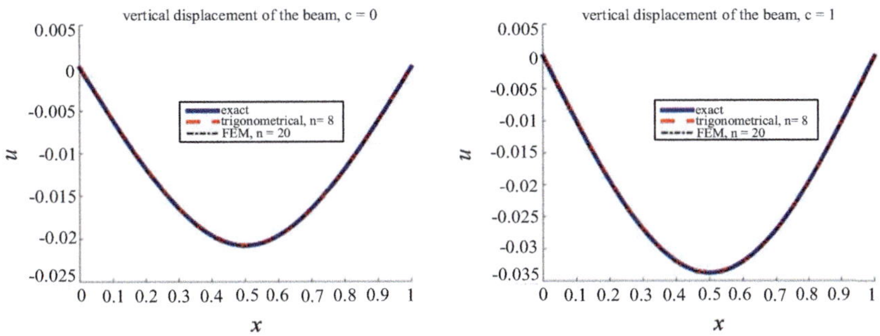

Figure 5.2. *Examples of solutions for beam under flexion. For a color version of the figure, see www.iste.co.uk/souzadecursi/variational.zip*

The results are analogous when $a(x)$ is not a constant. An example is shown in Figure 5.3, where we consider a linear growth of a:

$$a(x) = 1 + x\,, if\ x < \frac{1}{2};\quad a(x) = 2 - x, if\ x > \frac{1}{2}. \qquad \blacksquare$$

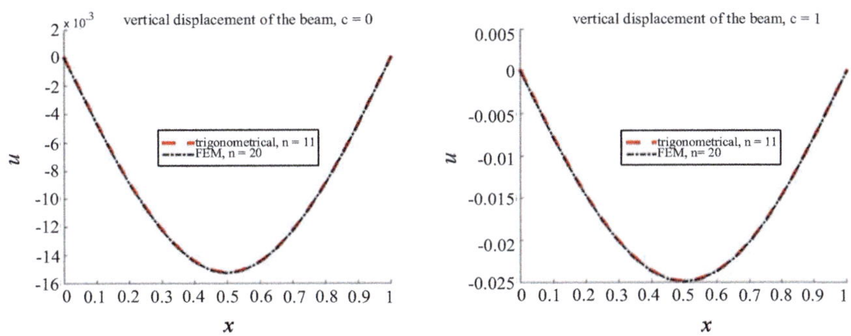

Figure 5.3. *Examples of solutions for beam under flexion (linear variation of a)*

EXAMPLE 5.6.– Let $\Omega \subset \mathbb{R}^2$ be a regular region such that $0 \in \Omega$. We consider the eigenvalue problem

$$u \neq 0, \Delta u = \lambda u \text{ on } \Omega, u = 0 \text{ on } \partial\Omega.$$

We set

$$V = \{v \in L^2(\Omega): \nabla v \in [L^2(\Omega)]^2 \text{ and } v = 0 \text{ on } \partial\Omega\}.$$

A variational formulation is given by

$$u \in V, u \neq 0, \int_\Omega |\nabla u|^2 = 1 \text{ and } \int_\Omega \nabla u . \nabla v = \lambda \int_\Omega uv, \forall v \in V,$$

Where $\lambda \in \mathbb{R}$ is an unknown real number. λ is an eigenvalue and u is the associated *mode*.

Let $F = \{\varphi_{ij}\}_{i,j \in \mathbb{N}} \subset V$ be a total family on V. We set

$$u = \sum_{i=0}^{+\infty} \sum_{j=0}^{+\infty} u_{ij}\varphi_{ij} \approx Pu = \sum_{i=0}^{n_1} \sum_{j=0}^{n_2} u_{ij}\varphi_{ij}$$

Then:

$$\sum_{m=0}^{n_1} \sum_{n=0}^{n_2} a_{ijmn}u_{mn} = \lambda \sum_{m=0}^{n_1} \sum_{n=0}^{n_2} b_{ijmn}u_{mn} , 0 \leq i \leq n_1, 0 \leq j \leq n_2 ,$$

where

$$a_{ijmn} = \int_\Omega \nabla\varphi_{ij} . \nabla\varphi_{mn} , \; b_{ijmn} = \int_\Omega \varphi_{ij}\varphi_{mn} .$$

These equations form a linear system for the determination of *Pu*. As previously observed, we may reduce the number of indexes to 2 by using a transformation

$$ind(i, j) = 1 + i + j(n_1 + 1).$$

Then,

$U_l = u_{mn}, A_{kl} = a_{ijmn}$ and $B_{kl} = b_{ijmn}, \ k = ind(i,j), \ l = ind(m,n)$,

and

$$\sum_{l=1}^{N} A_{kl}U_l = \lambda \sum_{l=1}^{N} B_{kl}U_l \, , 1 \leq i \leq N \, , N = (n_1+1)(n_1+1),$$

so that:

$$U \neq 0 \text{ and } AU = \lambda BU \, .$$

Thus $det\ (A - \lambda B) = 0$. We observe that $kU\ (and\ ku)$ are also solutions for any $k \neq 0$. Usually, k is chosen in order to normalize the eigenvectors.

We consider the situation where $\Omega = I_1 \times I_\alpha$, $I_s = s \times (-s \times) = (-s, s)$, $\{\psi_i\}_{i \in \mathbb{N}}$ is a total family on

$$H = \{v \in L^2(\mathrm{I}) \colon v' \in L^2(\mathrm{I}) \ and \ v(-1) = v(1) = 0 \, \}.$$

In this case, we may use $\varphi_{ij}(x_1, x_2) = \psi_i(x_1)\psi_j\left(\frac{x_2}{\alpha}\right)$.

Then,

$$a_{ijmn} = \alpha C_{im}D_{jn} + \frac{1}{\alpha} C_{jn}D_{im} \, , \ \ b_{ijmn} = \alpha D_{im}D_{jn}$$

where

$$C_{ij} = \int_{-1}^{1} \psi_i'(t)\psi_j'(t)dt \ , D_{ij} = \int_{-1}^{1} \psi_i(t)\psi_j(t)dt$$

We use $\alpha = 0.5$ and the trigonometrical family $\psi_k(t) = \begin{cases} sin\left(\frac{k+1}{2}\pi t\right), \ kodd \\ a + cos\left(\left(\frac{k}{2}+1\right)\pi t\right), \ keven \end{cases}, \ a = \begin{cases} 1, \ \left(\frac{k}{2}+1\right) odd \\ -1, \ \left(\frac{k}{2}+1\right) even \end{cases}.$

Results furnished by this family are shown in Figure 5.4.

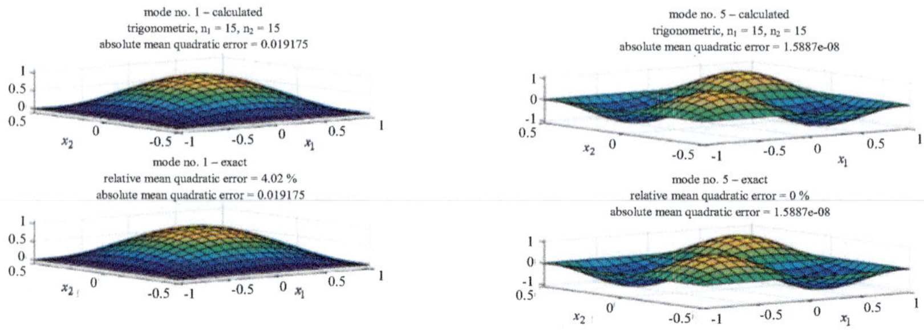

Figure 5.4. *Examples of modes furnished by the trigonometric family. For a color version of the figure, see www.iste.co.uk/souzadecursi/variational.zip*

We consider $\alpha = 0.5$ and the polynomial family $\psi_i(t) = t^{i+2} - at - b$, or

$$a = \begin{cases} 1, & i \ odd \\ 0, & i \ even \end{cases}, \quad b = \begin{cases} 0, & i \ odd \\ 1, & i \ even \end{cases}.$$

Results furnished by this family are exhibited in Figure 5.5.

Figure 5.5. *Examples of modes furnished by the polynomial family. For a color version of the figure, see www.iste.co.uk/souzadecursi/variational.zip*

For Q2 finite elements, $I_1$is partitioned in $2n_1$ subintervals corresponding to the points $(x_1)_i = -1 + (i-1)h_1, \ h_1 = 1/n_1$. Analogously, I_α is partitioned in $2n_2$ subintervals corresponding to the points $(x_2)_j = -\alpha + (j-1)h_2$, $h_2 = \alpha/n_2$. A node is a point $P_{ij} = ((x_1)_i, (x_2)_j)$. Element Q_{ij} is the rectangle limited by points P_{ij}, $P_{i+1,j}$, $P_{i+1,j+1}$, $P_{i,j+1}$. On each element, u is approximated by a polynomial function of degree 2 having as expression $a_0 + a_1x_1 + a_2x_2 + a_3x_1x_2$. Elementary mass and stiffness are evaluated as follows:

$$K_{el} = \begin{pmatrix} a & b & c & d \\ b & a & d & c \\ c & d & a & b \\ d & c & b & a \end{pmatrix}, \begin{matrix} a = (c_1 + c_2)/3 \\ b = (c_1 - 2c_2)/6 \\ c = -(c_1 + c_2)/6 \\ d = (c_2 - 2c_1)/6 \end{matrix}, \begin{matrix} c_1 = h_1/h_2 \\ c_2 = h_2/h_1 \end{matrix}$$

$$M_{el} = \frac{h_1h_2}{36}\begin{pmatrix} 4 & 2 & 1 & 2 \\ 2 & 4 & 2 & 1 \\ 1 & 2 & 4 & 2 \\ 2 & 1 & 2 & 4 \end{pmatrix}.$$

Results furnished by Q2 are shown in Figure 5.6.

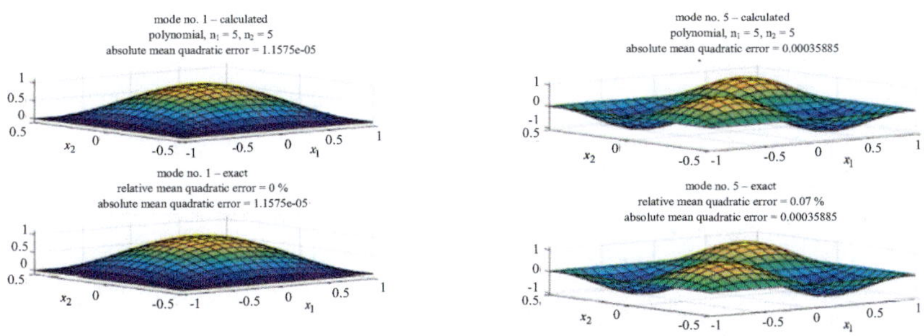

Figure 5.6. *Examples of modes furnished by the Q2 family. For a color version of the figure, see www.iste.co.uk/souzadecursi/variational.zip*

For T1 finite elements, rectangle Q_{ij} is split in two triangles by one of its diagonals. For instance, one triangle is limited by $P_1 = P_{ij}, P_2 = P_{i+1,j}, P_3 = P_{i+1,j+1}$, while another one is limited by $P_1 = P_{ij}$, $P_2 = P_{i+1,j+1}, P_3 = P_{i,j+1}$. On each element, u is approximated by a polynomial function of degree 1 having as expression $a_0 + a_1x_1 + a_2x_2$. Elementary mass and stiffness are usually evaluated as functions of $P_2 - P_1 = (x_{21}, y_{21})$, $P_3 - P_1 = (x_{31}, y_{31})$, $P_3 - P_2 = (x_{32}, y_{32})$;

$$K_{el} = \frac{d}{2} a^t a \,, \; a = \begin{pmatrix} -y_{32} & y_{31} & -y_{21} \\ x_{32} & -x_{31} & x_{21} \end{pmatrix} \,, \; d = x_{21}y_{31} - x_{31}y_{21}$$

$$M_{el} = \frac{d}{24} \begin{pmatrix} 2 & 1 & 1 \\ 1 & 2 & 1 \\ 1 & 1 & 2 \end{pmatrix}.$$

Results furnished by T1 are shown in Figure 5.7.

When α is small, the geometry of the rectangle may be approximated by a line and we expect that the eigenvalues end modes converge to the corresponding ones for a biclamped bar:

$$(u_{1d})_n = sin\left(\frac{n\pi}{2}(x_1 + 1)\right)$$

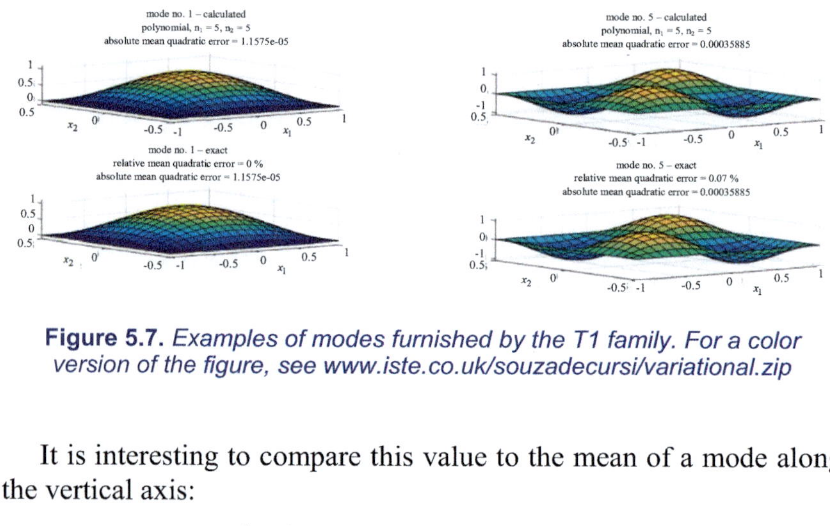

Figure 5.7. *Examples of modes furnished by the T1 family. For a color version of the figure, see www.iste.co.uk/souzadecursi/variational.zip*

It is interesting to compare this value to the mean of a mode along the vertical axis:

$$(u_{mean})_n = \frac{1}{2\alpha} \int_{-\alpha}^{\alpha} u_n(x_1, x_2) dx_2$$

As shown in Figure 5.8, u_{mean} and u_{1d} are close. ■

5.4. Evolution partial differential equations

Let us consider the situation where

$$\frac{du}{dt} + A(u) = f \ on\,\Omega, \quad u(0) = u_0$$

with

$$a(u,v) = (A(u),v); \ \ell(v) = (f,v); f, A(u) \in H.$$

In this case, we have

$$\left(\frac{du}{dt}, v\right) + a(u,v) = \ell(v), \ \forall v \in H.$$

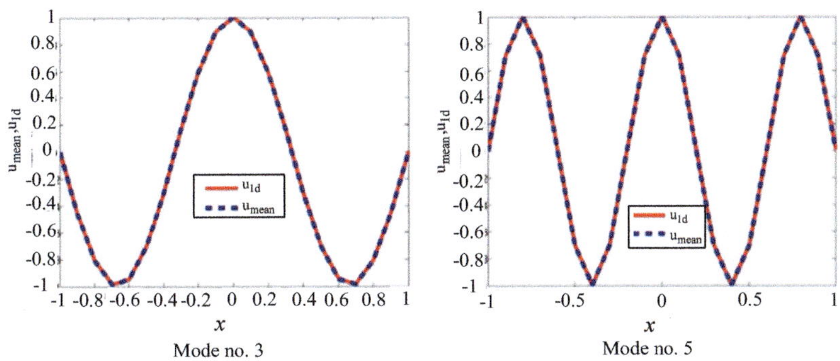

Figure 5.8. *Comparison between the mean along the vertical axis and a one-dimensional mode ($\alpha = 0.1$)*

5.4.1. *Linear equations*

Let us assume that $u \to a(u,v)$ is linear. Using the approximation introduced in sections 4.4, 4.7 and 4.8:

$$u \approx Pu = \sum_{j=1}^{k} u_j(t)\varphi_j(x),$$

and

$$\sum_{j=1}^{k} \left[(\varphi_j, v)\frac{du_j}{dt} + a(\varphi_j, v)u_j \right] = \ell(v), \quad \forall v \in H.$$

Taking $v = \varphi_i$, $\boldsymbol{U} = (u_1(t), \dots, u_k(t))$, we obtain

$$M\frac{dU}{dt} + KU = B,$$

where

$$M_{ij} = (\varphi_i, \varphi_j), \quad K_{ij} = a(\varphi_i, \varphi_j), \quad B_i = \ell(\varphi_i).$$

In addition, we may approximate

$$u_0 \approx Pu_0 = \sum_{j=1}^{k} u_j(0)\varphi_j(x).$$

In this case, we have

$$MU(0) = Y\,, \; Y_i = (u_0, \varphi_i)_0.$$

Thus,

$$U(0) = U_0 = M^{-1}Y.$$

These observations show that approximations of a linear differential equation using a total family or a Hilbert basis leads to the construction of:

1) a mass matrix M;

2) a stiffness matrix K;

3) a second member B, which may depends upon t;

4) an initial data U_0.

For instance, these matrices may be generated by finite element methods (FEM) or the continuous families presented.

The resulting equations may be integrated by intrinsic solvers from Matlab® (`ode45`, `ode23s`, etc.). For instance, we may implement a subprogram evaluating $M^{-1}(B - KU)$. Denoting by `invM` the inverse M^{-1} of M and `invMK` the value of $M^{-1}K$, we may use the code

```
function v = second_member(t,U,invM,invMK, B)
v = invM*B(t) - invMK*U;
end
```

Then, we call one of these solvers:

```
T = ...; % give the time interval (0,T)
solver = @solvername;  % solvername: ode45,ode23s,etc.
invM =  M \ eye(size(M)); % evaluate inv(M)
invMK = M \ K; % evaluate inv(M)*K
U0= M \ Y; % vector of initial conditions
B = @program_name;  % name of the subprogram evaluating
F.
secm = @(t,X) second_member(t,U,invM,invMK, B);
sol = solver(secm, [0,T], U0);
```

Returning structure *sol* contains:

1) `sol.x` : vector of time moments. The line `tc = sol.x;` copies the time moments \in vector `tc` – the solution has been evaluated at `tc(1), tc(2) – tc(end).`

2) `sol.y` : array of values of U. Each colum of `sol.y` contains the solution for one time moment: column `j` (`sol.y(:,j)`) contains the solution U at time `tc(j)`. Each line contains the solution for a given degree of freedom (dof) and all the times: line `i` (`sol.y(i,:)`) contains the evolution of U_i for all the values `tc`.

5.4.2. *Nonlinear equations*

Let us assume that $u \to a(u, v)$ is nonlinear. In this case,

$$\sum_{j=1}^{k} \left[(\varphi_j, v)\frac{du_j}{dt} + a(Pu, v)\right] = \ell(v), \quad \forall v \in H.$$

and we have

$$M\frac{dU}{dt} = F(t, U),$$

where

$$F_i(t,U) = B_i - a(Pu,\varphi_i).$$

Let us assume that `proj(U,basis)` furnishes Pu, while `a(pu,basis)` furnishes a vector $\boldsymbol{v}$ such that $v_i = a(PU,\varphi_i)$. In this case, the second member reads as

```
function v = second_member(t,U,invM,a, proj,basis,B)
pu = proj(U,basis);
aux = a(pu,basis) ;
v = invM*(B(t) -aux); % or M \ (B(t) - aux) ;
end
```

Then, the differential equation may be solved by an intrinsic solver from Matlab®.

5.4.3. *Motion equations*

Let us consider the equations describing the motion of an undamped system:

$$M\ddot{U} + KU = F$$

The model must be completed by initial conditions:

$$U(0) = U_0\,;\dot{U}_0 = \dot{U}_0.$$

The main assumption on this system concerns the existence of *normal modes*, i.e. the existence of vectors $\phi_i, i = 1,\ldots,n$ (n = dimension of the square matrices M, K and of vector U), such that the matrix $\Phi = [\phi_1,\ldots,\phi_n]$ having as columns the vectors simultaneously diagonalizes the matrices $M,\ K$:

$$\Phi^t M\Phi = \widetilde{M},\ \ \Phi^t K\Phi = \widetilde{K}\ ,$$

Where $\widetilde{M},\widetilde{K}$ are diagonal matrices. The vectors are determined by setting

$$U = \Phi\exp(i\omega t)$$

and solving

$$(K - \omega^2 M)\Phi = 0.$$

As an alternative, we may set

$$U = \mathrm{A}\Phi \sin(\omega t) + \mathrm{B}\Phi \cos(\omega t)$$

ω is called a *natural pulsation*. Φ is the associated eigenvector, called *normal mode*.

We set

$$X = \begin{pmatrix} U \\ \dot{U} \end{pmatrix}$$

and the problem is rewritten as:

$$\frac{dX}{dt} = \begin{pmatrix} 0 & Id \\ -M^{-1}K & 0 \end{pmatrix} X + \begin{pmatrix} 0 \\ M^{-1}F \end{pmatrix}, \quad X(0) = \begin{pmatrix} U_0 \\ \dot{U}_0 \end{pmatrix}$$

These equations correspond to evolution equations as treated in section 5.4.1. They may be integrated by intrinsic solvers from Matlab® (`ode45`, `ode23s`, etc.). The subprogram evaluating the right-hand side of this system reads as

```
function v = secm_matlab(t,X,invM, invMK,invMC,F)
n = length(X)/2;
v = zeros(size(X));
v(1: n) = X(n+1:2*n);
v(n+1:2*n) = invM*F(t) - invMK*X(1: n) -
invMC*X(n+1:2*n);
end
```

Subprogram `F` must be implemented independently and returns the force vector at time t.

Assuming that the values of $U_0, \dot{U}_0$ are defined in the Matlab vectors `U0` and `Up0`, respectively, then Matlab® solver is called as follows:

```
T = ...; % give the time interval (0,T)
```

```
solver = @solvername;  % solvername: ode45,ode23s,etc.
invM =  M \ eye(size(M)); % evaluate inv(M)
invMK = M \ K; % evaluate inv(M)*K
invMC = M \ C; % evaluate inv(M)*C
X0 = [U0; Up0]; % vector of initial conditions
F = @program_name;  % name of the subprogram evaluating
F.
secm = @(t,X) sec_matlab(t,X,invM, invMK,invMC, F);
sol = solver(secm, [0,T], X0);
```

5.4.3.1. *Newmark integration of motion equations*

One of the most popular numerical methods in dynamics is Newmark's scheme:

$$U(t) = U(t-\Delta t) + \Delta t\, \dot{U}(t-\Delta t) + \frac{(\Delta t)^2}{2}\left[(1-2\beta)\ddot{U}(t-\Delta t) + 2\beta\ddot{U}(\quad)\right]$$

$$\dot{U}(t) = \dot{U}(t-\Delta t) + \Delta t\left[(1-\gamma)\ddot{U}(t-\Delta t) + \gamma\ddot{U}(t)\right]$$

Here, β, γ are strictly positive parameters to be defined by the user. For $2\beta \geq \gamma \geq \frac{1}{2}$, the method is unconditionally stable. Typical values are $\beta = \frac{1}{2}, \gamma = \frac{1}{2}$ or $= \frac{1}{4}, \gamma = \frac{1}{2}$. For the system under consideration, we have:

$$U(t) = U(t-\Delta t) + \Delta t\, \dot{U}(t-\Delta t) + \frac{(\Delta t)^2}{2}[(1-2\beta)M^{-1}(F(t-\Delta t) - KU(t-\Delta t)) + 2\beta M^{-1}(F(t) - KU(t))],$$

$$\dot{U}(t) = \dot{U}(t-\Delta t) + \Delta t\,[(1-\gamma)M^{-1}(F(t-\Delta t) - KU(t-\Delta t)) + \gamma M^{-1}(F(t) - KU(t))],$$

or, equivalently,

$$MU(t) = MU(t-\Delta t) + \Delta t M\, \dot{U}(t-\Delta t) + \frac{(\Delta t)^2}{2}[(1-2\beta)(F(t-\Delta t) - KU(t-\Delta t)) + 2\beta(F(t) - KU(t))]$$

$$M\dot{U}(t) = M\dot{U}(t-\Delta t) + \Delta t\,[(1-\gamma)(F(t-\Delta t) - KU(t-\Delta t)) + \gamma(F(t) - KU(t))]$$

Let us introduce a time discretization involving nt subintervals of time:

$$\Delta t = \frac{T}{nt}, \quad t_i = (i-1)\Delta t, \quad 1 \leq i \leq nt$$

We denote

$$F^i = F(t_i), U^i = U(t_i), \dot{U}^i = \dot{U}(t_i), \ddot{U}^i = \ddot{U}(t_i).$$

Thus

$$MU^{i+1} = MU^i + \Delta t M\,\dot{U}^i + \frac{(\Delta t)^2}{2}[(1-2\beta)(F^i - KU^i) + 2\beta(F^{i+1} - KU^{i+1})]$$

$$M\dot{U}^{i+1} = M\,\dot{U}^i + \Delta t\left[(1-\gamma)(F^i - KU^i) + \gamma(F^{i+1} - KU^{i+1})\right]$$

The first equation reads as

$$(M + \beta(\Delta t)^2 K)U^{i+1} = MU^i + \Delta t M\,\dot{U}^i + \frac{(\Delta t)^2}{2}[(1-2\beta)(F^i - KU^i) + 2\beta F^{i+1}]$$

and determines U^{i+1}. Then, the second equation determines $\dot{U}^{i+1}$. A simple Matlab implementation is the following:

```
nt = ...; % number of time intervals
dt = T/nt; % time step
F = @(t) program_name(t, ...);  % subprogram evaluating F.
betta = ...; % value of beta.
gama = ... ; % value of gama.
A = MM + betta*dt*dt*KK;
t = 0;
Ui = U0;
Vi = Up0;
Fi = F(t);
Ai = M \ (Fi - K*Ui)
for npt = 1: nptmax
t = t + dt;
Fi1 = F(t);
    aux = M*Ui + dt*M*Vi + 0.5*dt*dt*((1-2*betta)*(Fi -
K*Ui) + 2*betta*Fi1);
    Ui1 = A \aux;
    Ai1 = M \ (Fi1 - K*Ui1);
Vi1 = Vi + dt*((1-gama)*Ai + gama*Ai1);
Ui = Ui1;
    Vi = Vi1;
    Ai = Ai1;
Fi = Fi1 ;
   end;
```

Newmark's scheme also reads as the construction of a sequence of vectors given by

$$\begin{pmatrix} U^{k+1} \\ \dot{U}^{k+1} \end{pmatrix} = \begin{pmatrix} Id & \Delta t \\ 0 & Id \end{pmatrix} \begin{pmatrix} U^{k} \\ \dot{U}^{k} \end{pmatrix} + \begin{pmatrix} \left(\frac{1}{2}-\beta\right)(\Delta t)^2 & \beta(\Delta t)^2 \\ (1-\gamma)\Delta t & \gamma\Delta t \end{pmatrix} \begin{pmatrix} \ddot{U}^{k} \\ \ddot{U}^{k+1} \end{pmatrix}$$

or

$$\begin{pmatrix} M & 0 \\ 0 & M \end{pmatrix} \begin{pmatrix} U^{k+1} \\ \dot{U}^{k+1} \end{pmatrix} =$$
$$\begin{pmatrix} M & \Delta t M \\ 0 & M \end{pmatrix} \begin{pmatrix} U^{k} \\ \dot{U}^{k} \end{pmatrix} + \begin{pmatrix} \left(\frac{1}{2}-\beta\right)(\Delta t)^2 & \beta(\Delta t)^2 \\ (1-\gamma)\Delta t & \gamma\Delta t \end{pmatrix} \begin{pmatrix} M\ddot{U}^{k} \\ M\ddot{U}^{k+1} \end{pmatrix}$$

Since

$$M\ddot{U} = F - KU,$$

we have

$$\begin{pmatrix} M\ddot{U}^{k} \\ M\ddot{U}^{k+1} \end{pmatrix} = \begin{pmatrix} F^{k} \\ F^{k+1} \end{pmatrix} - \begin{pmatrix} K & 0 \\ 0 & 0 \end{pmatrix} \begin{pmatrix} U^{k} \\ \dot{U}^{k} \end{pmatrix} - \begin{pmatrix} 0 & 0 \\ K & 0 \end{pmatrix} \begin{pmatrix} U^{k+1} \\ \dot{U}^{k+1} \end{pmatrix}$$

so that Newmark's scheme reads as

$$A_1 \begin{pmatrix} U^{k+1} \\ \dot{U}^{k+1} \end{pmatrix} = A_2 \begin{pmatrix} U^{k} \\ \dot{U}^{k} \end{pmatrix} + A_3 \begin{pmatrix} F^{k} \\ F^{k+1} \end{pmatrix},$$

with

$$A_1 = \begin{pmatrix} M + \beta(\Delta t)^2 K & 0 \\ \gamma\Delta t K & M \end{pmatrix},$$

$$A_2 = \begin{pmatrix} M - \left(\frac{1}{2}-\beta\right)(\Delta t)^2 K & \Delta t M \\ -(1-\gamma)\Delta t K & M \end{pmatrix};$$

$$A_3 = \begin{pmatrix} \left(\frac{1}{2}-\beta\right)(\Delta t)^2 Id & \beta(\Delta t)^2 Id \\ (1-\gamma)\Delta t Id & \gamma\Delta t Id \end{pmatrix}.$$

The first equation furnishes a linear system for U^{k+1}. Once U^{k+1} is determined, the second line furnishes $\dot{U}^{k+1}$. An alternative code is

```
nt = ...; % number of time intervals
dt = T/nt; % time step
F = @program_name;  % name of the subprogram evaluating
F.
betta = ...; % value of beta.
gama = ... ; % value of gama.
A1_11 =M + betta*dt*dt*K;
A1_21 = gama*dt*K;
A1_22 = M;
A2_11 = M - (0.5 - betta)*dt*dt*K;
A2_12 = dt*M;
A2_21 = -(1-gama)*dt*K;
A2_22 = M;
A3_11 = (0.5 - betta)*dt*dt;
A3_12 = betta*dt*dt ;
A3_21 = (1-gama)*dt;
A3_22 = (1-gama)*dt;
t = 0;
Uk = U0;
Upk = Up0;
Fk = F(t);
for npt = 1: nt
    t = t + dt;
Fk1 = F(t);
    B = A2_11*Uk + dt*M*Upk + A3_11*Fk + A3_12*Fk1;
    Uk1 = A1_11 \ B;
    B = - gama*dt*K*Uk1-(1-
gama)*dt*K*Uk+M*Upk+A3_21*Fk+A3_22*Fk1;
    Upk1 = M \ B;
    Uk = Uk1;
    Upk = Upk1;
    Fk = Fk1;
end;
```

5.4.3.2. *Fourier integration of motion equations*

If

$$F(t) = \bar{F} + \textstyle\sum_{i=1}^{N}[(F_A)_i \sin(\omega_i t) + (F_B)_i \cos(\omega_i t)];$$

then

$$U(x,t) = r(t) + \bar{U} + \textstyle\sum_{i=1}^{N}[(U_A)_i \sin(\omega_i t) + (U_B)_i \cos(\omega_i t)];$$

with

$$r(t) = \sum_{k=1}^{n}\left[a_k\phi_k \sin(\omega_{nat,i}t) + b_k\phi_k \cos(\omega_{nat,i}t)\right],$$

$$\left(K - \omega_{nat,i}^2 M\right)\phi_k = 0.$$

Here, ϕ_k are the eigenvectors and $\omega_{k,nat}$ are the natural pulsations associated to K . We have

$$M\ddot{r} + C\dot{r} + Kr = 0\ ,\ \forall\, A = \begin{pmatrix} a_1 \\ \dots \\ a_n \end{pmatrix}, B = \begin{pmatrix} b_1 \\ \dots \\ b_n \end{pmatrix}.$$

The coefficients $\overline{U}$, UA, UB are determined by

$$K\overline{U} = \overline{F}\ ;\ \left(K - \omega_i^2 M\right)(U_A)_i = (F_A)_i;\ \left(K - \omega_i^2 M\right)(U_B)_i = (F_B)_i.$$

Let us introduce the matrix $\Phi = [\phi_1, \dots, \phi_n]$ having as columns the eigenvectors, Ω_{nat} a diagonal matrix containing the natural frequencies in the diagonal, SIN_{nat} and COS_{nat}, diagonal matrices containing the terms $\sin(\omega_{nat,k}t)$, $\cos(\omega_{nat,k}t)$, respectively:

$$(\Omega_{\text{nat}})_{\text{ij}} = \delta_{ij}\omega_{i,nat}\ , (\text{SIN}_{\text{nat}})_{\text{ij}} = \delta_{ij}\sin(\omega_{nat,i}t)\ , (\text{COS}_{\text{nat}})_{\text{ij}} = \delta_{ij}\cos(\omega_{nat,i}t)$$

Then:

$$r(t) = (\Phi SIN_{nat})\, A + (\Phi COS_{nat})\, B$$

and

$$\dot{r}(t) = (\Phi\Omega_{\text{nat}}COS_{nat})\, A - (\Phi\Omega_{\text{nat}}COS_{nat})\, B.$$

Analogously, let us consider Ω a diagonal matrix containing the frequencies ω_i in the diagonal, SIN and COS, diagonal matrices containing the terms $\sin(\omega_i t)$, $\cos(\omega_i t)$, respectively:

$$\Omega_{\text{ij}} = \delta_{ij}\omega_i\ , \text{SIN}_{\text{ij}} = \delta_{ij}\sin(\omega_i t)\ , \text{COS}_{\text{ij}} = \delta_{ij}\cos(\omega_i t)$$

Then

$$U = r(t) + \overline{U} + SIN\, U_A + COS\, U_B \;, U_A = \begin{pmatrix} (U_A)_1 \\ \cdots \\ (U_A)_N \end{pmatrix}, U_B = \begin{pmatrix} (U_B)_1 \\ \cdots \\ (U_B)_N \end{pmatrix}$$

so that

$$\dot{U} = \dot{r}(t) + (\Omega\, COS)U_A - (\Omega SIN)U_B.$$

Thus

$$U(0) = \Phi\, B + \overline{U} + U_B \Longrightarrow \Phi\, B = U_0 - (\overline{U} + U_B);$$

$$\dot{U}(0) = (\Phi\Omega_{\text{nat}})\, A + \Omega U_A \Longrightarrow (\Phi\Omega_{\text{nat}})\, A = \dot{U}_0 - \Omega U_A;$$

which determines the coefficients A, B. A simple implementation of this method is the following one:

```
function [Ubar, U_A,U_B,a,b] =
solve_multichromatic(omega, M,K, Phi, om_nat, Fbar, F_A,
F_B, U0, V0)
U_A = cell(size(omega));
U_B = cell(size(omega));
Ubar = K\ Fbar;
for i = 1: length(omega)
om = omega(i);
    fA = F_A{i};
    fB = F_B{i};
    matsys = K - om^2*M;
uA = matsys \ fA;
    uB = matsys \ fB;
U_A{i} = uA;
    U_B{i} = uB;
end;
X_B = Ubar;
for i = 1:length(omega)
X_B = X_B + U_B{i};
end;
b = Phi \ (U0 - X_B);
X_A = zeros(size(U_A{1}));
for i = 1:length(omega)
    X_A = X_A + omega(i)*U_A{i};
```

```
end;
gama = Phi \ (V0 - X_A);
a = gama./om_nat;
return;
end
```

5.4.4. *Motion of a bar*

Let us consider a bar of length $\ell > 0$, submitted to a distributed load f, clamped at origin and force f_ℓ extremity ℓ. The motion's equations are

$$(ESu')' + f = \rho S\ddot{u}\ , u(x,0) = u_0, \dot{u}(x,0) = \dot{u_0}\ , u(0,t) = 0, u'(\ell,t) = \frac{f_\ell}{ES}.$$

where:

1) E = Young's modulus of the material;

2) S = surface of the right section;

3) ρ = volumetric density of the material.

FEM discretization starts by the generation of a set of vertices and elements, defined by a partition

$$0 = x_1 < x_2 < \cdots < x_n < x_{n+1} = \ell$$

Element number i is (x_i, x_{i+1}). When a uniform discretization is used, we have

$$x_i = (i-1)h\ ,\ h = \frac{\ell}{n}.$$

In this case, we may generate the discretization and the elements by using the following Matlab® code

```
ele = …; % give the value of ℓ
n = …; % give the number of elements
x = linspace(0, ele, n+1);
element = zeros(n,2);
for i = 1: n
    element(i,:) = [i i+1];
end;
```

The complete mass and stiffness cM, cK are generated by using integrals involving the shape functions on the elements, i.e. the *elementary matrices* m_e for mass and k_e for stiffness. For instance, let us consider the classical bar element involving 2 degrees of freedom by element. In this case,

$$k_e = \frac{ES}{h}\begin{pmatrix} 1 & -1 \\ -1 & 1 \end{pmatrix}; \quad m_e = \frac{\rho S h}{6}\begin{pmatrix} 2 & 1 \\ 1 & 2 \end{pmatrix}$$

For instance, we may use the Matlab code

```
elastmod = ...; % give the value of Young's modulus
surf = ...;  % give the value of the surface of the
right section
rho = ... ; % give the value of the volumetric density
of the material
m_e = rho*surf*h*[2 1 ; 1 2]/6;
k_e = elastmod*surf*[1 -1; -1 1]/h;
```

Matrices cM, cK are obtained by assembling the elementary matrices on the elements:

```
cM = zeros(n+1,n+1);
cK = zeros(n+1,n+1);
for i = 1:n
    ind = element(i,:);
    cM(ind, ind) = cM(ind, ind) + m_e;
    cK(ind, ind) = cK(ind, ind) + k_e;
end;
```

Boundary condition $u(0,t) = 0$ is taken into account by the elimination of the first line and column of cM, cK. For instance, we may use the code:

```
M = cM(2:n+1,2:n+1);
K = cK(2:n+1,2:n+1);
```

External forces are given by a time-dependent field $f(x,t)$. Force vector is obtained by assembling the elementary forces: on element i, corresponding to nodes `[i i+1]` and degrees of freedom `[i i+1]`, the forces are (f_i, f_{i+1}), $f_i = f(x_i, t)$. Assembling is made by using the vector of nodal forces $f = (f_1, \dots f_{n+1})^t$. Let us assume that the values of f , f_ℓ are furnished by subprograms `f`, `f_ele`, respectively. Then, an example of assembling is given by the subprogram

```
function cF = assembly_force(f,f_ele,x,n,t, elastmod,
surf)
cF = zeros(n+1,1);
for i = 1: n
dof = [i, i+1];
    f_el = [f(x(dof(1)),t), f(x(dof(2)),t)];
    cF(dof) =  cF(dof) + mean(f_el)*[1; 1];
end;
cF(n+1) = CF(n+1) +  f_ele(t)/(elastmod*surf);
return;
end
```

In this situation, assembling may be simplified:

$$cF = cMf + \delta_{n+1} f_\ell \ , \ \delta_{n+1} = (0, \dots, 0, 1)^t$$

At time t, this vector may be obtained by using the subprogram

```
function cF = assembly_force(f,f_ele, x,n,t, cM,
elastmod, surf)
f = zeros(n+1,1);
for i = 1: n+1
f(i) = f(x(i),t);
end;
cF = cM*f;
cF(n+1) = cF(n+1) + f_ele(t)/(elastmod*surf); );
return;
end
```

The elimination of the first degree of freedom is performed as follows:

```
function F = force(t,f,f_ele, x,n, cM, elastmod, surf)
cF = assembly_force(f,f_ele, x,n,t, cM, elastmod,
surf);
F = cF(2:end);
return;
end
```

Initial conditions are discretized as follows: assume that the values of $u_0(x), u_0(x)$ are furnished by subprograms `u_0`, `up_0`, respectively. Then, we may use the code

```
cU0 = zeros(n+1,1);
cUp0 = zeros(n+1,1);
for i = 1: n+1
    cU0(i) = u_0(x(i));
cUp0(i) = up_0(x(i));
end;
U0 = cU0(2:n+1);
Up0 = cUp0(2:n+1);
```

EXAMPLE 5.7.– A uniform monochromatic force

Let us consider the bar submitted to the oscillating forces

$$f(x,t) = p_A \sin(t) + p_B \cos(t) \ , f_\ell(t) = 0$$

For

```
omega = 1;
pA = 1;
pB = 2;
f = @(x,t) pA*sin(omega*t) + pB*cos(omega*t);
f_ele = @(t) 0;
```

we have the results below for the motion of the last node.

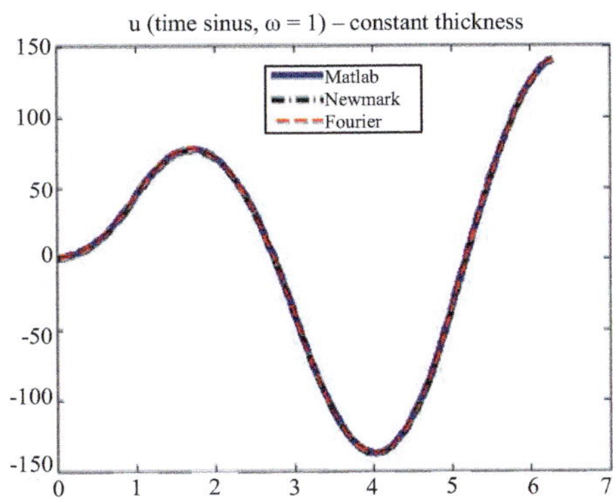

Figure 5.9. *Solution for a uniform monochromatic force*

EXAMPLE 5.8.– A uniform multichromatic force

Let us consider the bar submitted to the oscillating forces

$$f(x,t) = \sum_{i=1}^{N}\left(p_{A_i}\sin(\omega_i t) + p_{B_i}\cos(\omega_i t)\right),\ f_\ell(t) = 0$$

For

```
omega = [ 1 2 3 4];
pA = [ 4 3 2 1];
pB = [ 1 2 3 4];
pbar = 0;
f = @(x,t) somme_trigo(x,t, pbar, pA, pB, omega);
f_ele = @(t) 0;
```

we have the results below for the motion of the last node.

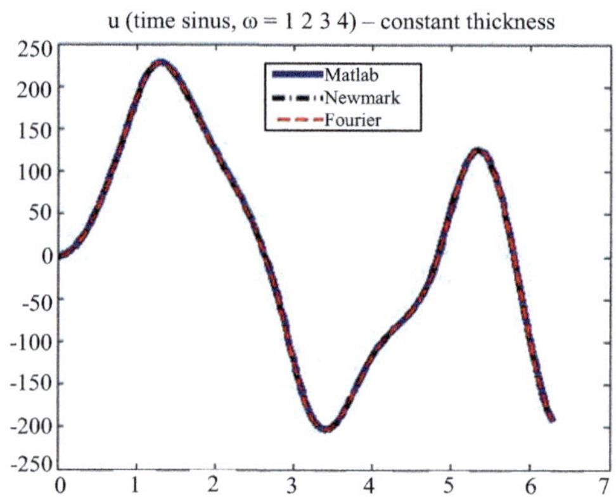

Figure 5.10. *Solution for a uniform multichromatic force*

5.4.5. *Motion of a beam under flexion*

Let us consider a beam of length $\ell > 0$, submitted to pure flexion under a distributed load f, clamped at origin and force f_ℓ extremity ℓ. The motion's equations are

$$(EIu'')'' + f = \rho S\ddot{u}\ , u(x,0) = u_0, \dot{u}(x,0) = \dot{u_0},$$

$$u(0,t) = 0, u'(0,t) = 0, u''(\ell,t) = 0, u'''(\ell,t) = \frac{f_\ell}{EI}.$$

where:

1) E = Young's modulus of the material;

2) S = surface of the right section;

3) I = inertia moment of the cross section;

4) ρ = volumetric density of the material.

FEM discretization is analogous to the preceding one. For instance, let us consider the classical beam element involving 4 degrees of freedom by element. In this case,

$$k_e = \frac{EI}{h^3}\begin{pmatrix} 12 & 6h & -12 & 6h \\ 6h & 4h^2 & -hh & 2h^2 \\ -hh & -6h & 12 & -22 \\ 6h & 2h^2 & -hh & 4h^2 \end{pmatrix};$$

$$m_e = \frac{\rho Sh}{420}\begin{pmatrix} 156 & 22h & 54 & -13h \\ 22h & 4h^2 & 13h & -3h^2 \\ 54 & 13h & 156 & -56h \\ -56h & -3h^2 & -36h & 4h^2 \end{pmatrix}$$

For instance, we may use the Matlab code

```
elastmod = ...; % give the value of Young's modulus
surf = ...;  % give the value of the surface of the right section
rho = ... ; % give the value of the volumetric density of the material
inertia = ... ; % give the value of the inertia of the right section
m_el = rho*surf*h*[
156     22*h      54    -13*h;
                    22*h    4*h^2    13*h   -3*h^2;
                     54      13*h     156     -22*h;
                   -13*h   -3*h^2   -22*h    4*h^2
            ]/420;
k_el = elastmod*inertia*[
                 12      6*h       -12          6*h;
                 6*h    4*h^2    -6*h       2*h^2;
                -12       -6*h       12         -6*h;
                 6*h    2*h^2    -6*h       4*h^2
]/h^3;
```

Analogously to the preceding situation, matrices cM, cK are obtained by assembling the elementary matrices on the elements:

```
ndof = 2*(n+1);
cM = zeros(ndof,ndof);
cK = zeros(ndof,ndof);
for  i = 1: n
ind = (2*i-1:2*i+2);
     cM(ind,ind) =  cM(ind,ind) + m_el;
     cK(ind,ind) =  cK(ind,ind) + k_el;
end;
```

Boundary conditions $u(0,t) = u'(0,t) = 0$ are taken into account by the elimination of the two first lines and columns of cM, cK. For instance, we may use the code:

```
M = cM(3:ndof,3:ndof);
K = cK(3:ndof,3:ndof);
```

External forces are given by a time-dependent field $f(x,t)$. The force vector is obtained by assembling the elementary forces: on element i, corresponding to nodes `[i  i+1]` and degrees of freedom `[2*i-1  2*i  2*i+1  2*i+2]`, the forces are (f_i, f_{i+1}), $f_i = f(x_i, t)$. An example of assembling is given by the subprogram

```
function cF = assembly_force(f,f_ele,x,n,t, elastmod,
inertia)
cF = zeros(2*n+2,1);
for i = 1: n
     dof = (2*i-1:2*i+2);
     f_el = [f(x(i),t); f(x(i+1),t)];
     cF(dof) = cF(dof) + mean(f_el)*h*[1; h/6; 1; -
h/6]/2;
end;
cF(2*n+1) = CF(2*n+1) +  f_ele(t)/(elastmod*inertia);
return;
end
```

The elimination of the two first degrees of freedom is performed as follows:

```
function F = force(t,f,f_ele, x,n, cM, elastmod,
inertia)
  cF = assembly_force(f,f_ele, x,n,t, cM, elastmod,
inertia);
  F = cF(3:end);
  return;
  end
```

Initial conditions are discretized analogously to the case of a bar.

EXAMPLE 5.9.– A uniform monochromatic force

Let us consider the beam submitted to the oscillating forces

$$f(x,t) = p_A \sin(t) + p_B \cos(t) \text{ , } f_\ell(t) = 0$$

For

```
omega = 1;
pA = 1;
pB = 2;
f = @(x,t) pA*sin(omega*t) + pB*cos(omega*t);
f_ele = @(t) 0;
```

we have the results below for the motion of the last node.

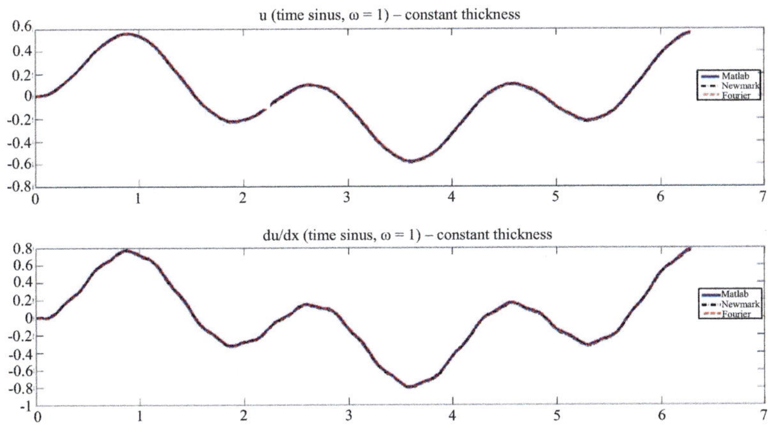

Figure 5.11. *Solution for a uniform monochromatic force*

EXAMPLE 5.10.– A uniform multichromatic force

Let us consider the beam submitted to the oscillating forces

$$f(x,t) = \sum_{i=1}^{N} \left(p_{A_i} \sin(\omega_i t) + p_{B_i} \cos(\omega_i t)\right) , f_\ell(t) = 0$$

For

```
omega = [ 1 2 3 4];
pA = [ 4 3 2 1];
pB = [ 1 2 3 4];
pbar = 0;
f = @(x,t) somme_trigo(x,t, pbar, pA, pB, omega);
f_ele = @(t) 0;
```

we have the results below for the motion of the last node.

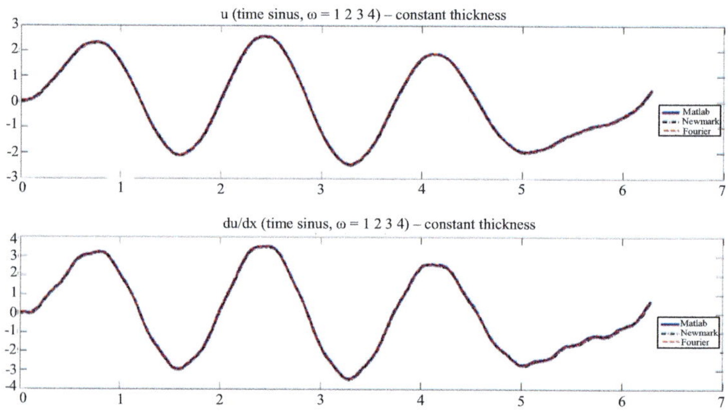

Figure 5.12. *Solution for a uniform multichromatic force*

5.5. Exercises

EXERCISE 5.1.– Let $V = H^1(\Omega)$.

1) Show that there exists $M \in \mathbb{R}$ such that: $\forall v \in V$: $|\ell(v)| \leq M\| \quad \|_1$.

2) Conclude that there exists one and only one $u \in V$ such that $(u,v)_0 = \ell(v), \forall v \in V$.

3) Show that $u(x) = 1$ *on* Ω.

EXERCISE 5.2.– Let $\Omega = (0,1)$, $a \in \Omega$, $P \in R$, $c \in L^2(\Omega)$. Let $F(v) = Pv(a) + (c,v)_0$; $(u,v)_0 = \int_\Omega uv$.

1) Show that

$$F(\varphi) = 0, \forall \varphi \in D(\Omega) \Leftrightarrow P = 0 \text{ and } c = 0.$$

2) Let $V = \{v \in H^1(\Omega) | v(0) = v'(0) = 0\}$ and the scalar product

$$(u,v)_V = (u'',v'')_0,\ \|v\|_V = \sqrt{(v,v)_V}.$$

3) Let $(v) = Pv(a) + (c,v)_0$. Use exercise 4.10 to show that there exists one and only one $u \in V$ such that $\forall\ v \in V$: $(u,v)_V + C(v) = 0$.

4) Assume that $u'''' \in L^2(\Omega)$.

a) Show that

$$\forall\ v \in V\text{: } u''(1)v'(1) - u'''(1)v(1) + Pv(a) + (u'''' + c, v)_0 = 0$$

b) Conclude that $P\varphi(a) + (u'''' + c, \varphi)_0 = 0,\ \forall \varphi \in D(\Omega)$.

c) Show that $P = 0$ and $u'''' + c = 0$ on Ω.

d) Show that $u(0) = u'(0) = u''(1) = u'''(1) = 0$.

5) Assume that $u'''' \in L^2(0,a), u'''' \in L^2(a,1)$, $[u'''(a)] = u(a+) - u(a-) \in \mathbb{R}$.

a) Show that, $\forall\ v \in V$:

$$u''(1)v'(1) - u'''(1)v(1) + (P + [u'''(a)])v(a) + (u'''' + c, v)_0 = 0$$

b) Conclude that $(P + [u'''(a)])\varphi(a) + (u'''' + c, \varphi)_0 = 0$, $\forall \varphi \in D(\Omega)$.

c) Show that $P + [u'''(a)] = 0$ and $u'''' + c = 0$ on Ω.

d) Show that $u(0) = u'(0) = u''(1) = u'''(1) = 0$.

6) Let $J: V \to R$ *be* given by $J(v) = \frac{1}{2}\|v\|_V^2 + C(v)$. *Let* $u^* = \underset{V}{ArgMin} J$.

1) Let $g(\lambda) = J(u^* + \lambda v)$, $\lambda \in \mathbb{R}$. Verify that, $\forall v \in V$, the minimum of g is attained at $\lambda = 0$.

2) Use that $g'(0) = 0$ in order to conclude that $(u^*, v)_V + C(v) = 0, \forall v \in V$.

3) Conclude that $u^* = u$. ■

EXERCISE 5.3.– Let $\Omega \subset R^2$ be a regular bounded domain,

$$V = \{v = (v_1, v_2) \in [H^1(\Omega)]^2 | \ v = 0 \ on \ \partial\Omega \ \};$$

$$S = \{v \in V | \ div(v) = 0 \ on \ \Omega\}.$$

The scalar product on V is defined as

$$(u, v)_V = (u, v)_{1,0} = (u_1, v_1)_{1,0} + (u_2, v_2)_{1,0} \, .$$

Let us introduce

$$(u.\nabla)(v) = (u_1\partial_1 v_1 + u_2\partial_2 v_1, u_1\partial_1 v_2 + u_2\partial_2 v_2);$$

1) Show that there exists a constant C such that $\forall v \in V$: $\|v\|_2 \leq C\|v\|_V$.

2) Verify that S is a vector subspace from V.

3) We aim to show that S is closed.

1) Verify that $v \in S \Leftrightarrow (v, \nabla\varphi)_{0,2} = 0, \ \forall \varphi \in D(\Omega)$.

2) Let $\{u_n : n \in \mathbb{N}\}_n \subset S$ satisfy $u_n \to u$ strongly in $(V, \|v\|_V)$. Verify that $\|u_n - u\|_2 \leq C\|u_n - u\|_V$. Show that $(u_n - u, \nabla\varphi)_0 \to 0$.

3) Conclude that S is closed.

4) Let $\eta \in V$ satisfy $(\eta, v)_0 = 0, \ \forall v \in S$. We aim to show that $\eta = \nabla\psi$: it suffices to show that $rot\ (v) = 0$.

1) Verify that $\forall \varphi \in D(\Omega) : v = (\partial_2\varphi, -\partial_1\varphi) \in S$.

2) Show that $(\partial_1 \eta_2 - \partial_2 \eta_1, \varphi)_{0,2} = 0, \ \forall \varphi \in D(\Omega)$.

3) Conclude.

5) Let $b(u, v, w) = \big((u.\nabla)(v), w\big)_0$.

a) Show that

$$\forall u, v \in V\colon\ b(u, v, v) = \frac{1}{2}\big(u, \nabla(|v|^2)\big)_{0,2} = \frac{1}{2}_{\Omega}\ u\nabla(|v|^2).$$

b) Verify that

$$\forall u, v \in V\colon\quad div\left(\frac{1}{2}|v|^2 u\right) = \frac{1}{2}u\nabla(|v|^2) + \frac{1}{2}|v|^2 div(u).$$

c) Show that

$$u \in S \Longrightarrow b(u, v, v) = 0, \quad \forall v \in V.$$

d) Show that

$$\forall u, v \in V\colon\quad b(v, u, u) = \frac{1}{2}\big(v, \nabla(|u|^2)\big)_{0,2} = \frac{1}{2}_{\Omega}\ v\nabla(|u|^2).$$

6) Let $f \in L^2(\Omega)$,$F(v) = (f, v)_0$.

a) Show that there exists a constant $K \in \mathbb{R}$ such that $\forall v \in V$: $|F(v)| \leq K\|v\|_V$.

b) Verify that $F \in V'$.

7) Let $a(u, v) = (u, v)_V + \ b(u, u, v)$.

a) Show that $\forall u \in V$: $a(u, u) = \|u\|_V^2$.

b) Verify that

$$\forall u, v \in V\colon\quad a(u, v) = \int_\Omega v\{-\Delta u + (u.\nabla)(u)\}$$

8) Let u be the solution of

$$u \in S;\quad a(u, v) = F(v), \quad \forall v \in S.$$

We aim to show that u verifies the Navier-Stokes' equations

$$u = 0 \text{ on } \partial\Omega,\ \ \text{div}(v) = 0 \text{ on } \Omega,$$

$$-\Delta u + (u.\nabla)(u) - \nabla p = f \text{ on } \Omega.$$

a) Show that $\forall u \in V$: $a(u,u) = \|u\|_V^2$.

b) Verify that

$$\int_\Omega v\{\Delta u - (u.\nabla)(u) + f\} = 0, \quad \forall v \in S.$$

4) Show that there exists $p: \Omega \to R$ such that $\Delta u - (u.\nabla)(u) + f = \nabla p$.

5) Conclude.

EXERCISE 5.4.– Let $\Omega \subset \mathrm{R}^2$ be a regular bounded domain,

$$V = \{v = (v_1, v_2) \in [H^1(\Omega)]^2 |\, v = 0 \; on \; \partial\Omega\},$$

with the scalar product

$$(u,v)_V = \lambda(u,v)_{1,0} + \mu\big(div(u), div(v)\big)_0.$$

We assume that $\lambda > 0$, $\lambda + 2\mu \geq 0$.

1) Verify that

$\partial_1 u_1 - \frac{1}{2}\mathrm{div}(u) = \frac{1}{2}(\partial_1 u_1 - \partial_2 u_2)$; $\partial_2 u_2 - \frac{1}{2}\mathrm{div}(u) = \frac{1}{2}(\partial_2 u_2 - \partial_1 u_1)$.

2) Let

$$A = \begin{pmatrix} \partial_1 u_1 - \frac{1}{2}\mathrm{div}(u) & \partial_2 u_1 \\ \partial_1 u_2 & \partial_2 u_2 - \frac{1}{2}\mathrm{div}(u) \end{pmatrix}$$

Show that $\sum_{i,j=1}^2 |A_{ij}|^2 + \frac{1}{2}|\mathrm{div}(u)|^2 \sum_{i,j=1}^2 |\partial_i u_j|^2 = \nabla u . \nabla u$.

3) Conclude that $A = 0$ and $\mathrm{div}(u) = 0 \Leftrightarrow \nabla u = 0$.

4) Show that

$$(u,u)_V =_\Omega \left\{\lambda_{i,j=1}^{2}\left|A_{ij}\right|^2 + \frac{1}{2}(\lambda + 2\mu)|\mathrm{div}(u)|^2\right\}$$

5) Conclude that $(\bullet,\bullet)_V$ is a scalar product on V .

6) Let $F \in V'$. Show that there exists a unique $u \in V$ such that

$$u \in V \text{ and } (u,v)_V = F(v), \quad \forall \quad v \in V.$$

7) Show that $\forall u, v \in V:(u,v)_{1,0} = \int_\Omega v.\Delta u\, u$

8) Show that $\forall u, v \in V:\left(div(u), div(v)\right)_0 = -\int_\Omega v.\nabla\left(div(u)\right).u$

9) Show that $\forall u, v \in V:\left(div(u), div(v)\right)_0 = -\int_\Omega v.\nabla\left(div(u)\right).u$

10) Let $f = (f_1, f_2) \in [L^2(\Omega)]^2$ and $F(v) = (f,v)_0$. Show that

$$\forall v \in V : \int_\Omega v.\left\{\lambda\Delta u + \mu\nabla\left(\mathrm{div}(v)\right) + f\right\} = 0.$$

11) Conclude that

$$\lambda\Delta u + \mu\nabla\left(div(v)\right) + f = 0 \;\; on \;\; \Omega; \; u = 0 \;\; on \;\; \partial\Omega. \qquad \blacksquare$$

6

Dirac's Delta

Dirac's delta δ_0 (sometimes denoted as $\delta(x)$) is a fundamental linear functional. It has been introduced in examples 3.10, 4.2, 4.3, 4.4, 4.5.

Dirac's delta is often used to denote instantaneous impulsions, concentrated forces, sources or other quantities having their action concentrated at a point or an instant of time. For instance, Dirac's delta is used to describe

1) An instantaneous variation of the velocity

2) A punctual source or a sink

3) A punctual charge

4) A punctual force

Formally, Dirac's delta δ_0 is a linear functional, i.e. a function that associates a real number to a function, i.e. $\delta_0: V \to \mathbb{R}$, where V is a convenient space of functions. For a function $\in V$, we have:

$$\delta_0(\varphi) = \varphi(0).$$

Usually, δ_0 is defined on the space formed by the indefinitely differentiable functions having compact support and taking its values on $\mathbb{R}$ (see below).

Since variations of energy are physically connected to the action of forces, it is usual to refer to Dirac's delta δ_0 as Dirac's delta function

$\delta(x)$ and, analogously, to use the notation $\int \delta(x)\varphi(x)dx$ to refer to $\delta_0(\varphi)$. These notations are not mathematically correct, but they are often useful in practice.

However, δ_{x_0} is given by $\delta_{x_0}(\varphi) = \varphi(x_0)$ and sometimes denoted by $\delta(x - x_0)$.

As often in Science, mentions to Dirac's delta appear prior to Dirac. For instance, Fourier [FOU 22] has considered equalities of the form.

$$F(x) = \frac{1}{\pi}\int_{-\pi}^{\pi} d\alpha \left[F(\alpha)\left(\frac{1}{2} + \sum_{i=1}^{+\infty} \cos\left(i(x - \alpha)\right)\right)\right].$$

Fourier also considered $f(x) = \frac{1}{\pi}\int_0^{+\infty} d\alpha\left[f(\alpha)\int_{-\infty}^{+\infty} \cos(p\alpha - px)\, dp\right]$,

which can be rewritten as:

$$f(x) = \frac{1}{\pi}\int_{-\infty}^{+\infty} \frac{f(\alpha)\sin(p\alpha - px)}{\alpha - x} dx\,,$$

where p is requested to take infinite values. Fourier wrote at page 546 of [FOU 22]:

> *We must give to p, in this expression, an infinite value; and this being so, the second member will express the value of f(x).*

The works of Fourier were completed by Poisson [POI 16, POI 23a, POI 23b], Cauchy [CAU 27b] and Dirichlet [DIR 29a, DIR 29b, DIR 37].

The next step is given by Oliver Heaviside with the operational calculus, which may be interpreted as a transformation of Analysis in Algebra.

The main idea of operational calculus is to define algebraic properties of each operator (derivation, integration, etc.) and use these properties in order to solve differential or integral equations. The first attempt to establish an operational calculus is due to Louis François Antoine Arbogast [ARB 00], followed by Francois-Joseph Servois [SER 14]. Works of Robert Bell Carmichael [CAR 55] and by George

Boole [BOO 59] tried to popularize this approach. But only the Heaviside works really generated interest and controversy.

Oliver Heaviside was never interested in academic subjects or academic recognition. School and academy were far from the interests of Heaviside, which were mostly to learn and apply knowledge. Charles Wheatstone was his uncle-in-law and helped him to become a telegrapher at the Danish Great Northern Telegraph Company. He came back to England in 1871 and started to publish papers in the journal Philosophy Magazine [HEA 73a, HEA 73b, HEA 73c, HEA 73d, HEA 74a, HEA 74b]. The works of Heaviside interested William Thomson (Lord Kelvin) and James Clerk Maxwell, who cited Heaviside's works in his famous [MAX 73]. With the creation of the Society of Telegraph Engineers, Heaviside asked for a membership, but his candidacy was rejected. He said that the reason was that "*they didn't want telegraph clerks*". He was eventually accepted as a member, with the help of Thomson (Lord Kelvin). Heaviside made many contributions to telecommunications and electromagnetism. Namely, he is the real author of the actual form of Maxwell's equations: Maxwell produced 20 equations, which were simplified by Heaviside in order to produce the set of four equations nowadays known as Maxwell's equations. He also worked on the *Telegraph's equation* and patented the coaxial cable. Among the ideas attributed to Heaviside, we will find the Lorentz force, the Cherenkov radiation, the electromagnetic mass, the Kennelly-Heaviside layer in the ionosphere, the use of inductors in telephone and telegraph lines and, last but not the least, the first complete and unified presentation of vector algebra.

Heaviside also developed an operational calculus involving impulsions and derivative of step functions – thus, Dirac's delta. Heaviside's operational calculus is particularly efficient in practical situations and became popular in Engineering. But Heaviside was not interested in formal proofs and presented just a set of rules for the algebraical manipulations of integrals and derivatives. One of the most popular citations of Heaviside is [HEA 93]:

> *Mathematics is an experimental science and definitions do not come first, but later on.*

Heaviside published in the *Proceedings of the Royal Society* two articles containing a presentation of operational calculus [HEA 93, HEA 94]. A third paper completing the presentation was rejected by the referee William Burnside. Heaviside's operational calculus remained informal until the works of Thomas John I'Anson Bromwich [BRO 11], John Renshaw Carson [CAR 26], Norbert Wiener [WIE 26]. Namely, these works connected Heaviside's operational calculus to Laplace Transforms [EUL 44, EUL 50a, EUL 68, LAG 70, LAP 10, LAP 12] (see [DEA 81, DEA 82, DEA 92] for a complete history of Laplace Transforms), giving a solid mathematical background to the results of Heaviside.

Gustav Robert Kirchhoff [KIR 83] and Hermann von Helmholtz [HEL 97] considered the limit of Gaussian distributions when the standard deviation goes to zero. Notice that Kirchhoff's integral formula may be interpreted on fundamental solutions generated by Dirac's delta. It is also interesting to notice that Cauchy mentions in [CAU 27a] that, for $\alpha, \varepsilon \rightarrow 0+$,

$$f(a) \approx \frac{1}{\pi} \int_{\alpha-\varepsilon}^{\alpha+\varepsilon} \frac{\alpha f(\mu)}{\alpha^2+(\mu-\alpha)^2} d\mu ,$$

which may be interpreted in the same way (limit of the mean of a parameterized Cauchy distribution) – see [VDP 50, KAT 13] for more information.

Paul Adrien Maurice Dirac has introduced δ for operational reasons. In [DIR 58], he wrote (see also [BUE 05, BAL 03]):

> *To get a convenient method of handling this question a new mathematical notation is required, which will be given in the next section.*

In fact, Dirac desired to write the conditions of orthonormality for families indexed by continuous parameters. For a discrete family $F = \{\varphi_i : i \in \mathbb{N}^*\}$, orthonormality are written using Kronecker's delta δ_{ij} :

$$(\varphi_i, \varphi_j) = \delta_{ij} = \begin{cases} 0, i \neq j \\ 1, i = j \end{cases},$$

but Dirac developed a "convenient method" for the orthonormalization of a continuous family $F = \{\varphi_\lambda : \lambda \in \Lambda\}$, so that he needed a "new mathematical notation". The next section is called "*The delta function*" and introduces Dirac's delta as a function such that:

$$\int_{-\infty}^{+\infty} \delta(x)dx = 1 \, ; \, \delta(x) = 0 \, for \, x \neq 0 \, .$$

Dirac has immediately recognized that [DIR 58]:

> *$\delta(x)$ is not a function of x according to the usual definition of a function ... Thus, $\delta(x)$ is not a quantity which can be generally used in mathematical analysis like an ordinary function, but its use must be confined to certain simple types of expression for which it is obvious that no inconsistency can arise.*

A consistent mathematical framework was furnished by Laurent Schwarz and the Theory of Distributions in 1950 [SCH 50, SCH 51]. This framework is summarized in the following.

6.1. A simple example

Let us consider a mass m at rest at the position $x = 0$ for $t < 0$. At $t = 0$, the mass receives a *kick* and passes suddenly from rest to motion with a velocity v_0. The velocity $v(t)$ of the mass as function of the time is shown in Figure 6.1.

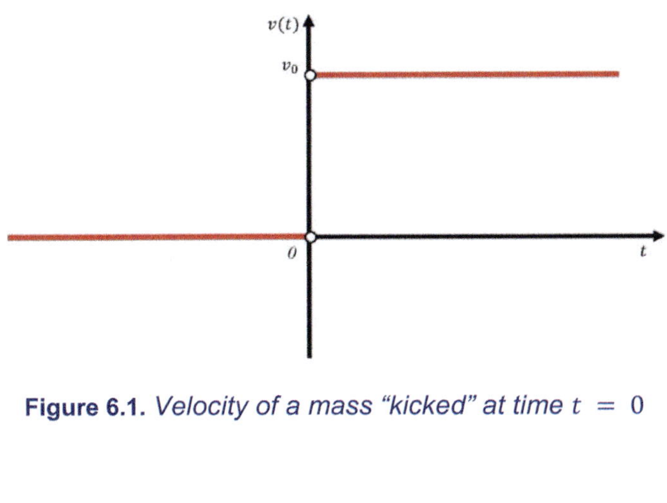

Figure 6.1. *Velocity of a mass "kicked" at time* $t = 0$

Let us try to use Newton's formulation: in this case, we write the classical law force = change in motion= change in [mass × velocity], i.e.,

$$\frac{d}{dt}\left(m\frac{dx}{dt}\right) = \frac{d}{dt}(mv) = F\ .$$

Thus,

$$t < 0: mv(t) = 0 \Longrightarrow \frac{d}{dt}(mv) = 0 \Longrightarrow F(t) = 0;$$

while

$$t > 0: mv(t) = mv_0 \Longrightarrow \frac{d}{dt}(mv) = 0 \Longrightarrow F(t) = 0;$$

so that:

$$F(t) = 0, \forall\, t \neq 0\ .$$

This equality shows that F is not an usual function, since it generates a change in the energy of the system, but it is equal to zero everywhere, except, eventually, $t = 0$. For a usual function, this implies that its work (or power) on any displacement u (or any velocity v) is equal to zero, since, for any other function φ and $a > 0$,

$$F\ usual\ function\ \Rightarrow \int_{-a}^{a} F(t)\varphi(t)dt = \int_{-a}^{0} F(t)\varphi(t)dt + \int_{0}^{a} F(t)\varphi(t)dt,$$

so that:

$$F\ usual\ function\ \Rightarrow \int_{-a}^{a} F(t)\varphi(t)dt = 0 + 0 = 0\ .$$

In fact, this simple situation shows that it is necessary to introduce a *new tool* for the evaluation of the variation of the energy of the system: indeed, the energy J of the mass is:

$$J(t) = 0, t < 0\ ;\ J(t) = \frac{1}{2}mv_0^2, t > 0\ .$$

Thus,

$$\frac{dJ}{dt} = 0, \forall\, t \neq 0\,.$$

Again,

$$\frac{dJ}{dt}\ is\ not\ an\ usual\ function.$$

As previously observed, the construction of a mathematical model for this situation involves Dirac's delta: according to example 3.10, the derivative of J is a multiple of Dirac's delta. We come back to this example in section 7.6.

6.2. Functional definition of Dirac's delta

6.2.1. *Compact support functions*

Let $\Omega \subset \mathbb{R}^n$ be an open domain and $\varphi: \Omega \rightarrow \mathbb{R}$ be a function. Let $S \subset \Omega$: we say that φ is supported by S if and only if there exists $x \in \Omega$ such that $\varphi(x) \neq 0$. The *support* of φ, denoted by $supp(\varphi)$, is the minimal closed set that supports φ:

$$supp(\varphi) = \cap\{S: S\ is\ closed\ and\ \varphi \text{ is supported by } S\,\}$$

$supp(\varphi)$ is the closure of the set of the points where φ is not zero-valued, i.e. the closure of the set $A = \{x|\varphi(x) \neq 0\}$: $supp(\varphi) = \bar{A}$.

We say that φ has a bounded support if and only if $supp(\varphi)$ is bounded. Thus, φ has a bounded support if and only if there is a bounded set B such that $A \subset B$.

We say that φ has a compact support if and only if $supp(\varphi)$ is compact. Thus, φ has a compact support if and only if there is a compact set B such that $A \subset B$.

6.2.2. *Infinitely differentiable functions having a compact support*

Let $\Omega \subset \mathbb{R}^n$ be an open domain. We set:

$$\mathcal{D}(\Omega) = \{\varphi: \Omega \to \mathbb{R}: \varphi \text{ is indefinitely differentiable and } supp(\varphi) \text{ is compact}\}.$$

i.e.

$$\mathcal{D}(\Omega) = \{\varphi: \Omega \to \mathbb{R}: \varphi \in C^\infty(\Omega) \text{ and } supp(\varphi) \text{ is compact}\}.$$

Thus, there is a compact subset $A \subset \Omega$ such that φ vanishes outside A. Since Ω is open, this implies that $\boldsymbol{\varphi}$ and all its derivatives vanish on the boundary of Ω.

$\mathcal{D}(\Omega)$ is not empty: let us consider the function:

$$\mu(s) = \begin{cases} 0, if\ s \le 0 \\ \exp\left(-\frac{1}{s}\right), if\ s > 0 \end{cases}.$$

We have $\mu \in C^\infty(\mathbb{R})$. Thus, the function:

$$\varphi(x) = \mu(r^2 - \|x - x_0\|^2)$$

verifies

$$\varphi \in C^\infty(\Omega);\ \varphi(x) = 0, \text{if } \|x - x_0\| \ge r\,;\ \varphi(x) > 0, if\ \|x - x_0\| < r$$

so that:

$$\varphi \in C^\infty(\Omega);\ supp(\varphi) = \{x \in \mathbb{R}^n: \|x - x_0\| \le r\}.$$

Since Ω is open, there is a value of $r > 0$ such that $supp(\varphi) \subset \Omega$, so that$\varphi \in \mathcal{D}(\Omega)$.

6.2.3. *Formal definition of Dirac's delta*

$\delta_0: \mathcal{D}(\Omega) \longrightarrow \mathbb{R}$ is given by $\delta_0(\varphi) = \varphi(0)$. However, for $x_0 \in \Omega$, $\delta_{x_0}(\varphi) = \varphi(x_0)$. Let us consider a norm $\|\blacksquare\|$ on $\mathcal{D}(\Omega)$ such that: there exists $C \in \mathbb{R}$, independent of φ, verifying

$$\left|\delta_{x_0}(\varphi)\right| \leq C\|\varphi\|, \forall\ \varphi \in \mathcal{D}(\Omega),$$

i.e. δ_{x_0} is continuous on the space $(\mathcal{D}(\Omega), \|\blacksquare\|)$. By the extension principle (theorem 3.23), δ_{x_0} extends to any space V such that $(\mathcal{D}(\Omega), \|\blacksquare\|)$ is dense on V, i.e. such that $V \subset \overline{\mathcal{D}(\Omega)}$ (for the norm under consideration). In other terms, δ_{x_0} extends to the space V formed by the limits of sequences $\{\varphi_n : n \in \mathbb{N}\} \subset \mathcal{D}(\Omega)$: $V = \{v : \exists\ \{\varphi_n : n \in \mathbb{N}\} \subset \mathcal{D}(\Omega), \|\varphi_n - v\| \to 0\ for\ n \to +\infty\}$. For instance:

- $V = C^0(\Omega)$ and $\|\varphi\| = sup\{|\varphi(x)| : x \in \Omega\}$;
- $V = H^1(\Omega)$ and $\|\varphi\| = \left(\int_\Omega (|\varphi(x)|^2 + |\nabla\varphi(x)|^2)\, dx\right)^{1/2}$;
- $V = H_0^1(\Omega)$ and $\|\varphi\| = \left(\int_\Omega |\nabla\varphi(x)|^2\, dx\right)$.$^{1/2}$

6.2.4. *Dirac's delta as a probability*

Dirac's delta may be seen as a probability: assume that P is a probability defined on Ω, such that, for any $S \subset \Omega : P(S) = 1, if\ x_0 \in S$; $P(S) = 0, otherwise$ (i.e. $if\ x_0 \notin S$). In other words, the single point of non-null probability is x_0: $P(\{x_0\}) = 1$, while $P(\Omega - \{x_0\}) = 0$. In this case, the mean of an element $\varphi \in \mathcal{D}(\Omega)$ is:

$$E(\varphi) = \varphi(x_0) = \delta_{x_0}(\varphi).$$

Thus, δ_{x_0} may be interpreted as the "density of probability" associated with the point x_0. The following notations are often used in order to recall this property:

$$\varphi(x) = \int \delta(y - x)\varphi(y)dy = \int \delta_x(y)\varphi(y)dy.$$

In these notations, Dirac's delta appears as a probability density. In fact, the probability corresponding to Dirac's delta *does not* possess a probability density in the standard sense, but these notations are more comprehensive and allow manipulation of Dirac's delta as standard functions.

6.3. Approximations of Dirac's delta

The natural approximation for a Dirac's delta δ_{x_0} is furnished by a sequence of probabilities converging to the one that is concentrated at x_0. For instance, let us consider a n-dimensional Gaussian variable having as density

$$f(\xi, h) = \frac{1}{h^n (2\pi)^{n/2}} exp\left(-\frac{1}{2h^2} \|\xi\|^2\right). \qquad [6.1]$$

This variable has mean equal to zero and variance equal to h^2. Thus, for $h \to 0+$, the distribution tends to a Dirac's delta δ_0. By setting:

$$g(y, x, h) = f(y - x, h), \qquad [6.2]$$

We obtain a distribution such that:

$$E(u) = \int g(y, x, h) u(y) dy \to \int \delta(y - x) \varphi(y) dy = u(x) \; for \; h \to 0+.$$

In these expressions, the integrals are evaluated on the whole space $\mathbb{R}^n$. In practice, they are evaluated on a finite part $\Omega \subset \mathbb{R}^n$. In addition, a numerical evaluation has to be performed: the simplest way consists of considering a discretization of Ω in np subsets $\omega_1, \ldots, \omega_{np}$: a point $p_j \in \omega_j$ is selected in each subset and we set:

$$u(x) = E(u) \approx \sum_{j=1}^{np} u_j V_j g(p_j, x, h) \; ; \; V_j = \int_{\omega_j} dy \; ; \; u_j = u(p_j). \qquad [6.3]$$

As alternative, we may consider a truncation of the Gaussian distribution: for instance, its restriction to the subset $\Omega \subset \mathbb{R}^n$. In this case,

$$E(u) = \frac{\int_\Omega g(y,x,h) u(y) dy}{\int_\Omega g(y,x,h) dy}.$$

The preceding formulation extends straightly to the new formulation: we have:

$$u(x) \approx \frac{\sum_{j=1}^{np} V_j g(p_j, x, h) u_j}{\sum_{j=1}^{np} V_j g(p_j, x, h)} \ ; \ u_j = u(p_j). \qquad [6.4]$$

6.4. Smoothed particle approximations of Dirac's delta

We may consider different densities f, other than the Gaussian one defined in equation [6.1]. Considering other functions f – called *kernels* in this context – leads to an approximation method usually referred to as *smoothed particle approximations*. This approach is based on the general approximation given in equation [6.3], which has the general form:

$$u(x) \approx \sum_{j=1}^{np} V_j g(p_j, x, h) u_j \ ; \ u_j = u(p_j). \qquad [6.5]$$

In the context of smoothed particle approximations, p_j is called a *particle*, V_j is the *volume* associated with the particle and g is the *kernel*. By analogy with the usual physical quantities, we consider that $V_j = m_j / \rho_j$, where m_j is the *mass* of the particle and ρ_j is its volumetric mass density – briefly referred as its *density*. Thus,

$$u(x) \approx \sum_{j=1}^{np} \frac{m_j}{\rho_j} g(p_j, x, h) u_j \ ; \ u_j = u(p_j). \qquad [6.6]$$

Taking $x = p_i$ and $u = \rho$, we have:

$$\rho(p_i) = \rho_i = \sum_{j=1}^{np} \frac{m_j}{\rho_j} g(p_j, p_i, h) \rho_j = \sum_{j=1}^{np} m_j g(p_j, p_i, h) \qquad [6.7]$$

Equations [6.6] and [6.7] define the *smoothed particle approximation* of $u(x)$. Usually, masses are taken as all equal, for instance $m_j = 1, \forall j$ – in such a case, we neglect the particular case of boundary points, which correspond to a fraction of the standard volume of a particle. Different kernels may be found in the literature, namely about smoothed particle hydrodynamics (SPH). The most popular approaches use:

$$g(y,x,h) = \alpha(h)\psi\left(\frac{\|y-x\|}{h}\right),$$

where $\alpha(h)$ is a normalizing factor such that $\int_{\mathbb{R}^n} g(y,x,h)dy = 1$. For instance:

– Cubic kernel: $\alpha(h) = \frac{2}{3}, \frac{10}{7\pi}, \frac{1}{\pi}$ (dimension 1, 2 or 3, respectively)

$$\psi(r) = \begin{cases} 1 - \frac{3r^2}{2} + \frac{3r^3}{4}, if\ 0 \le r \le 1 \\ \frac{1}{4}(2-r)^3, if\ 1 \le r \le 2 \\ 0, otherwise \end{cases}$$

– Quartic kernel: $\alpha(h) = \frac{1}{24}, \frac{96}{1199\pi}, \frac{1}{20\pi}$ (dimension 1, 2 or 3, respectively)

$$\psi(r) = \begin{cases} \left(\frac{5}{2}-r\right)^4 - 5\left(\frac{3}{2}-r\right)^4 + 10\left(\frac{1}{2}-r\right)^4, if\ 0 \le r \le \frac{1}{2} \\ \left(\frac{5}{2}-r\right)^4 - 5\left(\frac{3}{2}-r\right)^4, if\ \frac{1}{2} \le r \le \frac{3}{2} \\ \left(\frac{5}{2}-r\right)^4, if\ \frac{3}{2} \le r \le \frac{5}{2}, \\ 0, otherwise \end{cases}$$

– Quintic kernel: $\alpha(h) = \frac{1}{120}, \frac{7}{478\pi}, \frac{1}{120\pi}$ (dimension 1, 2 or 3, respectively)

$$\psi(r) = \begin{cases} (3-r)^5 - 6(2-r)^5 + 15(1-r)^5, if\ 0 \le r \le 1 \\ (3-r)^5 - 6(2-r)^5, if\ 1 \le r \le 2 \\ (3-r)^5, if\ 2 \le r \le 3, \\ 0, otherwise \end{cases}$$

– Gaussian kernel: $\alpha(h) = \frac{1}{h\sqrt{2\pi}}, \frac{7}{2\pi h^2}, \frac{1}{h^3(2\pi)^{3/2}}$ (dimension 1, 2 or 3, respectively)

$$\psi(r) = e^{-r^2/2}\_$$

6.5. Derivation using Dirac's delta approximations

Let us initially consider the one-dimensional (1D) situation: since

$$u(x) = \int \delta(y-x)u(y)dy\,,$$

we have:

$$\frac{d}{dx}u(x) = u'(x) = -\int u(y)\frac{d}{ds}\delta(s)\Big|_{s=y-x} dy\,.$$

However, as previously observed, δ *is not* a usual function and its derivative does not have a meaning in the traditional framework of standard functions. We must work in a variational framework: $p = \delta'(s)$ is the solution of the variational equality

$$\int p(s)\eta(s)ds = -\int \delta(s)\eta'(s)ds\ ,\forall\eta \in \mathcal{D}(\mathbb{R})\,.$$

Of course, we may consider using the approximation:

$$\delta(y-x) \approx g(y,x,h)$$

and

$$u'(x) \approx \int u(y)\frac{d}{dx}g(y,x,h)dy,$$

which leads to the approximation:

$$u'^{(x)} \approx \sum_{j=1}^{np} V_j u_j \frac{d}{dx} g\left(p_j,x,h\right)\ ;\ V_j = \int_{\omega_j} dy\ ;\ u_j = u\left(p_j\right). \qquad [6.8]$$

A second form of approximation is based on the equation:

$$u'(x) = \frac{\int \delta\prime(y-x)\left(u(y)-u(x)\right)dy}{\int (y-x)\delta\prime(y-x)dy}$$

which leads to:

$$u'(x) \approx \frac{\sum_{j=1}^{np} V_j(u_j - u(x))\frac{d}{dx}g(p_j,x,h)}{\sum_{j=1}^{np} V_j(p_j - x)\frac{d}{dx}g(p_j,x,h)} \; ; \; V_j = \int_{\omega_j} dy \; ; \; u_j = u(p_j). \qquad [6.9]$$

Into an analogous way, smoothed particle approximations may be derived: equation [6.8] reads as:

$$u'(x) \approx \sum_{j=1}^{np} \frac{m_j}{\rho_j} u_j \frac{d}{dx} g(p_j, x, h) \; ; \; u_j = u(p_j). \qquad [6.10]$$

while equation [6.9] reads as:

$$u'(x) \approx \frac{\sum_{j=1}^{np} \frac{m_j}{\rho_j}(u_j - u(x))\frac{d}{dx}g(p_j,x,h)}{\sum_{j=1}^{np} \frac{m_j}{\rho_j}(p_j - x)\frac{d}{dx}g(p_j,x,h)} \; ; \; u_j = u(p_j). \qquad [6.11]$$

When considering partial derivatives, the approach is analogous: we may consider the approximations ($u_j = u(p_j)$ in all the equations [6.12], [6.13], [6.14] and [6.15])

$$\frac{\partial u}{\partial x_i}(x) \approx \sum_{j=1}^{np} V_j u_j \frac{\partial}{\partial x_i} g(p_j, x, h) \; ; \; V_j = \int_{\omega_j} dy \; ; \qquad [6.12]$$

$$\frac{\partial u}{\partial x_i}(x) \approx \sum_{j=1}^{np} \frac{m_j}{\rho_j} u_j \frac{\partial}{\partial x_i} g(p_j, x, h) \; ; \qquad [6.13]$$

$$\frac{\partial u}{\partial x_i}(x) \approx \frac{\sum_{j=1}^{np} V_j(u_j - u(x))\frac{\partial}{\partial x_i}g(p_j,x,h)}{\sum_{j=1}^{np} V_j(p_j - x)\frac{\partial}{\partial x_i}g(p_j,x,h)} \; ; \; V_j = \int_{\omega_j} dy \; ; \qquad [6.14]$$

and

$$\frac{\partial u}{\partial x_i}(x) \approx \frac{\sum_{j=1}^{np} \frac{m_j}{\rho_j}(u_j - u(x))\frac{\partial}{\partial x_i}g(p_j,x,h)}{\sum_{j=1}^{np} \frac{m_j}{\rho_j}(p_j - x)\frac{\partial}{\partial x_i}g(p_j,x,h)}. \qquad [6.15]$$

6.6. A Matlab® class for smoothed particle approximations

As previously observed, smoothed particle approximations involve kernels. We define the kernels in a separated class:

```
classdef spkernel
    methods (Static)
        function v = gaussian(h)
            v.value = @(y,x) val(y,x,h);
            v.d1 = @(y,x) der(y,x,h);
            function v1 = val(y,x,h)
                r = norm(y-x)/h;
                v1 = exp(-0.5*(r^2))/(h*sqrt(2*pi));
            end
            function v1 = der(y,x,h)
                r = norm(y-x)/h;
                dr = sign(y-x)/h;
                v1 = -dr*r*exp(-
0.5*(r^2))/(h*sqrt(2*pi));
            end
        end;
        function v = cubic(h)
            v.value = @(y,x) val(y,x,h);
            v.d1 = @(y,x) der(y,x,h);
            function v1 = val(y,x,h)
                r = norm((x-y)./h);
                if r < 1
                    v1 = 1 - 1.5*r^2 + 0.75*r^3;
                elseif r < 2
                    v1 = 0.25*(2-r)^3;
                else
                    v1 = 0;
                end
                v1 = 2*v1/3;
            end
            function v1 = der(y,x,h)
                r = norm((x-y)./h);
                dr = sign(y-x)/h;
                if r < 1
                    v1 = -3*r + 2.25*r^2;
                elseif r < 2
                    v1 = -0.75*(2-r)^2;
                else
                    v1 = 0;
                end
                v1 = 2*v1*dr/3;
            end
        end;
        function v = quartic(h)
            v.value = @(y,x) val(y,x,h);
            v.d1 = @(y,x) der(y,x,h);
            function v1 = val(y,x,h)
                r = norm((x-y)./h);
                if r < 0.5
```

```
                    v1 = (2.5 - r)^4 - 5*(1.5-r)^4 +
10*(0.5-r)^4;
                elseif r < 1.5
                    v1 = (2.5 - r)^4 - 5*(1.5-r)^4 ;
                elseif r < 2.5
                    v1 = (2.5 - r)^4;
                else
                    v1 = 0;
                end
                v1 = v1/24;
            end
            function v1 = der(y,x,h)
                r = norm((x-y)./h);
                dr = sign(y-x)/h;
                if r < 0.5
                    v1 = -4*(2.5 - r)^3 + 20*(1.5-
r)^3 - 40*(0.5-r)^3;
                elseif r < 1.5
                    v1 = -4*(2.5 - r)^3 + 20*(1.5-
r)^3 ;
                elseif r < 2.5
                    v1 = -4*(2.5 - r)^3;
                else
                    v1 = 0;
                end
                v1 = v1*dr/24;
            end
        end;
        function v = quintic(h)
            v.value = @(y,x) val(y,x,h);
            v.d1 = @(y,x) der(y,x,h);
            function v1 = val(y,x,h)
                r = norm((x-y)./h);
                if r < 1
                    v1 = (3 - r)^5 - 6*(2-r)^5 +
15*(1-r)^5;
                elseif r < 2
                    v1 = (3 - r)^5 - 6*(2-r)^5;
                elseif r < 3
                    v1 = (3 - r)^5;
                else
                    v1 = 0;
                end
                v1 = v1/120;
            end
            function v1 = der(y,x,h)
                r = norm((x-y)./h);
                dr = sign(y-x)/h;
                if r < 1
```

```
                    v1 = -5*(3 - r)^4 + 30*(2-r)^4 +
75*(1-r)^4;
                elseif r < 2
                    v1 = -5*(3 - r)^4 + 30*(2-r)^4;
                elseif r < 3
                    v1 = -5*(3 - r)^4;
                else
                    v1 = 0;
                end
                v1 = v1*dr/120;
            end
        end;
    end
end
```

Program 6.1. *A class for kernels*

Approximations are evaluated by the class below:

```
classdef spa
    properties
        kernel
        xp
        np
        m
    end
    methods (Access = public)
        %
        %  constructor
        %
        function obj = spa(kernel,h,xp,m)
            obj.kernel = kernel;
            obj.xp = xp;
            obj.np = length(xp);
            obj.m = m;
        end
        function nb = particle_density(obj)
            nb = zeros(size(obj.xp));
            for i = 1: obj.np
                s1 = 0;
                for j = 1: obj.np
                    aux =
obj.kernel.value(obj.xp(j),obj.xp(i));
                    s1 = s1 + aux;
                end;
                nb(i) = s1;
            end;
```

```
        end
        function dnb = particle_density_der(obj)
            dnb = zeros(size(obj.xp));
            for i = 1: obj.np
                s1 = 0;
                for j = 1: obj.np
 aux = obj.kernel.d1(obj.xp(j),obj.xp(i))*(obj.xp(j) -
obj.xp(i));
                    s1 = s1 + aux;
                end;
                dnb(i) = s1;
            end;
        end
        function v = approx(obj,x,up,nb)
            v = zeros(size(x));
            for i = 1: length(x)
                s1 = 0;
                for j = 1: obj.np
              aux = obj.m(j)*obj.kernel.value
(obj.xp(j),x(i))/nb(j);
                 s1 = s1 + aux*up(j);
                end;
                v(i) = s1;
            end;
        end
        function v = approx_der(x,ux, up,nb)
            v = zeros(size(x));
            for i = 1: length(x)
                s1 = 0;
                s2 = 0;
                for j = 1: obj.np
                aux = obj.m(j)*obj.kernel.d1
(obj.xp(j),x)/nb(j);
                    s1 = s1 + aux*(up(j)-ux);
                    s2 = s2 + aux*(obj.xp(j) - x);
                end;
                v(i) = s1/s2;
            end
        end
    end
end
```

Program 6.2. *A class for kernels*

Let us illustrate the use of these classes: we consider the function $u(x) = \sin(x)$ on $\Omega = (0,2\pi)$. The function is defined as follows:

```
u = @(x) sin(x);
a = 0;
b = 2*pi;
```

We generate the particles, the masses and the kernel:

```
np = 200;
hp = (b-a)/np;
xp = a:hp:b;
kp = 3;
h = kp*hp;
kernel = spkernel.quintic(h);
m = ones(size(xp));
m(1) = 0.5;
m(np+1) = 0.5;
```

Then, we create the data corresponding to particles:

```
up = zeros(size(xp));
for i = 1: np+1
 up(i) = u(xp(i));
end;
```

and we evaluate the approximation:

```
sp = spa(kernel,h,xp,m);
nb = sp.particle_density();
x = xp;
uc = sp.approx(x,up, nb);
```

The results obtained are exhibited in Figure 6.2.

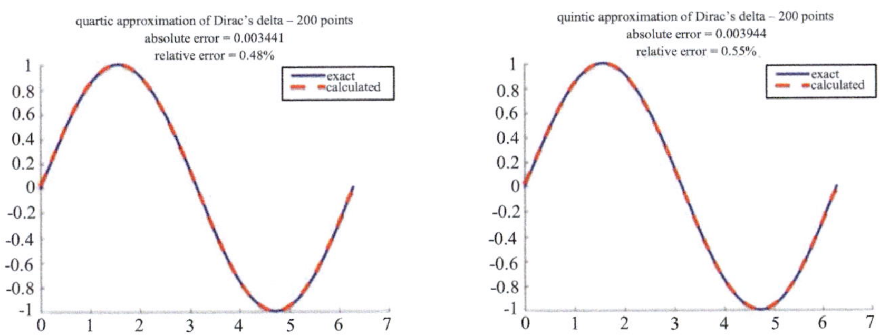

Figure 6.2. *An example of smoothed particle approximation. For a color version of this figure, see www.iste.co.uk/souzadecursi/variational.zip*

Derivatives may be evaluated analogously. For instance, the code

```
ux = up;
duc = sp.approx_der(x,ux,up,nb);
```

evaluates an approximation of the derivatives. The results obtained are exhibited in Figure 6.3.

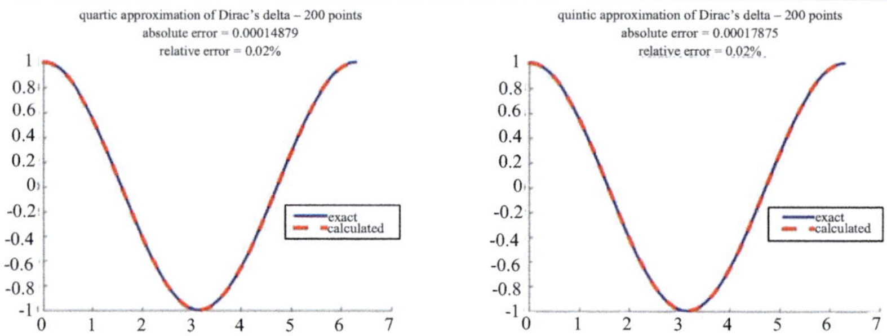

Figure 6.3. *An example of numerical derivation using smoothed particle approximation. For a color version of this figure, see www.iste.co.uk/souzadecursi/variational.zip*

6.7. Green's functions

Green's functions are a powerful tool, namely for the solution of differential equations. Many methods such as, for instance, vortex methods, particle methods, boundary element methods are based on Green's functions. The richness and possibilities of this method have not been entirely exploited at this date.

George Green (1793–1841) was a miller in a small village called Sneinton, near Nottingham (now, it is part of the city). Green had no formal education in mathematics, was not considered as a researcher and was totally unknown by the scientific establishment. He published his first work at the age of 35 (in 1828) [GRE 28]. The publication was paid for by a subscription of 51 people, among whom was the duke of Newcastle. Some of them considered what the potentialities of Green would be with a proper mathematical education and sought for his enrolment in a university.

In particular, Sir Edward French Bromhead repeatedly insisted at Cambridge until success: in 1832, at the age of 40, Green started university studies at Cambridge – with a significant lack of knowledge in almost all important things at his time: luckily, he was awarded the first prize in Math in the first year, which offset his catastrophic results in Greek and Latin. He graduated in 1838, at the age of 45, as fourth of his group of students. In 1845, four years after his death, a young researcher aged 21, called William Thomson – later known as Lord Kelvin – discovered Green's book. He understood the possibilities offered for the solution of differential equations and popularized the works of Green.

6.7.1. *Adjoint operators*

Operators have been introduced in section 3.10, as a convenient way for the generation of Hilbert basis. We have manipulated operators in theorem 4.6 (but the word "operator" was not used, since it was unnecessary).

Let us recall that an operator A on a Hilbert space V is an application $A: V \to V$. In a general situation, the values of A may be undefined for some elements of V: we call domain of A – denoted by $D(A)$ – the part of V for which A has a meaning:

$$D(A) = \{v \in V: A(v) \in V\} = \{v \in V: \|A(v)\| < \infty\}$$

The *adjoint* of A is the operator $A^*: V \to V$ such that:

$$(A(u), v) = (u, A^*(v)), \forall u \in D(A), v \in D(A^*).$$

It is usual to define A^* (and often A too) on a dense subset of $W \subset V$, such as, for instance, $\mathcal{D}(\Omega)$. A is said to be *selfadjoint* if and only if $A^* = A$.

EXAMPLE 6.1.– Let $\Omega = (0,1)$, $V = L^2(\Omega)$, $(u, v) = \int_\Omega uv$, $\|u\| = \sqrt{(u, u)}$ and $A(u) = u'$. We have:

$$D(A) = \{v \in V: \|v'\| < \infty\} = H^1(\Omega).$$

Let us consider $\in \mathcal{D}(\Omega)$. Then;

$$(A(u),v) = \int_0^1 u'v = u(1)\underbrace{\varphi(1)}_{=0} - u(0)\underbrace{\varphi(0)}_{=0} - \int_0^1 u\varphi' ,$$

so that:

$$(A(u),\varphi) = -\int_0^1 u\varphi' = (u,-A(\varphi)) \, , \forall \varphi \in \mathcal{D}(\Omega)$$

Thus, $A^* = -A$ on $\mathcal{D}(\Omega)$- the equality holds on any subspace W such that $\mathcal{D}(\Omega)$ is dense on W (for instance, $W = H_0^1(\Omega)$). For a more general function $v \in D(A)$:

$$(A(u),v) = \int_0^1 u'v = u(1)v(1) - u(0)v(0) - \int_0^1 uv' ,$$

which shows that:

$$(A(u),v) - (u,A^*(v)) = \underbrace{u(1)v(1) - u(0)v(0)}_{boundary\ terms}.$$

This equality may be exploited in order to solve differential equations of the form $u' = f$. ■

EXAMPLE 6.2.– Let $\Omega = (0,1)$, $V = L^2(\Omega)$, $(u,v) = \int_\Omega uv$, $\|u\| = \sqrt{(u,u)}$ and $A(u) = u''$. We have:

$$D(A) = \{v \in V\colon \|v''\| < \infty\} \subset H^2(\Omega).$$

We have :

$$(A(u),\varphi) = \int_0^1 u\varphi'' = (u,A(\varphi)) \, , \forall \varphi \in \mathcal{D}(\Omega)$$

Thus, $A^* = A$on $\mathcal{D}(\Omega)$- the equality holds on any subspace W such that $\mathcal{D}(\Omega)$ is dense on W. Since $A^* = A$, A is selfadjoint. We have, for a general v

$$(A(u),v) - \big(u,A^*(v)\big) = \underbrace{u'(1)v(1) - u'(0)v(0) - u(1)v'(1) + u(0)v'(0)}_{boundary\ terms}.$$ ■

6.7.2. *Green's functions*

Let us consider a differential operator $D: V \rightarrow V$. Current examples of D are the following:

– For a system having mass matrix M, damping matrix C and stiffness matrix K, The motion equations read as $M\ddot{x} + C\dot{x} + Kx = F$ and;

$$Dx = M\ddot{x} + C\dot{x} + Kx$$

– For Laplace's equation - $\Delta u = f$, $Du = - \Delta u$;

– For the linear heat equation $\frac{\partial \theta}{\partial x} - \text{div}(c\nabla\theta) = f$, $D\theta = - \frac{\partial \theta}{\partial x} - div(c\nabla\theta)$.

DEFINITION 6.1.– $G(y, x)$ is a Green function or fundamental solution associated with the differential operator D if and only if $D_y^* G(x, y) = \delta(y - x)$, where D_y^* is D^* applied to the variable y. ■

G is not uniquely determined: no boundary conditions are specified. In fact, analogously to standard differential equations, we may consider any function g satisfying $D_y^* g = 0$ and $G + g$ is a Green's function. Namely, particular choices of g may be performed in order to obtain Green's functions adapted to particular situations.

EXAMPLE 6.3.– Let us consider the simple differential operator $Du = \frac{du}{dx}$. We have $D_y^* G(x, y) = - \frac{d}{dy} G(x, y)$. The equation $D_y^* G(x, y) = \delta(y - x)$ corresponds to

$$\frac{d}{dy} G(x, y) = - \delta(y - x)$$

A particular solution is

$$G(x, y) = -\frac{1}{2} - \frac{1}{2} sign(y - x).$$

Indeed, let us consider a function $\varphi \in \mathcal{D}(\Omega)$ (see section 6.2.2 for the definition of this set): we have $\varphi(y) = 0$ for $\|x\|$ large enough (since $supp(\varphi)$ is compact). Thus,

$$-\int_{\mathbb{R}} G(x,y)\varphi'(y)dy = \int_{x}^{+\infty} \varphi'(y)dy = -\varphi(x) =$$
$$-\int_{\mathbb{R}} \delta(y-x)\varphi(y)dy\,,$$

so that $D_y^* G(x,y) = \delta(y-x)$ (in the variational sense – see section 6.2). The solution of $D_y^* g = 0$ is $g = g(x)$. Thus, the general form of Green's function is:

$$G(x,y) = -\frac{1}{2} sign(y-x) + g(x).$$ ■

6.7.3. *Using fundamental solutions to solve differential equations*

6.7.3.1. *The value of Dirac's delta at a smooth boundary point*

In this section, we are interested in the value of integrals of the form $\int_{\Omega} u(y)\,\delta(y-x)dy$ when x is a boundary point of Ω, i.e. for $x \in \partial\Omega$.

Let us consider the situation in dimension 1: $\Omega = (a,b)$ is an interval. Recall that Dirac's delta may be approximated by a sequence of Gaussian distributions having densities (see section 6.2)

$$\varphi(y) = \psi(y-x,\varepsilon), \psi(s,\varepsilon) = \frac{1}{\varepsilon\sqrt{2\pi}} \exp\left(-\frac{1}{2}\left(\frac{s}{\varepsilon}\right)^2\right)$$

Let $x \in \overline{\Omega}$ and $= (y-x)/\varepsilon$. We set $A = min\{a-x, 0\}, B = max\{b-x, 0\}$. We have:

$$\int_{\Omega} u(y)\,\varphi(y)dy = \int_{a}^{b} u(y)\varphi(y)dy = \int_{A/\varepsilon}^{B/\varepsilon} u(x+$$
$$\varepsilon z)\,\psi(\varepsilon z,\varepsilon)\varepsilon dz$$

Thus,

$$\int_{\Omega} u(y)\,\varphi(y)dy = \int_{A/\varepsilon}^{B/\varepsilon} u(x+\varepsilon z)\frac{1}{\sqrt{2\pi}} \exp\left(-\frac{z^2}{2}\right) dz\,.$$

For a regular function u, $u(x+\varepsilon z) = u(x) + \varepsilon z u'(x) + \cdots$. Since, $\forall n > 0$,

$$\int_0^{+\infty} z^n \frac{1}{\sqrt{2\pi}} exp\left(-\frac{z^2}{2}\right) dz < \infty \; ; \; \int_{-\infty}^{0} z^n \frac{1}{\sqrt{2\pi}} exp\left(-\frac{z^2}{2}\right) dz < \infty$$

we have:

$$\varepsilon^n \int_{A/\varepsilon}^{B/\varepsilon} z^n \frac{1}{\sqrt{2\pi}} exp\left(-\frac{z^2}{2}\right) dz \to 0 \; for \; \varepsilon \to 0+$$

and

$$\int_\Omega u(y)\, \varphi(y) dy \approx u(x) \int_{A/\varepsilon}^{B/\varepsilon} \frac{1}{\sqrt{2\pi}} exp\left(-\frac{z^2}{2}\right) dz$$

For $x \in \Omega$, we have $a < x < b$, so that $A < 0$ and $B > 0$: thus,

$$\int_{A/\varepsilon}^{B/\varepsilon} \frac{1}{\sqrt{2\pi}} exp\left(-\frac{z^2}{2}\right) dz \to \int_{-\infty}^{+\infty} \frac{1}{\sqrt{2\pi}} exp\left(-\frac{z^2}{2}\right) dz = 1$$

and

$$\int_\Omega u(y)\, \varphi(y) dy \to u(x)$$

For $x = a$, we have $A = 0$and $B > 0$: thus,

$$\int_{A/\varepsilon}^{B/\varepsilon} \frac{1}{\sqrt{2\pi}} exp\left(-\frac{z^2}{2}\right) dz \to \int_{0}^{+\infty} \frac{1}{\sqrt{2\pi}} exp\left(-\frac{z^2}{2}\right) dz = \frac{1}{2}$$

and

$$\int_\Omega u(y)\, \varphi(y) dy \to \frac{1}{2} u(x)$$

For $x = b$, we have $A < 0$and $B = 0$: thus,

$$\int_{A/\varepsilon}^{B/\varepsilon} \frac{1}{\sqrt{2\pi}} exp\left(-\frac{z^2}{2}\right) dz \to \int_{-\infty}^{0} \frac{1}{\sqrt{2\pi}} exp\left(-\frac{z^2}{2}\right) dz = \frac{1}{2}$$

and

$$\int_\Omega u(y)\, \varphi(y) dy \to \frac{1}{2} u(x)$$

Finally,

$$\int_{\Omega} u(y)\,\delta(y-x)dy = \begin{cases} u(x), if\ x \in \Omega \\ \frac{1}{2}u(x), if\ x \in \partial\Omega. \end{cases}$$

In a multidimensional situation, the results are analogous, but we must consider multivariate Gaussian distributions: in dimension n the density is:

$$\varphi(y) = \psi(\|y-x\|, \varepsilon), \psi(s,\varepsilon) = \frac{1}{\varepsilon^n\sqrt{(2\pi)^n}}\, exp\left(-\frac{1}{2}\left(\frac{s}{\varepsilon}\right)^2\right)$$

Let us consider the unit ball S given by:

$$S = \{\xi \in \mathbb{R}^n : \|\xi\| \leq 1\}$$

and $S_\delta = x + \delta S = \{y : \|y-x\| < \delta\}$. We introduce the set $\Omega_\delta = \Omega \cup S_\delta$: then,

$$\int_{\Omega} u(y)\,\varphi(y)dy = \int_{\Omega_\delta} u(y)\,\varphi(y)dy - \int_{S_\delta^e} u(y)\,\varphi(y)dy,$$

where $S_\delta^e = S_\delta - S_\delta^i$, $S_\delta^i = S_\delta \cap \Omega$ (see Figure 6.1). We have $S_\delta^e = \{y \in S_\delta : y \notin \Omega\}$ and $S_\delta^i = \{y \in S_\delta : y \in \Omega\}$. Assuming that x is a smooth point, the unitary normal n is defined and we have the situation shown in Figure 6.4:

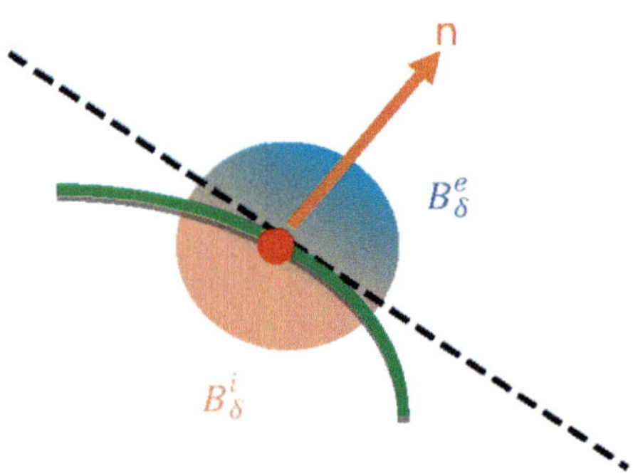

Figure 6.4. *Approximation of the set Ω by Ω_δ. For a color version of this figure, see www.iste.co.uk/souzadecursi/variational.zip*

Taking $z = (y - x)/\varepsilon$, we have $y = x + \ \varepsilon z$ and

$$\varphi(y)dy = \psi(z)dz\ ;\ \psi(z) = \frac{1}{\sqrt{(2\pi)^n}}\, exp\left(-\frac{1}{2}\|z\|^2\right)$$

Let us consider a reduced Gaussian n-dimensional vector Z having the density ψ. Then;

$$\int_{S_\delta} \varphi(y)dy = \int_{\delta S/\varepsilon} \psi(z)dz = P\left(\|Z\| < \frac{\delta}{\varepsilon}\right)$$

Thus, for $\varepsilon > 0$ small enough, the Gaussian distribution concentrates on S_δ:

$$\int_{S_\delta} \varphi(y)dy = \int_{\delta S/\varepsilon} \psi(z)dz \to 1\ for\ \varepsilon \to 0+$$

and

$$\int_{S_\delta} \varphi(y)dy \to \int_{\Omega_\delta} \varphi(y)dy\ \ for\ \varepsilon \to 0+.$$

Moreover,

$$u(y) = u(x + \varepsilon z) = u(x) + \ \varepsilon z^t \nabla u(x) + \cdots,$$

so that:

$$\int_{\Omega_\delta} u(y)\,\varphi(y)dy =$$
$$u(x) \int_{S_\delta} \varphi(y)dy + \ \varepsilon \nabla u(x)^t \int_{S_\delta} z\varphi(y)dy + \cdots.$$

Thus,

$$\int_{\Omega_\delta} u(y)\,\varphi(y)dy\ \ \to u(x) for\ \varepsilon \to 0+.$$

However,

$$\int_{S_\delta^e} \varphi(y)dy = P\left(\|Z\| < \frac{\delta}{\varepsilon}\ and\ x + \varepsilon Z \notin \Omega\right)$$

Let n be the unitary normal at point x, pointing outwards Ω. By approximating

$$\int_{S_\delta^e} \varphi(y)dy \approx P\left(\|Z\| < \frac{\delta}{\varepsilon}\ and\ n^t Z \geq 0\right),$$

we have:

$$\int_{S_\delta^e} u(y)\,\varphi(y)dy = u(x)P\left(\|Z\| < \frac{\delta}{\varepsilon} \text{ and } n^t Z \geq 0\right) + \varepsilon \nabla u(x)^t \int_{S_\delta^e} z\varphi(y)dy + \cdots .$$

and

$$\int_{S_\delta^e} u(y)\,\varphi(y)dy \approx u(x)P(n^t Z \geq 0) for\ \varepsilon \to 0+.$$

Since the inequality $n^t Z \geq 0$ defines a half space, we have $P(n^t Z \geq 0) = 1/2$ and

$$\int_\Omega u(y)\,\varphi(y)dy \approx \frac{1}{2}u(x).$$

Thus, we have:

$$\int_\Omega u(y)\,\delta(y-x)dy = \begin{cases} u(x), if\ x \in \Omega \\ \frac{1}{2}u(x), if\ x \in \partial\Omega. \end{cases}$$

REMARK 6.1.– If x is not a smooth point, a correction must be applied, corresponding to the fraction of 2π limited by the left and right tangents.

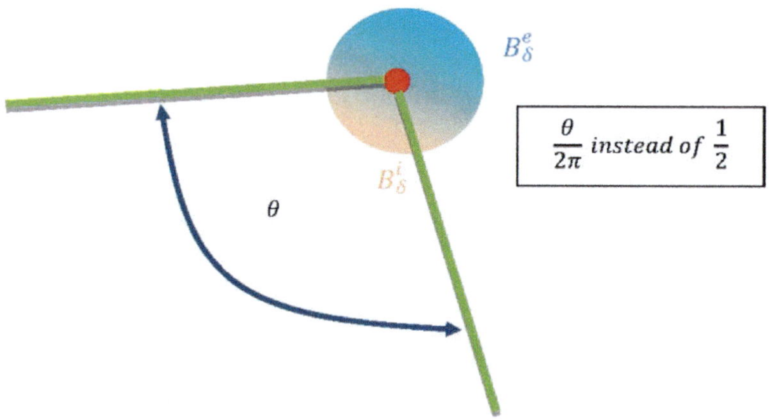

Figure 6.5. *Approximation of the set Ω in the non-smooth case. For a color version of this figure, see www.iste.co.uk/souzadecursi/variational.zip*

6.7.3.2. *Reducing the problem to the boundary values*

The main idea is to use the fact that

$$(A(u),v)-\left(u,A^{*}(v)\right)=boundary\ terms$$

in order to reduce the problem to the determination of the boundary values. For instance, let us consider a function u such that:

$$Du=f\ on\ \Omega$$

and the fundamental solution

$$D_y^{*}G(x,y)=\delta(y-x)\,.$$

Let $x\in\partial\Omega$ be a smooth point. Then;

$$\left(u,D_y^{*}G(x,y)\right)=\int_\Omega u(y)\,\delta(y-x)dy=\frac{1}{2}u(x)$$

while

$$\left(D_y u,G(x,y)\right)=\left(f,G(x,y)\right)=\int_\Omega f(y)\,G(x,y)dy$$

Thus,

$$\underbrace{\left(u,D_y^{*}G(x,y)\right)-\left(D_y u,G(x,y)\right)}_{boundary\ terms}=\frac{1}{2}\,u(x)-\int_\Omega f(y)\,G(x,y)dy\,,\ \boxed{\forall x\in\partial\Omega.}$$

This equation may be used in order to determine the values of u on the boundary $\partial\Omega$. Once these values are determined, we may find values in Ω by:

$$\underbrace{\left(u,D_y^{*}G(x,y)\right)-\left(D_y u,G(x,y)\right)}_{boundary\ terms}=u(x)-\int_\Omega f(y)\,G(x,y)dy\,,\ \boxed{\forall x\in\Omega}$$

EXAMPLE 6.4.– Let us consider the boundary value problem

$$u'' = f on (0,1)\,, u(0) = 0, u'(1) = 1, f(x) = x\,.$$

Since;

$$\int_0^1 u''(y)\,G(x,y)dy = \underbrace{u'(1)}_{=1}G(x,1) - u'(0)G(x,0) - \int_0^1 u'(y)\,G'_y(x,y)dy\,,$$

and

$$\int_0^1 u'(y)\,G'_y(x,y)dy = u(1)G'_y(x,1) - \underbrace{u(0)}_{=0}\,G'_y(x,0) -$$

$$\int_0^1 u(y)\,G''_y(x,y)dy\,,$$

we have:

$$\int_0^1 f(y)\,G(x,y)dy = G(x,1) - u'(0)G(x,0)$$

$$-u(1)G'_y(x,1) + \int_0^1 u(y)\,\delta(y-x)dy\,,$$

For $x = 0 \in \partial\Omega$, $\int_0^1 u(y)\,\delta(y-x)dy = u(0)/2$ and

$$\int_0^1 f(y)\,G(0,y)dy = G(0,1) - u'(0)G(0,0) - u(1)G'_y(0,1) + \frac{1}{2}\underbrace{u(0)}_{=0}$$

i.e.

$$u'(0)G(0,0) + u(1)G'_y(0,1) = G(0,1) - \int_0^1 f(y)\,G(0,y)dy$$

For $x = 1 \in \partial\Omega$, $\int_0^1 u(y)\,\delta(y-x)dy = u(1)/2$ and

$$\int_0^1 f(y)\,G(1,y)dy = G(1,1) - u'(0)G(1,0) - u(1)G'_y(1,1) + \frac{1}{2}u(1)$$

i.e.

$$u'(0)G(1,0) + u(1)\left[G'_y(1,1) - \frac{1}{2}\right] = G(1,1) - \int_0^1 f(y)\,G(1,y)dy$$

These equations form a linear system which determines $u'(0)$ and $u(1)$. An example of a Green function for this problem is $G(x,y) = g(y-x), g(s) = \frac{1}{2}|s|$. Indeed,

$$\int_0^1 G(x,y)\varphi''(y)dy = \frac{1}{2}\int_0^x (x-y)\varphi''(y)dy + \frac{1}{2}\int_x^1 (y-x)\varphi''(y)dy$$

Thus, for $\varphi \in \mathcal{D}(\Omega)$,

$$\int_0^1 G(x,y)\varphi''(y)dy = -\frac{x}{2}\underbrace{\varphi'(0)}_{=0} + \frac{1}{2}\int_0^x \varphi'(y)dy + (1-x)\underbrace{\varphi'(1)}_{=0} - \frac{1}{2}\int_x^1 \varphi'(y)dy$$

and

$$\int_0^1 G(x,y)\varphi''(y)dy = \frac{1}{2}\int_0^x \varphi'(y)dy - \frac{1}{2}\int_x^1 \varphi'(y)dy\,,$$

which shows that:

$$\frac{d}{dy}G(x,y) = \begin{cases} -1/2\,, if\ y < x \\ 1/2, if\ y > x. \end{cases}$$

By using a second integration by parts, we get, for $\varphi \in \mathcal{D}(\Omega)$,

$$\int_0^1 G(x,y)\varphi''(y)dy = \frac{1}{2}\left(\varphi(x) - \underbrace{\varphi(0)}_{=0}\right) - \frac{1}{2}\left(\underbrace{\varphi(1)}_{=0} - \varphi(x)\right) = \varphi(x),$$

which shows that:

$$\frac{d^2}{dy^2}G(x,y) = \delta(y-x)$$

Thus, the equations for the boundary values $u'(0)$ and $u(1)$ read as:

$$\frac{1}{2}u(1) = \frac{1}{3}$$

$$-\frac{1}{2}u'(0) + \frac{1}{2}u(1) = \frac{1}{12}$$

Consequently,

$$u(1) = \frac{2}{3}\ ;\ u'(0) = \frac{1}{2}.$$

For $x \in \Omega$:

$$\int_0^1 f(y)\, G(x,y)dy = G(x,1) - u'(0)G(x,0) - u(1)G_y'(x,1) + u(x)$$

and

$$u(x) = \int_0^1 f(y)\, G(x,y)dy - G(x,1) + \frac{1}{2}G(x,0) + \frac{2}{3}G_y'(x,1))$$

i.e.

$$u(x) = \frac{1}{2}x + \frac{1}{6}x^3 .$$

■

6.7.3.3. *Green's functions for Laplace's operator*

Let us consider the situation where;

$$\Delta u = f \text{ on } \Omega \,, u = u_0 \text{ on } \Gamma_0 \,; \frac{\partial u}{\partial n} = h \text{ on } \partial\Omega - \Gamma_0$$

The main tool for the use of Green functions in problems involving the Laplace's operator is the second Green identity:

$$\int_\Omega \left[u(y)\Delta_y G(x,y) - G(x,y)\Delta_y u(y)\right]dy = \int_{\partial\Omega} \begin{bmatrix} u(y)\dfrac{\partial G}{\partial n_y}(x,y) \\ -G(x,y)\dfrac{\partial u}{\partial n_y}(y) \end{bmatrix} d\Gamma$$

Green's function verifies

$$\Delta_y G(x,y) = \delta(y-x),$$

so that:

$$\int_\Omega u(y)\Delta_y G(x,y)dy = \begin{cases} u(x), if\ x \in \Omega \\ \frac{1}{2}u(x), if\ x \in \partial\Omega. \end{cases}$$

Thus, for $x \in \partial\Omega$, we have:

$$\frac{1}{2}u(x) - \int_{\Omega} G(x,y)f(y)dy = \int_{\partial\Omega}\left[u(y)\frac{\partial G}{\partial n_y}(x,y) - G(x,y)\frac{\partial u}{\partial n_y}(y)\right]d\Gamma$$

and, for $x \in \Gamma_0$,

$$\frac{1}{2}u_0(x) - \int_{\Omega} G(x,y)f(y)dy = \int_{\partial\Omega}\left[u_0(y)\frac{\partial G}{\partial n_y}(x,y) - G(x,y)\frac{\partial u}{\partial n_y}(y)\right]d\Gamma\,,$$

what determines $\frac{\partial \mathrm{u}}{\partial \mathrm{n}}$ on Γ_0.

However, for $x \in \partial\Omega - \Gamma_0$,

$$\frac{1}{2}u(x) - \int_{\Omega} G(x,y)f(y)dy = \int_{\partial\Omega}\left[u(y)\frac{\partial G}{\partial n_y}(x,y) - G(x,y)h(y)\right]d\Gamma\,,$$

which determines u on $\partial\Omega - \Gamma_0$.

Then, for $x \in \Omega$,

$$u(x) = \int_{\Omega} G(x,y)f(y)dy + \int_{\partial\Omega}\left[u(y)\frac{\partial G}{\partial n_y}(x,y) - G(x,y)\frac{\partial u}{\partial n_y}(y)\right]d\Gamma$$

what determines u on Ω.

This method may be applied once we have a Green function for the Laplace's operator: such a function may be found in the literature. For instance,

$$G(x,y) = g(\|y - x\|),$$

where g depends upon the dimension:

Dimension	g
1	$g(r) = \frac{1}{2} r$
2	$g(r) = \frac{1}{2\pi} \log(r)$
3	$g(r) = \frac{-1}{4\pi r}$

Green's functions for other problems – such as the Heat equation, Helmholtz equation, etc. – may be found in the literature.

6.7.3.4. *An example of implementation in 2D*

In this section, we give an example of numerical use of Green's functions for the computer solution of partial differential equations. Let us recall that, for $x \in \partial\Omega$, we have:

$$\frac{1}{2}u(x) - \int_\Omega G(x,y)f(y)dy = \int_{\partial\Omega} \left[u(y)\frac{\partial G}{\partial n_y}(x,y) - G(x,y)\frac{\partial u}{\partial n_y}(y) \right] d\Gamma$$

When discretized, these equations furnish a linear system for the determination of the boundary values $u(x)$ and $\frac{\partial u}{\partial n_y}$. Indeed, if the boundary is discretized in segments and at each segment, either $u(x)$ or $\frac{\partial u}{\partial n_y}$ is given (and assumed to be constant), we may solve the equation in order to determine the unknown value. We give below an example of implementation inspired from [KTS 02, ANG 07]. We take into account three types of boundary conditions:

$$u(x) = u_0 \; ; \; \frac{\partial u}{\partial n_y} = g \, , \frac{\partial u}{\partial n_y} - \alpha^2 u = 0.$$

We have:

$$\frac{\partial G}{\partial y_i}(x,y) = \frac{y_i - x_i}{2\pi\|y-x\|^2}$$

Thus,

$$\frac{\partial G}{\partial n_y}(x,y) = \frac{(y-x).n}{2\pi\|y-x\|^2}$$

For a 2D implementation, the boundary $\partial\Omega$ is discretized in $nseg$ segments (P_i, P_{i+1}). A segment (P_i, P_{i+1}) has a unitary normal n_i, a length $\ell_i = \|P_{i+1} - P_i\|$ and it is described by $P(t) = P_i + t\ell_i\tau_i$, $t \in (0,1)$ and $\tau_i = (-n_{i,2}, n_{i,1})$. For each segment, we must evaluate integrals as:

$$\int_{P_i}^{P_{i+1}} u(y)\frac{\partial G}{\partial n_y}(x,y)d\Gamma \approx \frac{\ell_i u_i}{2\pi} n_i.(P_i - x)\int_0^1 \frac{dt}{\|P(t)-x\|^2},$$

$$\int_{P_i}^{P_{i+1}} G(x,y)\frac{\partial u}{\partial n_y}(y)d\Gamma = \frac{\ell_i}{2\pi}\left(\frac{\partial u}{\partial n}\right)_i \int_0^1 log(\|P(t)-x\|)dt\,.$$

In these approximations, we assume that both u and $\frac{\partial u}{\partial n}$ are constant on the segment. These integrals may be evaluated explicitly or numerically. For instance, we may observe that $\|P(t) - x\|$ is the square root of a quadratic polynomial in t. Thus, $\|P(t) - x\|^2$ is a quadratic polynomial in t and $log(\|P(t) - x\|)$ is the log of a quadratic polynomial in t. Thus, an example of Matlab® code is the following:

```
function v = int_dg(a,b,c,t, chouia)
% returns the numerical value of
%  int(dt/(at^2 + bt + c))
aa = 2*a*t + b;
if abs(aa) > chouia
    aux1 = 4*a*c - b*b;
    if aux1 > chouia
        aux = sqrt(aux1);
        v = 2*atan(aa/aux)/aux;
    else
        v = -2/aa;
```

```
    end;
else
    v = 0;
end;
return;
end

function v = int_g(a,b,c,t,chouia)
% returns the value of
%  int(log(at^2 + bt + c) dt)
aux1 = 4*a*c - b*b;
if aux1 > chouia
    aux = sqrt(aux1);
    aa = t + b/(2*a);
    aaa = abs(aa);
    a1 = (log(a)-2)*t;
    if aaa > chouia
        a2 = aa*log(abs(t*t + (b/a)*t + (c/a)));
        a3 = aux*atan((2*a*t + b)/aux)/a;
    else
        a2 = 0;
        a3 = 0;
    end;
    v = a1 + a2 + a3;
else
    aa = t + b/(2*a);
    a1 = -2*t;
    aaa = abs(aa);
    if aaa > chouia
        a2 = 2*aa*log(aaa);
    else
        a2 = 0;
    end;
    a3 = log(a)*t;
    v = a1 + a2 + a3;
end;
return;
end
```

The integrals over each segment must be assembled in order to form a linear system. At each segment, the type boundary condition is defined by a value bcd, which takes the value 0 when u is given, 1 when $\frac{\partial u}{\partial n_y}$ is given and 2 for the condition $\frac{\partial u}{\partial n_y} - \alpha^2 u = 0$. The corresponding value is stocked in a table bcd. An example of assembling of the matrices of the system is the following (`coef` contains the value `of` α^2):

```
function [aa, bb] =
system_matrices(xms,xs,ds,ts,ns,bcd,bcv,nel,chouia,coef)
aa = zeros(nel,nel);
bb = zeros(nel,1);
for i = 1: nel
  x = xms(i,:);
  s = 0;
  for j = 1:nel
    h = ds(j);
    a = h*h;
    dx = xs(j,:) - x;
    b = 2*h*(dx(1)*ts(j,1) + dx(2)*ts(j,2));
    c = dx(1)*dx(1) + dx(2)*dx(2);
    c1 = int_g(a,b,c,1,chouia);
    c2 = int_g(a,b,c,0,chouia);
    aux_g = c1-c2;
    cd1 = int_dg(a,b,c,1,chouia);
    cd2 = int_dg(a,b,c,0,chouia);
    aux_dg = cd1 - cd2;
    c_g = h*aux_g/(4*pi);
    ns_dot_dx = ns(j,1)*dx(1) + ns(j,2)*dx(2);
    c_dg = h*ns_dot_dx*aux_dg/(2*pi);
    if i == j
      c_id = 1;
    else
      c_id = 0;
    end;
    if bcd(j) == 0
      aa(i,j) = -c_g;
      s = s + bcv(j)*(0.5*c_id - c_dg);
    elseif bcd(j) == 1
      aa(i,j) = c_dg - 0.5*c_id;
      s = s + c_g*bcv(j);
    else
      aa(i,j) = c_dg - 0.5*c_id - coef*c_g;
    end;
  end;
  bb(i) = s;
end;
```

The solution is performed as follows:

```
function [phi, dphidn] = solution(bcd,bcv,aa,bb,nel,coef)
%
```

```
% solution of the linear system
%
sol = aa \ bb;
phi = zeros(nel, 1);
dphidn = zeros(nel,1);
for i = 1: nel
  if bcd(i) == 0
    phi(i) = bcv(i);
    dphidn(i) = sol(i);
  elseif bcd(i) == 1
    phi(i) = sol(i);
    dphidn(i) = bcv(i);
  else
    phi(i) = sol(i);
    dphidn(i) = coef*sol(i);
 end;
end;
```

Tangents and normals are generated by:

```
function [ts,ns,ds,xms] = vecteurs(xs,element,nel)
%
% generation of tangents et normals
%
% INPUT:
% xs : points - type double
% element : definition of the elements - type integer
% nel : number of elements - type integer
%
% OUTPUT:
% ts : tangents - type double
% ns : normals - type double
% ds : lengths - type double
% xms : middle points - type double
%
ds = zeros(nel,1);
xms = zeros(nel,2);
ns = zeros(nel, 2);
ts = zeros(nel, 2);
for i = 1: nel
ind = element(i,:);
    dx = xs(ind(2),:) - xs(ind(1),:);
    ds(i) = norm(dx);
    xms(i,:) = 0.5*(xs(ind(2),:) + xs(ind(1),:));
    ts(i,:) = (xs(i+1,:) - xs(i,:))/ds(i);
    ns(i,1) = ts(i,2);
```

```
    ns(i,2) = -ts(i,1);
end;
%
return;
end
```

EXAMPLE 6.5.– Let us consider $\Omega = (0,1) \times (0,1)$ and the solution of

$$\Delta u = 0 \; on \; \Omega ,$$

$$for \; x \in \partial\Omega : \; u(x) = \; x_1^2 \; on \; x_2 = 0; \; \frac{\partial u}{\partial n} = \begin{cases} 2 \, , x_1 = 1 \\ 0, x_1 = 0 \\ -2, x_2 = 1 \end{cases}$$

The exact solution is $u(x) = x_1^2 - x_2^2$. We present the discretization in Figure 6.6 and the results obtained in Figure 6.7.

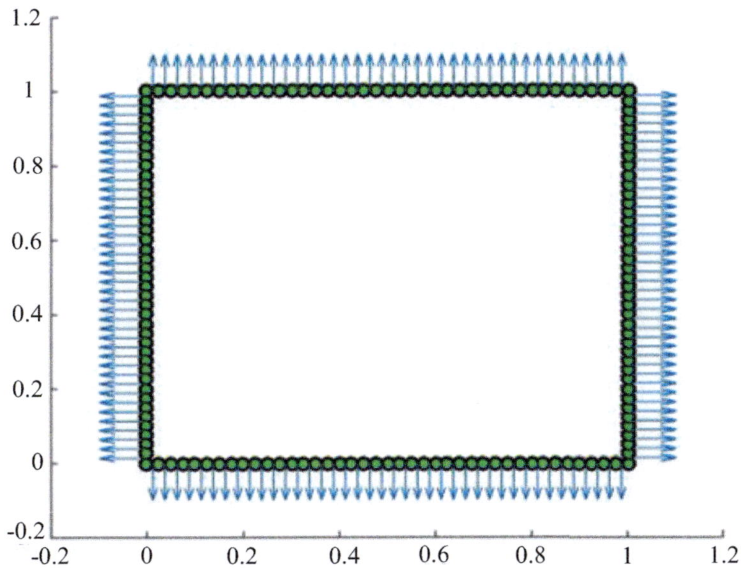

Figure 6.6. *The discretization used in example 6.5. For a color version of this figure, see www.iste.co.uk/souzadecursi/variational.zip*

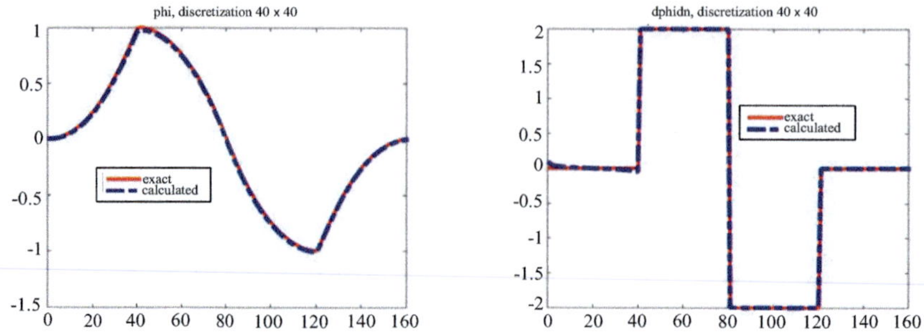

Figure 6.7. *Example of numerical solution by Green's function (example 6.5). For a color version of this figure, see www.iste.co.uk/souzadecursi/variational.zip*

7

Functionals and Calculus of Variations

For people interested in ancient myths, calculus of variations may be dated from the creation of Carthage by Elissa (also called Dido). This mythical phenycian queen which "bought all the land she could enclosed with a bull's hide" ("*mercatique solum, facti de nomine Byrsam, taurine quantum possent circumdare tergo*" [VIR 19]), delimited a circle using a string made of tires cut in a bull's hide. In fact, philosophers have long considered the connections between area and perimeter and looked for the maximal area for a given length (namely Zenodorus the Greek and Pappus of Alexandria) and Hero of Alexandria formulated a principle of economy, which corresponds to a variational principle.

Other people date calculus of variations from the famous problem stated by Johann Bernouilli about the brachistochrone ([BER 96], even if the same problem had already been proposed by Galilei Galileo precedently [GAL 38, ERL 88] – in addition, Fermat and Huyghens previously formulated physical principles leading to problems of Calculus of variations.

The basic problem of calculus of variations consists of determining a field that realizes the minimum or the maximum of a real quantity associated with it, such as, for instance, the energy, the area, the length, a characteristic time, etc. In the case of a real function $f: \mathbb{R} \to \mathbb{R}$, such a point x verifies $f'(x) = 0$, meaning that the tangent to the curve $(x, f(x))$ is a horizontal straight line. The extension of such a result, to the situation where f becomes energy and x becomes a field, requests the definition of derivatives for functions of functions.

A first analysis was made by Vito Volterra, which introduced the first definition of functional derivatives [VOL 87]. Jacques Hadamard remarked the works of Volterra and met him at conferences such as the International Congress of Mathematics, in 1897 in Zurich (then 1900 in Paris and 1904 in Heildelberg). Hadamard was interested in the ideas of Volterra on functional calculus and introduced the terminology *functional* in order to describe the "*functions of lines*" of Volterra [HAD 02, HAD 03a, HAD 10]. Hadamard had a student – Maurice Fréchet – who rapidly developed a new theory about differentials of functions of functions [FRE 04, FRE 11a, FRE 11b, FRE 12, FRE 14]. Fréchet also published the notes of a course by Hadamard about this subject [HAD 10].

The works of Fréchet did not end the discussion about differentials of functions. A young mathematician, René Eugène Gâteaux, proposed a new point of view, based on the extension of the notion of directional derivative. In fact, René Gâteaux was interested in extensions of Riezs's theorems to nonlinear situations, i.e. in the representation of nonlinear functionals [GAT 13a, GAT 13b, GAT 14a, GAT 14b]. A simple approach in order to generate such a representation consists of the use of expansions analogous to the Taylor series. These questions were investigated, namely by Volterra, Hadamard and Fréchet, and led to the theory of functional derivatives. Some of Gâteaux's results [GAT 19a, GAT 19b, GAT 22] were published after his death in October 1914, during a battle in World War I, but his approach has been shown to be very convenient in practical situations, which explains the popularity of Gâteaux derivatives.

7.1. Differentials

Let us consider a regular function $f: \mathbb{R} \longrightarrow \mathbb{R}$. In fact, the value df/dt *does not represent the variations* of f, but gives *the slope of the tangent line* to the graphics of the function at the point $(x, f(x))$.

The variations of f are represented by its *differential* df, which is a mathematical object of a different kind: df is a *linear function* that locally approximates $f(x + \delta x) - f(x)$. In other terms, we have $df(x) = T$, where T is a *linear function* such that $f(x + \delta x) - f(x) \approx T(\delta x)$. So,

$$f(x + \delta x) - f(x) \approx T(\delta x) = df(x)(\delta x).$$

$df(x)(\delta x)$ is the first-order approximation of $f(x + \delta x) - f(x)$:

$$f(x + \delta x) - f(x) = df(x)(\delta x) + O(\delta x^2).$$

These elements are illustrated in Figure 7.1.

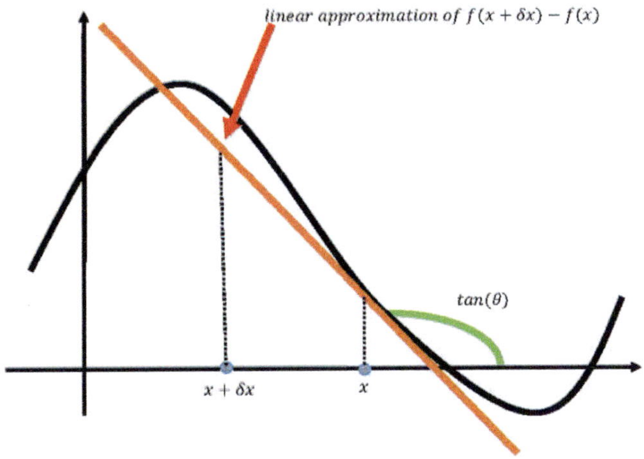

Figure 7.1. *Differential of a functional. For a color version of the figure, see www.iste.co.uk/souzadecursi/variational.zip*

7.2. Gâteaux derivatives of functionals

An interesting property of df is the following:

$$df(x)(\delta x) = \lim_{\lambda \to 0} \frac{f(x+\lambda\delta x)-f(x)}{\lambda}$$

The value $df(x)(\delta x)$ is called the *Gâteaux derivative* of f at the point x in the direction δx. The linear application $\delta x \longrightarrow df(x)(\delta x)$

is the *Gâteaux differential* of f at x. These notions extend to energies defined for physical fields. Let us recall a fundamental definition:

DEFINITION 7.1.– Let V be a vector space. A *functional* is an application $J: V \longrightarrow \mathbb{R}$. ■

Thus, a functional is simply a function defined on V, taking its values on the real line $\mathbb{R}$. Functionals are used in order to represent energies, distances between desired states and effective ones, or quantities defined in order to give a numerical measurement of some qualitative criteria. Namely, energies, variations of energy and virtual works are represented by functionals. In section 3.9.1, we have considered a particular class formed by *linear functionals,* i.e. *functionals* $\ell: V \longrightarrow \mathbb{R}$ verifying $\ell(u + \lambda v) = \ell(u) + \lambda\ell(v), \forall\, u, v \in V, \lambda \in \mathbb{R}$. For instance, virtual works are linear functionals in the space V formed by the virtual motions.

DEFINITION 7.2.– Let V be a vector space and $J: V \longrightarrow \mathbb{R}$ be a functional. The *Gâteaux derivative* of J at the point $u \in V$in the direction δu is:

$$DJ(u)(\delta u) = \lim_{\lambda \to 0} \frac{J(u+\lambda\delta u)-J(u)}{\lambda}$$ ■

The linear functional $V \ni \delta u \longrightarrow dJ(u)(\delta u) \in \mathbb{R}$ is the *Gâteaux differential* of J at the point u. ■

We have our first result:

THEOREM 7.1.– Let V be a vector space and $\ell: V \longrightarrow \mathbb{R}$ be a linear functional. Then:

$$D\ell(u)(\delta u) = \ell(\delta u)\,.$$ ■

PROOF.– We have $\ell(u + \lambda\delta u) = \ell(u) + \lambda\ell(\delta u)$, so that:

$$\frac{\ell(u+\lambda\delta u)-\ell(u)}{\lambda} = \frac{\lambda\ell(\delta u)}{\lambda} = \ell(\delta u)$$

and we have $D\ell(u)(\delta u) = \ell(\delta u)$. ■

Let us recall that $u \in V$ achieves the minimum of J on $C \subset V$ – or that u is the optimal point of J on C – if and only if $u \in C$ and $J(u) \leq J(v), \forall\, v \in V$. In this case, we write:

$$u = \arg\min_C J = \arg\min\{J(v): v \in C\}.$$

The second result is the following:

THEOREM 7.2.– Let V be a vector space and $J: V \longrightarrow \mathbb{R}$ be a Gâteaux-differentiable functional. Let,

$$u = \arg\min \{J(v): v \in V\}.$$

Then, $DJ(u)(\delta u) = 0, \forall\, \delta u \in V$. ■

PROOF.– Let $\delta u \in V$ and $\lambda > 0$. Then, we have $u + \lambda\delta u \in V$ and $u - \lambda\delta u \in V$, so that:

$$J(u + \lambda\delta u) \geq J(u) \quad and \quad J(u - \lambda\delta u) \geq J(u)$$

Since $\lambda > 0$,

$$\frac{J(u+\lambda\delta u)-J(u)}{\lambda} \geq 0 \quad and \quad \frac{J(u-\lambda\delta u)-J(u)}{\lambda} \geq 0.$$

Thus, by taking the limit for $\lambda \to 0+$:

$$DJ(u)(\delta u) \geq 0 \; and \; DJ(u)(-\delta u) \geq 0\,.$$

Since $\delta u \to DJ(u)(\delta u)$ is linear, we have $DJ(u)(-\delta u) = -DJ(u)(\delta u)$, so that $DJ(u)(\delta u) \geq 0 \; and \; -DJ(u)(\delta u) \geq 0$, what shows that $DJ(u)(\delta u) = 0$. ■

The third result is the following:

THEOREM 7.3.– Let $C \subset V$ be a non-empty convex subset (see section 3.7.2) and

$$u = \arg\min \{J(v): v \in C\}\,.$$

Then, $DJ(u)(\delta u) \geq 0, \forall\, \delta u$ such that $u + \delta u \in C$ or, equivalently,

$$DJ(u)(v - u) \geq 0, \forall\, v \in C.$$ ■

PROOF.– Let $\delta u \in V$ verify $u + \delta u \in C$ and $\lambda \in (0,1)$. Then, convexity of C implies that $u + \lambda\delta u = (1-\lambda)u + \lambda(u+\delta u) \in C$, so that:

$$J(u+\lambda\delta u) \geq J(u) \Longrightarrow \frac{J(u+\lambda\delta u)-J(u)}{\lambda} \geq 0$$

Thus, by taking the limit for $\lambda \to 0+$, we obtain $DJ(u)(\delta u) \geq 0$. Finally, we observe that $v \in C \implies \delta u = v - u$ verifies $v = u + \delta u \in C$, so that we have the result. ■

7.3. Convex functionals

DEFINITION 7.3.– Let V be a vector space and $J: V \longrightarrow \mathbb{R}$ be a functional. We say that J is *convex* if and only if

$$J(\theta u + (1-\theta)v) \leq \theta J(u) + (1-\theta)J(v), \forall\, u, v \in V, \theta \in (0,1).$$ ■

Convexity of a functional may be interpreted as shown in Figure 7.2: let

$$f(\theta) = J(\theta u + (1-\theta)v).$$

The curve formed by the points $(\theta, f(\theta))$ remains inferior to the segment of line connecting the points $(0, J(v))$ and $(1, J(u))$.

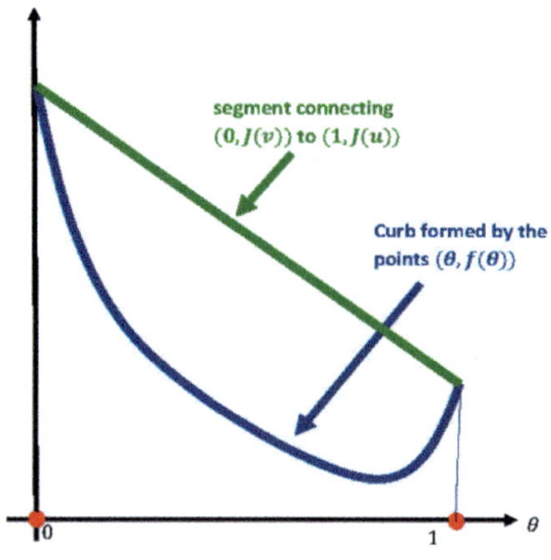

Figure 7.2. *A convex functional*

We have:

THEOREM 7.4.– Let V be a vector space and $J: V \longrightarrow \mathbb{R}$ be a Gâteaux-differentiable functional. Assume that J is convex. Then:

$$J(v) - J(u) \geq DJ(u)(v-u), \forall u, v \in V. \quad \blacksquare$$

PROOF.– Let $u, v \in V$ and $\theta \in (0,1)$. Then,

$$J(\theta v + (1-\theta)u) \leq \theta J(v) + (1-\theta)J(u).$$

Since

$$J(\theta v + (1-\theta)u) = J\big(u + \theta(v-u)\big),$$

We have

$$J\big(u + \theta(v-u)\big) - J(u) \leq \theta\big(J(v) - J(u)\big),$$

so that:

$$\frac{J\big(u+\theta(v-u)\big)-J(u)}{\theta} \leq J(v) - J(u),$$

Taking the limit for $\theta \to 0+$: $DJ(u)(v-u) \leq J(v) - J(u)$ ■

THEOREM 7.5.– Let V be a vector space and $J: V \longrightarrow \mathbb{R}$ be a Gâteaux-differentiable functional. Assume that J is convex and $u \in V$ verifies $DJ(u)(\delta u) = 0, \forall\, \delta u \in V$. Then $u = \arg\min \{J(v): v \in V\}$ ■

PROOF.– Let $v \in V$. From theorem 7.4: $J(v) - J(u) \geq DJ(u)(v-u) = 0$. Thus, $J(v) \geq J(u)$. Since v is arbitrary, we have $u = \arg\min \{J(v): v \in V\}$. ■

THEOREM 7.6.– Let V be a vector space and $J: V \longrightarrow \mathbb{R}$ be a Gâteaux-differentiable convex functional. Then

$$\big(DJ(v) - DJ(u)\big)(v-u) \geq 0, \forall u, v \in V. \quad \blacksquare$$

PROOF.– We have:

$$J(v) - J(u) \geq DJ(u)(v - u) \ \ and\ J(u) - J(v) \geq DJ(v)(u - v)$$

Since $DJ(v)(u - v) = -DJ(v)(v - u)$, we have

$$J(v) - J(u) \geq DJ(u)(v - u) \ \ and\ J(u) - J(v) \geq -DJ(v)(v - u)$$

Adding these two inequalities,

$$0 \geq DJ(u)(v - u) - DJ(v)(v - u) = -\big(DJ(v) - DJ(u)\big)(v - u),$$

so that:

$$\big(DJ(v) - DJ(u)\big)(v - u) \geq 0\,, \forall u, v \in V.$$

THEOREM 7.7.– Let $C \subset V$ be a not empty convex set and $J: V \longrightarrow \mathbb{R}$ be a Gâteaux-differentiable functional. Assume that J is convex and $u \in C$ verifies $DJ(u)(\delta u) \geq 0$, $\forall\ \delta u$ such that $u + \delta u \in C$ (or, equivalently, $DJ(u)(v - u) \geq 0$, $\forall\ v \in C$). Then $u = \arg\min\ \{J(v): v \in V\}$.

PROOF.– We have:

$$J(v) - J(u) \geq DJ(u)(v - u) \geq 0$$

so that $J(v) \geq J(u), \forall u \in C$ and we obtain the result.

7.4. Standard methods for the determination of Gâteaux derivatives

7.4.1. *Methods for practical calculation*

The practical evaluation of Gâteaux differentials is performed by using one of the following methods:

1) Derivatives of a real function: let us consider a point $u \in V$, a virtual motion $\delta u \in V$. We set, for $\lambda \in \mathbb{R}$, $f(\lambda) = J(u + \lambda\delta u)$. Then

$$DJ(u)(\delta u) = \lim_{\lambda \to 0} \frac{J(u+\lambda\delta u) - J(u)}{\lambda} = \lim_{\lambda \to 0} \frac{f(\lambda) - f(0)}{\lambda} = f'(0).$$

2) Local expansions limited to first-order: let us consider a point $u \in V$, a virtual motion $\delta u \in V$. We have for $\lambda \in \mathbb{R}$, $f(\lambda) = J(u + \lambda\delta u)$. Then

$$J(u + \lambda\delta u) = J(u) + \lambda DJ(u)(\delta u) + O(\lambda^2).$$

However, we may take $f(\lambda) = J(u + \lambda\delta u)$ and approximate

$$f(\lambda) = f(0) + \lambda f'(0) + O(\lambda^2).$$

Thus, we may consider the expansion:

$$f(\lambda) = A + B\lambda + O(\lambda^2)$$

and we have $DJ(u)(\delta u) = B$.

EXAMPLE 7.1.– For instance, let us consider the simple system formed by a mass m and a spring having rigidity k and natural length x_0, under the action of a force f.

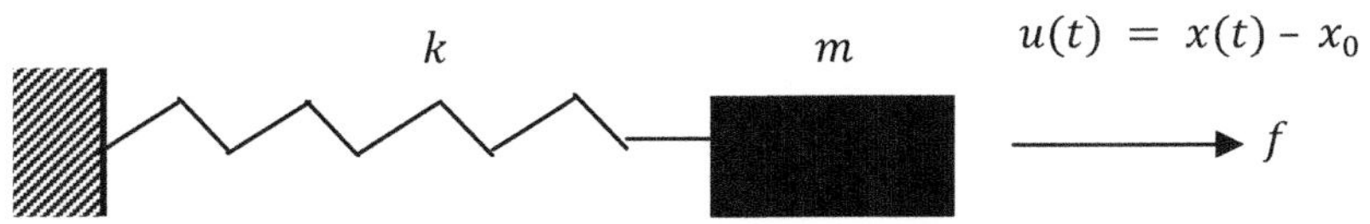

Figure 7.3. *A simple system*

The position of the mass at time t is $x(t)$ and its initial position is $x(0) = x_0$. The displacement of the mass is $u(t) = x(t) - x_0$. The kinetic energy of the mass is:

$$K(u) = \frac{1}{2}\int_0^T m\dot{u}^2 dt$$

We have:

$$f(\lambda) = K(u + \lambda\delta u) = A + B\lambda + C\lambda^2,$$

where,

$$A = \frac{1}{2}\int_0^T m\dot{u}^2 dt\,, B = \int_0^T m\dot{u}(\delta\dot{u})dt\,, C = \frac{1}{2}\int_0^T m(\delta\dot{u})^2 dt\,.$$

Thus,

$$DK(u)(\delta u) = f'(0) = B = \int_0^T m\dot{u}(\delta\dot{u})dt.$$

However,

$$f(\lambda) = A + B\lambda + O(\lambda^2)$$

and we have $DK(u)(\delta u) = B$ again.

However, the elastic energy of the spring is:

$$W(u) = \frac{1}{2}\int_0^T ku^2 dt$$

In this case,

$$f(\lambda) = W(u + \lambda\delta u) = A + B\lambda + C\lambda^2,$$

where:

$$A = \frac{1}{2}\int_0^T ku^2 dt\,, B = \int_0^T ku(\delta u)dt\,, C = \frac{1}{2}\int_0^T k(\delta u)^2 dt.$$

Thus,

$$DW(u)(\delta u) = f'(0) = B = \int_0^T ku(\delta u)dt.$$

However,

$$f(\lambda) = A + B\lambda + O(\lambda^2)$$

and we have yet $DW(u)(\delta u) = B$.

Finally, the potential due to the force f is;

$$F(u) = -\int_0^T f u dt$$

This expression corresponds to a linear functional. Thus,

$$DF(u)(\delta u) = F(\delta u) = -\int_0^T f\delta u dt.$$ ■

7.4.2. Gâteaux derivatives and equations of the motion of a system

In Lagrange's formalism, the motion of a system is described by using Gâteaux differentials: let $q = (q_1, \dots, q_n)$ be the degrees of freedom of the system. Assume that the Kinetic energy is $K(q, \dot{q})$, while the potential energy is $V(q)$. The *Lagrangian* of the system is $L = K - V$ and the *action* of the system is:

$$A(q) = \int_0^T L dt.$$

Let us consider δq such that $\delta q(0) = \delta q(T) = 0$. Then, the motion is described by:

$$DA(q)(\delta q) = 0, \forall\ \delta q \text{ such that } \delta q(0) = \delta q(T) = 0.$$

This variational equation corresponds to the following system of differential equations:

$$\frac{d}{dt}\left(\frac{\partial L}{\partial \dot{q}_i}\right) = \frac{\partial L}{\partial q_i}\ , i = 1, \dots, n.$$

These last equations are known as Euler–Lagrange's equations associated with the Lagrangian L.

EXAMPLE 7.2.– As an example, let us consider the system shown in Figure 7.4. In this case, $n = 1$ and $q_1 = u$. We have;

$$L = \frac{1}{2} m\dot{u}^2 - \frac{1}{2} k u^2 + f u,$$

so that:

$$A = K(u) - W(u) - F(u)$$

and

$$DA(q)(\delta q) = DK(u)(\delta u) - DW(u)(\delta u) - DF(u)(\delta u),$$

i.e.

$$\int_0^T m\dot{u}(\delta\dot{u})dt - \int_0^T ku(\delta u)dt + \int_0^T f\delta u dt.$$

Integrating by parts, we obtain:

$$\int_0^T (f - ku - m\ddot{u})(\delta u)dt = 0, \forall\ \delta u \textit{ such that } \delta u(0) = \delta u(T) = 0,$$

which implies that:

$$f - ku - m\ddot{u} = 0 \textit{ on } (0,T).$$

Thus, $m\ddot{u} = f - ku$ on $(0,T)$. We observe that:

$$\frac{\partial L}{\partial \dot{u}} = m\dot{u}, \frac{\partial L}{\partial u} = f - ku.$$

So, Euler–Lagrange's equations read as:

$$m\ddot{u} = \frac{d}{dt}(m\dot{u}) = f - ku.$$ ■

EXAMPLE 7.3.– Let us consider again the situation shown in section 6.1. Let us introduce a slight modification in the standard Newton's law: the original one states that force = change in motion = change in [mass x velocity]. From the variational standpoint, only changes in energy have to be considered and Newton's law becomes work for a variation δu on a time interval (t_1, t_2) = change in energy for δu on (t_1, t_2) = change in [mass x velocity^2/2] for δu on (t_1, t_2). Thus, we consider:

$$K(u) = \frac{1}{2}\int_{t_1}^{t_2} m\dot{u}^2 dt$$

We look for a linear approximation T such that:

$$K(u + \delta u) - K(u) = T(\delta u) + O(|\delta u|^2) \approx T(\delta u), \forall \delta u$$

As previously observed, we may determine the approximation by evaluating the Gâteaux differential of K:

$$T(\delta u) = DK(u)(\delta u) = \lim_{\lambda\to 0}\frac{K(u+\lambda\delta u)-K(u)}{\lambda}.$$

As previously, indicated, we may consider $\Omega = (t_1, t_2)$ and the set of test functions,

$D(\Omega)=\{\varphi:\Omega\to R : \exists\ \omega\subset\Omega,\ \omega$ closed and bounded, $\varphi(x)=0$ if $x\notin\omega\}$.

All the elements of $\mathcal{D}(\Omega)$ vanish at the ends of the interval: $\varphi(t_1) = \varphi(t_2) = 0, \forall\varphi \in \mathcal{D}(\Omega)$. We evaluate $T(\varphi)$for $\varphi \in \mathcal{D}(\Omega)$: whenever,

$$T(\varphi) = -\int_{t_1}^{t_2} f\varphi dt\ ,\forall\varphi \in \mathcal{D}(\Omega),$$

f is the force generating the change in the kinetic energy of the mass. As previously established, we have:

$$T(\delta u) = \int_{t_1}^{t_2} m\dot{u}(\delta\dot{u})dt\ \Rightarrow T(\varphi) = \int_{t_1}^{t_2} m\dot{u}\dot{\varphi}dt.$$

If $t_1 < t_2 < 0$, we have:

$$T(\varphi) = 0,\ \ \forall\varphi \in \mathcal{D}(\Omega) \Rightarrow f = 0, since\ \int_{t_1}^{t_2} f\varphi dt = 0, \forall\varphi, for\ f = 0.$$

If $0 < t_1 < t_2$, we have:

$$T(\varphi) = mv_0\int_{t_1}^{t_2}\dot{\varphi}dt = mv_0\left(\underbrace{\varphi(t_2)}_{=0} - \underbrace{\varphi(t_1)}_{=0}\right) = 0,\ \ \forall\varphi \in \mathcal{D}(\Omega)$$

and we have $f = 0$ again. If $t_1 < 0 < t_2$, we have:

$$T(\varphi) = mv_0\int_0^{t_2}\dot{\varphi}dt = -mv_0\varphi(0),\ \ \forall\varphi \in \mathcal{D}(\Omega)$$

and $T = mv_0\delta_0$, where δ_0 is the *Dirac's mass* supported by the origin $x = 0$. As previously observed, δ_0 *is not a usual function*. It is a *distribution*; i.e., a linear functional T defined on $\mathcal{D}(\Omega)$. ■

7.4.3. Gâteaux derivatives of Lagrangians

A particular situation often found concerns functionals $J(u)$ defined by integrals involving u and its derivatives. This situation corresponds to Lagrangians.

1) If

$$J(u) = \int_\Omega f(x, u(x), u'(x))\, dx,$$

Then

$$DJ(u)(v) = \int_\Omega \left[\frac{\partial f}{\partial u}(x, u(x), u'(x))v(x) + \frac{\partial f}{\partial u'}(x, u(x), u'(x))v'(x)\right] dx$$

2) More generally, let $U(x) = \left(x, u(x), u'(x), \dots, u^{(n)}(x)\right)$ and $J(u) = \int_\Omega f\left(x, U(x)\right) dx$.

Then

$$DJ(u)(v) = \int_\Omega \left[\frac{\partial f}{\partial u}\left(U(x)\right)v(x) + \sum_{i=1}^{n} \frac{\partial f}{\partial u^{(i)}}\left(U(x)\right)v^{(i)}(x)\right] dx$$

3) Analogously, let

$$u_{,i} = \partial u/\partial x_i \text{ and } U(x) = \left(x, u(x), u_{,1}(x), \dots, u_{,n}(x)\right).$$

If

$$J(u) = \int_\Omega f\left(x, U(x)\right) dx,$$

Then

$$DJ(u)(v) = \int_\Omega \left[\frac{\partial f}{\partial u}\left(U(x)\right)v(x) + \sum_{i=1}^{n} \frac{\partial f}{\partial u_{,i}}\left(x, U(x)\right)v_{,i}(x)\right] dx.$$

7.4.4. Gâteaux derivatives of fields

The notion of Gâteaux derivative extends straightly to the situation where a field X is a function of u: $w = W(u)$. In this case, we may consider

$$Dw(u)(\delta u) = \lim_{\lambda \to 0} \frac{W(u+\lambda\delta u)-W(u)}{\lambda}.$$

In practice, w is often defined by an equation involving w, u and their derivatives. For instance, let us assume that:

$$F(w, w', u) = 0 \quad on \ \Omega.$$

Let us introduce

$$\delta w_\lambda = \frac{w(u+\lambda\delta u)-w(u)}{\lambda}, \delta w_\lambda' = \frac{w'(u+\lambda\delta u)-w'(u)}{\lambda}$$

and assume that:

$$\delta w_\lambda \longrightarrow \delta w = Dw(u)(\delta u) \ \ and \ \ \delta w_\lambda' \longrightarrow \delta w' = Dw'(u)(\delta u) \ \ in\ L^2(\Omega)$$

Since, for $\varphi \in \mathcal{D}(\Omega)$,

$$\int_\Omega \delta w_\lambda \varphi' dx = -\int_\Omega \delta w_\lambda' \varphi dx,$$

we have:

$$\int_\Omega \delta w \varphi' dx = -\int_\Omega \delta w' \varphi dx.$$

Thus, $\delta w' = (\delta w)'$. Moreover,

$$F(w(u+\lambda\delta u), w'(u+\lambda\delta u), u+\lambda\delta u) = 0.$$

Let us assume that F is regular enough in order to have an expansion:

$$F(\xi_1 + \delta\xi_1, \xi_2 + \delta\xi_2, \xi_3 + \delta\xi_3) = F(\xi_1, \xi_2, \xi_3) + \sum_{i=1}^{3} A_i(\xi_1, \xi_2, \xi_3)\delta\xi_i + o\left(\sum_{i=1}^{3} \delta\xi_i^2\right).$$

Then, the equality:

$$\frac{F(w(u+\lambda\delta u),w'(u+\lambda\delta u),u+\lambda\delta u)-F(w,w',u)}{\lambda}=0$$

yields that

$$A_1(w,w',u)\delta w_\lambda + A_2(w,w',u)\delta w_\lambda' + A_3(w,w',u)\delta u + o\big(\lambda(w^2+(w')^2+u^2)\big)=0,$$

and we have:

$$A_1(w,w',u)\delta w + A_2(w,w',u)\delta w' + A_3(w,w',u)\delta u = 0,$$

Thus, δw is the solution of a differential equation. Under similar assumptions, the case where,

$$F(w,w',w'',u)=0 \quad on\ \Omega,$$

$$F(\xi_1+\delta\xi_1,\xi_2+\delta\xi_2,\xi_3+\delta\xi_3,,\xi_4+\delta\xi_4)=F(\xi_1,\xi_2,\xi_3,\xi_4)+\sum_{i=1}^{4}A_i(\xi_1,\xi_2,\xi_3,\xi_4)\delta\xi_i+o\big(\sum_{i=1}^{4}\delta\xi_i^2\big).$$

leads to the differential equation:

$$A_1(w,w',w'',u)\delta w + A_2(w,w',w'',u)\delta w' + A_3(w,w',w'',u)\delta w'' + A_4(w,w',w'',u)\delta u = 0$$

7.5. Numerical evaluation and use of Gâteaux differentials

Gâteaux differentials may be used in conjunction with Hilbertian basis in order to obtain numerical values, to implement descent methods, or to determine optimal points associated with Lagrangians. The basic use consists of either determining a derivative or solving an equation concerning the Gâteaux differential. In the following, $\{\varphi_n\}_{n\in\mathbb{N}}$ is a total family (such as, for instance, a Finite Element one) or a Hilbertian basis (such as, for instance, Legendre Polynomials).

7.5.1. *Numerical evaluation of a functional*

As observed above, the numerical evaluation of a functional may be performed by using a total family or a Hilbert basis. For instance, we may use the families defined in section 2.2.6.

A situation often found in practice consists of functionals defined by *Lagrangian densities*:

$$J(u) = \int_a^b f_\ell\big(t, u(t), u'(t)\big)dt.$$

f_ℓ is the Lagrangian density. If u and u' are furnished as subprograms, we may evaluate $J(u)$ by using numerical integration, as in section 1.5. If they are given as tables of data, we may use the methods presented in section 1.3. As an example, let us consider the situation where,

$$f_\ell\big(t, u(t), u'(t)\big) = \sqrt{[u_1'(t)]^2 + [u_2'(t)]^2}$$

Let $a = 0, b = 2\pi, u_1(t) = 2\cos(t), u_2(t) = 2\sin(t)$. The exact value is $J(u) = 4\pi$. The code

```
f_ell = @(t,u,du) sqrt(du(1)^2 + du(2)^2);
a = 0;
b = 2*pi;
%
u = @(t) 2*[cos(t), sin(t)];
du = @(t) 2*[-sin(t), cos(t)];
%
xlim.lower.x = a;
xlim.upper.x = b;
xlim.dim = 1;
g = @(t) f_ell(t, u(t), du(t));
sp = subprogram_integration(g,xlim);
v1 = sp.result ;
```

produces `v1 = 12.5664`$\approx 4\pi$. If the data is furnished as a table of values, we may use the class Riemann:

```
ns = 500;
h = (b-a)/ns;
```

```
x   = a:h:b;
F = spam.points('vector',g,x);
p.x = x;
p.dim   = 1;
p.measure = b-a;
data.points = p;
data.values = F;
v2 = riemann.trpzd(data);
```

This code produces the result `v2 = 12.5664`$\approx 4\pi$.

Assuming that an approximation $u \approx Pu = \sum_{i=1}^{n} u_i \varphi_i$ is given, we may approximate,

$$J(u) \approx J(Pu) = J(\sum_{i=1}^{n} u_i \varphi_i).$$

For instance, let us consider the approximation of vector $u(t) = 2(\cos(t), \sin(t))$ in a polynomial basis $\varphi_i(t) = t^{i-1}$ given by its orthogonal projection Pu on this basis. For $n = 10$, we obtain `v1 = 12.5663`$\approx 4\pi$.

Notice that $J(Pu)$ is a function of the vector of coefficients $U = (u_1, \ldots, u_n)$.

7.5.2. Determination of a Gâteaux derivative

Assume that a point $u \approx \sum_{i=1}^{n} u_i \varphi_i$ is given. The Gâteaux derivative $J'(u) \in V$ is approximated as follows: $J'(u) \approx p = \sum_{i=i}^{n} p_i \varphi_i$ and we have:

$DJ(u)(v) = (J'(u), v), \forall\, v \in V \Longrightarrow DJ(u)(\varphi_i) = (J'(u), \varphi_i), for\ 0 \leq i \leq n.$

Thus,

$$\left(\sum_{j=1}^{n} p_j \varphi_j\,, \varphi_i\right) = DJ(u)(\varphi_i), for\ 1 \leq i \leq n\,,$$

which leads to a linear system $A.P = B$, where,

$$A_{ij} = \left(\varphi_i, \varphi_j\right);\ \ B_i = DJ(u)(\varphi_i);\ \ P_j = p_j \ \ with\ \ 1 \leq i,j \leq n\,.$$

Once this system is solved for the coefficients P, the approximated value of $J'(u) \approx \sum_{i=1}^{n} p_i \varphi_i$ is determined.

As an example, let us consider $\Omega = (a, b)$ and the Lagrangian density:

$$f_\ell(t, u(t), u'(t)) = [u'(t)]^2 + [u(t)]^2$$

We have:

$$DJ(u)(\delta u) = \int_a^b (u'(t)\delta u'(t) + u(t)\delta u(t))dt.$$

Functional J has a meaning in Sobolev space $H^1(\Omega)$, i.e. for the scalar product:

$$(p, v)_1 = \int_\Omega (pv + p'v')dt$$

This scalar product may be defined as follows:

```
xlim.lower.x = a;
xlim.upper.x = b;
xlim.dim = 1;
sp0 = @(u,v) subprogram_integration(@(t)
u(t)*v(t),xlim).result;
sp1 = @(u,v,du,dv) sp0(u,v) + sp0(du,dv);
```

We have $DJ(u)(v) = (u, v)_1$, so that, when using this scalar product, $p = u$. Functional J has also a meaning in Sobolev space $V = \{v \in H^1(\Omega): v(a) = 0\}$ with the scalar product:

$$(p, v)_{1,0} = \int_\Omega p'v'dt.$$

This scalar product may be defined by the code:

```
sp10 = @(u,v,du,dv) sp0(du,dv);
```

In this case, the equality

$$(p, \delta u)_{1,0} = DJ(u)(\delta u), \forall \delta u \in V$$

is integrated by parts, which leads to:

$$p \in V \ and \ \left(p'(b) - u'(b)\right)\delta u(b) - \int_a^b (p'' + u'' - u)\delta u \, dt = 0, \ \ \forall \delta u \in V.$$

Thus,

$$p'' = u - u'' \ on \ (a,b) \ , p(a) = 0, \ p'(b) = u'(b).$$

Let us consider $u(t) = \sin(t), a = 0, b = 2\pi$. For scalar product $(\blacksquare, \blacksquare)_1$, we have $p(t) = \sin(t)$, while scalar product $(\blacksquare, \blacksquare)_{1,0}$ corresponds to $p(t) = 2\sin(t) - t$. When using $(\blacksquare, \blacksquare)_1$, the family is $\varphi_i(t) = t^{i-1}$ with $n = 10$, which may be created as follows:

```
n = 10;
phi = polynomial_basis(0,0,1,n-1);
bp = basis(phi);
```

The coefficients may be determined by using

```
A = zeros(n,n);
B = zeros(n,1);
for i = 1: n
    phi1 = @(t)  bp.phi.value(i,t);
    dphi1 = @(t) bp.phi.d1(i,t);
    A(i,i) = sp1(phi1,phi1,dphi1,dphi1);
    fff = @(t) df_ell(t,u(t),du(t),phi1(t),dphi1(t)) ;
    aux = subprogram_integration(fff,xlim);
    B(i) = aux.result;
    for j = i+1: n
    phi2 = @(t)  bp.phi.value(j,t);
    dphi2 = @(t) bp.phi.d1(j,t);
        aux = sp1(phi1,phi2,dphi1,dphi2);
        A(i,j) = aux;
        A(j,i) = aux;
    end;
end;
U = A \ B;
```

Then, `p = bp.projection(U,tc)` produces a table of values `p(j)`$= p($`tc(j)`$)$, for a given table of time moments `tc`.

For the scalar product $(\blacksquare, \blacksquare)_{1,0}$, the family is $\varphi_i(t) = t^i$ with $n = 10$. The results are exhibited in the Figure 7.4.

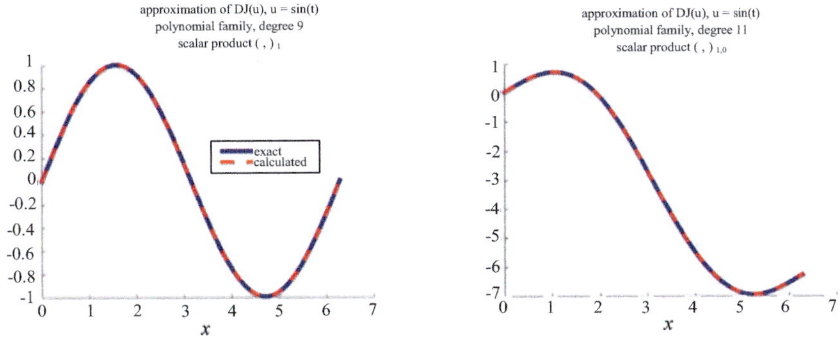

Figure 7.4. *Determination of a Gâteaux derivative for two different scalar products*

It must be noticed that an attempt to use the scalar product $(\blacksquare, \blacksquare)_0$ fails, except in particular situations: indeed, the equation

$$p \in L^2(\Omega) \ \ and \ \ (p, \delta u)_0 = DJ(u)(\delta u), \forall \delta u \in L^2(\Omega)$$

leads to:

$$u'(a)\delta u(a) - u'(b)\delta u(b) - \int_a^b (p + u'' - u)\delta u \, dt = 0, \ \ \forall \delta u \, .$$

This variational equation does not have any solution, except in the case where $u'(b) = u'(a) = 0$. In this case, the solution is $p = u - u''$. For instance, when considering $u(t) = \sin(t), a = 0, b = 2\pi$, these conditions are not satisfied, since $u'(b) = u'(a) = 1$ and p does not exist.

7.5.3. *Determination of the derivatives with respect to the Hilbertian coefficients*

In some situations, we are interested in the evaluation of $\partial J / \partial u_i$, where $u \approx Pu = \sum_{i=0}^{n} u_i \varphi_i$. In this case, we consider $U = (u_1, \ldots, u_n)$ and the function $J_n \colon \mathbb{R}^n \longrightarrow \mathbb{R}$ defined by:

$$J_n(U) = J\left(\sum_{j=1}^{n} u_j \varphi_j\right).$$

The aim is to determine $\partial J_n / \partial u_i$. By considering the function:

$$g(\alpha) = J\left(\alpha\varphi_i + \sum_{j=1}^{n} u_j\varphi_j\right),$$

it may be easily verified that:

$$\frac{\partial J_n}{\partial u_i} = g'(0) = DJ\left(\sum_{j=1}^{n} u_j\varphi_j\right)(\varphi_i).$$

As an example, let us consider $\Omega = (0,1)$ and the Lagrangian density

$$f_\ell\big(t, u(t), u'(t)\big) = [u'(t)]^2 + [u(t)]^2,$$

for which

$$DJ(u)(\delta u) = \int_a^b (u'(t)\delta u'(t) + u(t)\delta u(t))dt.$$

Let us consider the trigonometric family $\varphi_1(t) = 1$, $\varphi_{2i}(t) = \cos(2i\pi t)$, $\varphi_{2i+1}(t) = \sin(2i\pi t)$. This family is orthogonal and $\varphi'_{2i} = -2i\pi\varphi_{2i+1}$, $\varphi'_{2i+1} = 2i\pi\varphi_{2i}$, so that:

$$DJ\left(\sum_{j=1}^{n} u_j\varphi_j\right)(\varphi_i) = \int_0^1 (\varphi'_i(t)\varphi'_i(t) + \varphi_i(t)\varphi_i(t))dt$$

And we have:

$$\frac{\partial J_n}{\partial u_1} = u_1 \ ; \ \frac{\partial J_n}{\partial u_{2i}} = \left[\frac{1}{2} + 2(i\pi)^2\right] u_{2i} \ ; \ \frac{\partial J_n}{\partial u_{2i+1}} = \left[\frac{1}{2} + 2(i\pi)^2\right] u_{2i+1}.$$

We may obtain a numerical evaluation by using the class

```
classdef periodic_trigonometrical_basis
    properties
        zero
        xmin
        xmax
        nel
    end
    properties(Dependent)
```

```
        spmatrix
        intmatrix
    end
    methods
        function obj =
periodic_trigonometrical_basis(zero,xmin,xmax,order)
            obj.xmin = xmin;
            obj.xmax = xmax;
            obj.zero = zero;
            obj.nel = 2*order + 1;
        end
        function v = value(obj,i,t)
            if i == 1
                v = 1;
            else
                s = (t-obj.xmin)/(obj.xmax -
obj.xmin);
                if mod(i,2) == 0
                    if abs(s) < obj.zero
                        v = 1;
                    else
                        aux = i*pi*s;
                        v = cos(aux);
                    end;
                else
                    if abs(s) < obj.zero
                        v = 0;
                    else
                        aux =(i-1)*pi*s;
                        v = sin(aux);
                    end;
                end;
            end;
        end
        function v = d1(obj,i,t)
            if i == 1
                v = 0;
            else
                s = (t-obj.xmin)/(obj.xmax -
obj.xmin);
                if mod(i,2) == 0
```

```
                    if abs(s) < obj.zero
                        v = 0;
                    else
                        aux = i*pi*s;
                        v = -i*pi*sin(aux);
                    end;
                else
                    if abs(s) < obj.zero
                        v = (i-1)*pi;
                    else
                        aux = (i-1)*pi*s;
                        v = (i-1)*pi*cos(aux);
                    end;
                end;
                v = v/(obj.xmax - obj.xmin);
            end;
        end
        function v = d2(obj,i,t)
            if i == 1
                v = 0;
            else
                s = (t-obj.xmin)/(obj.xmax -
obj.xmin);
                if mod(i,2) == 0
                    if abs(s) < obj.zero
                        v = -1;
                    else
                        aux = i*pi*s;
                        v = -(i*pi)^2*cos(aux);
                    end;
                else
                    if abs(s) < obj.zero
                        v = 0;
                    else
                        aux = (i-1)*s;
                        v = -((i-1)*pi)^2*sin(aux);
                    end;
                end;
                v = v/(obj.xmax - obj.xmin)^2;
            end;
        end
    end
end
```

Program 7.1. *A class for periodic trigonometrical basis*

Then, we generate the family by:

```
n = 3;
phi = periodic_trigonometrical_basis(0,0,1,n);
bp = basis(phi);
```

We generate the gradient of J_n at the point $U = (u_1, \dots, u_n)$ as follows:

```
gradjn = zeros(size(U));
for i = 1: bp.nel
    phi1 = @(t)  bp.phi.value(i,t);
dphi1 = @(t) bp.phi.d1(i,t);
u = @(t) bp.projection(U,t);
    du = @(t) bp.derive(U,t);
    fff = @(t)
df_ell(t,u(t),du(t),phi1(t),dphi1(t)) ;
aux = subprogram_integration(fff,xlim);
    gradjn(i) = aux.result;
  end;
```

For $n = 3$ and `U=ones(2*n+1,1)`, we have:

```
gradjn=[1.0000  20.2392  20.2392  79.4568  79.4568
178.1529 178.1529],
```

which corresponds to the exact values.

7.5.4. *Solving an equation involving the Gâteaux differential*

Let $F: V \longrightarrow \mathbb{R}$ be a linear functional. The numerical determination of u such that:

$$u \in V \text{ and } DJ(u)(v) = F(v), \forall\, v \in V$$

is performed by setting $u \approx \sum_{i=1}^{n} u_i \varphi_i$ and considering the set of n equations for the n unknowns $U = (u_1, \dots, u_n)$:

$$E_i(U) = DJ\left(\sum_{j=1}^{n} u_j \varphi_j\right)(\varphi_i) - F(\varphi_i) = 0, for\ 1 \le i \le n.$$

This set of equations corresponds to:

$$E_i(U) = \frac{\partial J_n}{\partial u_i}(U) - F(\varphi_i) = 0, for\ 1 \le i \le n.$$

It must be solved for the unknown U by appropriate numerical methods, such as, for instance, Newton-Raphson, fixed point, relaxation, least squares etc. Once the solution is determined, the approximation $u \approx \sum_{i=1}^{n} u_i \varphi_i$ may be calculated.

As an example, let us consider $\Omega = (0,1)$, $V = H^1(\Omega)$ and the Lagrangian density:

$$f_\ell(t, u(t), u'(t)) = [u'(t)]^2 + [u(t)]^2,$$

for which:

$$DJ(u)(\delta u) = \int_0^1 (u'(t)\delta u'(t) + u(t)\delta u(t))dt.$$

Let us consider $F(v) = (f, v)_0$. Then, the equality $DJ(u)(\delta u) = F(\delta u)$ reads as:

$$u'(a)\delta u(a) - u'(b)\delta u(b) - \int_0^1 (f + u'' - u)\delta u\, dt = 0, \quad \forall \delta u \in V.$$

Thus,

$$u'' - u = -f\ on\ (0,1), \quad u'(0) = u'(1) = 0$$

For instance, if $f(t) = (1 + \pi^2)\sin(\pi t)$, then $u(t) = \sin(\pi t)$. Let us consider the trigonometric family $\varphi_1(t) = 1$, $\varphi_{2i}(t) = \cos(2i\pi t)$, $\varphi_{2i+1}(t) = \sin(2i\pi t)$. As previously remarked, this family is orthogonal and verifies,

$$\frac{\partial J_n}{\partial u_1} = u_1; \ \frac{\partial J_n}{\partial u_{2i}} = \left[\frac{1}{2} + 2(i\pi)^2\right] u_{2i}; \ \frac{\partial J_n}{\partial u_{2i+1}} = \left[\frac{1}{2} + 2(i\pi)^2\right] u_{2i+1}.$$

Moreover,

$$F(\varphi_1) = \frac{2(1+\pi^2)}{\pi}, F(\varphi_{2i}) = \frac{2(1+\pi^2)}{\pi(1-4i^2)}, F(\varphi_{2i+1}) = 0.$$

which determines the coefficients. Numerically, we may evaluate the vector $E(U) = (E_1(U), \dots, E_n(U))$ by the subprogram.

```
function eq = eqsdergat(U,basis,df_ell,xlim,f)
eq = zeros(basis.nel,1);
for i = 1: basis.nel
    phi1 = @(t)  basis.phi.value(i,t);
    dphi1 = @(t) basis.phi.d1(i,t);
    u = @(t) basis.projection(U,t);
    du = @(t) basis.derive(U,t);
    fff = @(t) df_ell(t,u(t),du(t),phi1(t),dphi1(t)) ;
    aux1 = subprogram_integration(fff,xlim);
    fff = @(t) f(t)*phi1(t) ;
    aux2 = subprogram_integration(fff,xlim);
    eq(i) = aux1.result - aux2.result;
end;
end
```

Then, we generate a function of U as follows:

```
f = @(t) (1+pi^2)*sin(pi*t);
eq = @(U) eqsdergat(U,bp,df_ell,xlim,f);
```

Then we may solve the system `eq = 0` by an adequated method. For instance, we may minimize `fobj = @(U) norm(eq(U))`. The point

```
U = [6.9198 -0.1140 0 -0.0058 0 -0.0011 0]
```

corresponds to `fobj (U)` $\approx 3.1 \times 10^{-15}$. It is very close to the exact value.

7.5.5. *Determining an optimal point*

The numerical determination of u such that:

$$u = \arg\min_V J$$

may be performed by different approaches:

1) Optimization approach: consider the function

$$J_n(u_1, \dots, u_n) = J\left(\sum_{j=1}^{n} u_j \varphi_j\right)$$

previously introduced and determined a numerical solution of the problem

$$U = \arg\min_{\mathbb{R}^n} J_n.$$

Analogously to the preceding situations, we have $u \approx \sum_{i=1}^{n} u_i \varphi_i$.

2) *Optimization approach involving Gâteaux derivative*: from the numerical standpoint, an alternative consists of using a descent method where the gradients are evaluated by using the Gâteaux derivative. For instance, the gradient descent with a fixed step reads as:

$$u^{(k+1)} = u^{(k)} - \mu J'\left(u^{(k)}\right).$$

More generally, a descent direction $d^{(k)}$ must be determined by using information furnished by the preceding iterations and $J'\left(u^{(k)}\right)$. Classical methods for the determination of $d^{(k)}$ are gradient, Fletcher–Reeves, DFP, BFGS, etc. Once $d^{(k)}$ is known, we must determine a step $\mu^{(k)} \geq 0$ in order to generate a new point by

$$u^{(k+1)} = u^{(k)} + \mu^{(k)} d^{(k)},$$

Usual methods for the determination of $\mu^{(k)}$ may be found in the literature. For instance, optimal step, fixed step, Armijo's, Wolfe's, and Goldstein's approaches may be used. In practice, this approach may request finite-dimensional approximations $J'(u) \approx \sum_{i=0}^{n} p_i \varphi_i$, $u \approx \sum_{i=0}^{n} u_i \varphi_i$.

3) *Equation approach:* solve the equation $DJ(u) = 0$, i.e. the n equations for the n unknowns $U = (u_1, \dots, u_n)$:

$$E_i(U) = DJ\left(\sum_{j=1}^{n} u_j \varphi_j\right)(\varphi_i) = 0, for\ 1 \leq i \leq n.$$

Then, approach $u \approx \sum_{i=1}^{n} u_i \varphi_i$. Notice that the equation approach is not generally equivalent to the optimization one: convexity assumptions are requested for the equivalence.

7.6. Minimum of the energy

In physics, one of the most useful principles is the *principle of the minimum of the energy*, which may be seen as a consequence of the *second principle of thermodynamics* – as formulated by Ludwig Boltzmann, the entropy can never decrease for an adiabatic system: thus, if S is the entropy such a system, we have $dS \geq 0$ (otherwise, we have the Clausius-Durhem's inequality). As a consequence, for any isolated system, equilibrium corresponds to both a maximum of the entropy and a minimum of the energy – we say that these two principles are *dual*: there is a duality between maximum of entropy and minimum of energy. In mathematical terms, we must consider:

1) the space of the parameters: a Hilbert space V;

2) the energy: a functional $J: V \to \mathbb{R}$.

Then, the principle of the minimum of the energy states that equilibrium parameters u correspond to (recall that $DJ(u)$ is the Gâteaux differential of J at the point u).

$$u = \arg\min\{J(v): v \in V\} \Longrightarrow DJ(u) = 0.$$

EXAMPLE 7.4.– Let us consider a system formed by a mass m, a spring having rigidity k under the action of the gravity $f = mg$ (Figure 7.5)

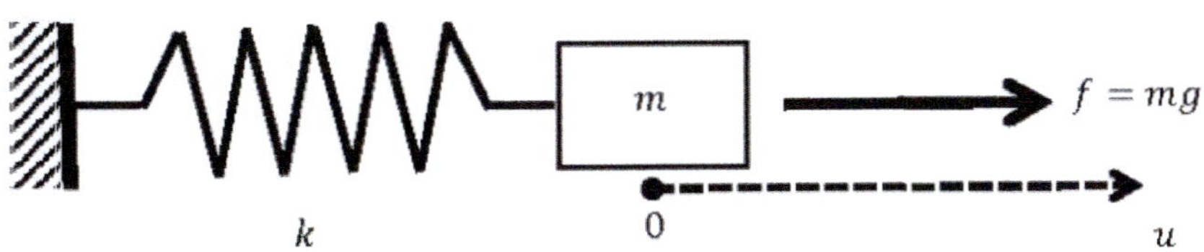

Figure 7.5. *Situation of example 7.4*

Let us denote by u the displacement of the mass. Here, $u \in V = \mathbb{R}$. The energy of the system is:

$$J(u) = \underbrace{\frac{k}{2}u^2}_{\substack{elastic\\ energy}} \underbrace{- mgu}_{\substack{gravity's\\ potential}}.$$

Thus,

$$DJ(u) = ku - mg = ku - f$$

and the equilibrium corresponds to $ku = mg = f$. ■

If the system is submitted to some restrictions, the space of the parameters is not the whole space V: we must consider the admissible set C, formed by the parameters satisfying the conditions:

$$u = \arg\min\{J(v): v \in C\}.$$

Here, we don't have $DJ(u) = 0$, since the minimum is not over the whole space.

EXAMPLE 7.5.– Let us consider again the system of example 7.4, but let us introduce an obstacle at distance $d > 0$ of the mass.

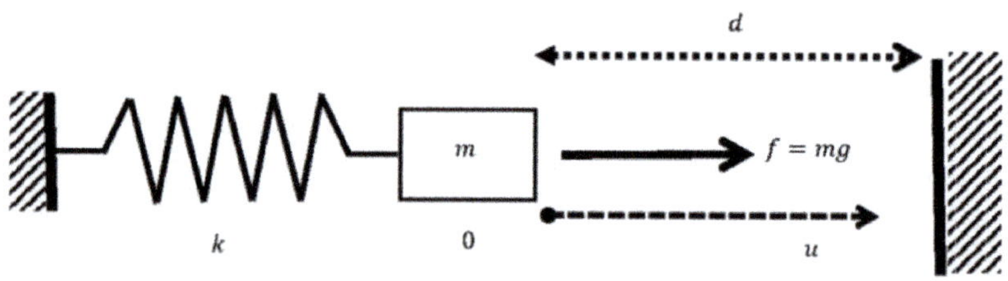

Figure 7.6. *Situation of example 7.5*

In this case,

$$C = \{u \in \mathbb{R} : u \leq d\}.$$

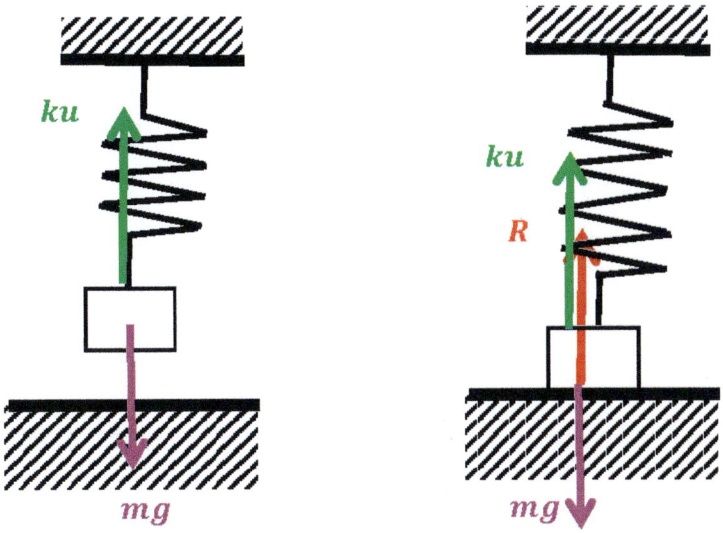

Figure 7.7. *Analysis of the situation of example 7.5*

Let $\overline{u} = mg/k$. Two situations are possible: if $\overline{u} < d$, the external force f is small enough in order that the spring's force equilibrates f before the mass touches the obstacle and we have $u = mg/k$, so that $DJ(u) = 0$. If $\overline{u} > d$, the external force f is large enough in order to bring the mass to the contact with the obstacle. In this case, we have $u = d$, so that $DJ(u) = kd - f < 0$: in fact, the contact with the obstacle provokes a contact reaction $R = mg - kd$, which equilibrates the resulting force to zero. ■

7.7. Lagrange's multipliers

As shown in example 7.5, when constraints have to be taken into account, we have not $DJ(u) = 0$. In fact, we must introduce Lagrange's multipliers λ in a *penalty functional* $P(u;\lambda)$ and construct the Lagrangian $L(u;\lambda) = J(u) + P(u;\ \lambda)$. These elements are defined according to the case under consideration:

constraint	Lagrange's Multiplier $\boldsymbol{\lambda}$	Properties of λ	Penalty $P(u;\lambda)$
$\psi(x,u(x)) \leq 0$, $\forall x \in \Omega$	$\lambda(x)$, $\lambda: \Omega \to \mathbb{R}$	$\lambda(x) \in \Lambda = \mathbb{R}_+$: $\lambda(x) \geq 0, \forall x \in \Omega$; $\lambda(x)\psi(x,u(x)) = 0 \; on \; \Omega$; $\lambda(x) = 0$ if $\psi(x,u(x)) < 0$	$\int_\Omega \lambda(x)\psi(x,u(x))dx$
$\psi(x,u(x)) = 0$, $\forall x \in \Omega$	$\lambda(x)$, $\lambda: \Omega \to \mathbb{R}$	$\lambda(x)$ *is arbitrary*: $\lambda(x) \in \Lambda = \mathbb{R}, \forall x \in \Omega$	$\int_\Omega \lambda(x)\psi(x,u(x))dx$
$I(u) \leq 0$, $I: V \to \mathbb{R}$	$\lambda \in \mathbb{R}$	$\lambda \in \Lambda = \mathbb{R}_+ : \lambda \geq 0$; $\lambda I(u) = 0$; $\lambda = 0$ if $I(u) < 0$	$\lambda I(u)$
$I(u) = 0$, $I: V \to \mathbb{R}$	$\lambda \in \mathbb{R}$	λ *is arbitrary*: $\lambda \in \Lambda = \mathbb{R}$	$\lambda I(u)$

Table 7.1. *Lagrange's Multipliers and penalties*

We have:

$$L(u;\eta) \leq L(u;\lambda) \leq L(v;\lambda), \forall v \in V, \eta \in \Lambda. \quad [7.1]$$

Thus, $(u;\lambda)$ is a saddle-point of Lon $V \times \Lambda$. Moreover, we have:

$$D_u L(u;\lambda)(\delta u) = 0, \forall \delta u \in V \; ; \; D_\lambda L(u;\lambda)(\delta\lambda) \leq 0, \forall \delta\lambda \text{ such that } \lambda + \delta\lambda \in \Lambda.$$

EXAMPLE 7.6.– Let us consider the situation of example 7.5. The constraint is $I(u) = u - d \leq 0$. Thus, $P(u;\lambda) = \lambda I(u) = \lambda(u - d)$ and $L(u;\lambda) = J(u) + \lambda(u - d)$. We have:

$$D_u L(u;\lambda) = ku - mg + \lambda = 0.$$

Moreover,

$$\lambda \geq 0, \lambda = 0 \; if \; u - d < 0.$$

Thus,

$$u < d \Rightarrow \lambda = 0 \Rightarrow mg - ku = \lambda = 0 \Rightarrow \frac{mg}{k} < d$$

and

$$u = d \Rightarrow mg - kd = \lambda \geq 0 \Rightarrow \frac{mg}{k} \geq d.$$

So, the solution is given by:

$$u = d, \lambda = mg - kd, if \; \frac{mg}{k} \geq d; u = \frac{mg}{k}, if \; \frac{mg}{k} < d.$$ ■

EXAMPLE 7.7.– Let us consider an inextensible string of length $\ell > 0$, linear mass density $\rho > 0$. Such a continuous medium is described by $a \in (0, \ell)$, where a is a particle of the string. Let us denote by $u(a)$ the position of the particle a of the string: in a bidimensional approach, the position is a field $u: (0, \ell) \to \mathbb{R}^2$ and $V = \{v: (0, \ell) \to \mathbb{R}^2\}$. Since the string is inextensible, we have ($u' = du/da$)

$$\psi(a, u(a)) = |u'(a)| - 1 = 0, \forall a \in (0, \ell).$$

Assume that the string is moored at the origin of the system of coordinates. We have also $I_1(u) = u_1(0) = 0, I_2(u) = u_2(0) = 0$. The energy of the string is the gravitational one:

$$J(u) = \int_0^\ell \rho g \, u_2(a) da.$$

Let us denote $\lambda = (\lambda_1, \lambda_2, \lambda_3), I(u) = (I_1(u), I_2(u))$ and $\lambda_2 = (\lambda_2, \lambda_3)$. Observe that $\lambda_1 = \lambda_1(a)$. Then,

$$P(u; \lambda) = \lambda_2 I(u) + \int_0^\ell \lambda_1(a)(|u'(a)| - 1) da.$$

Thus,

$$L(u; \lambda) = \lambda_2 u(0) + \int_0^\ell [\rho g \, u_2(a) + 1(a)(|u'(a)| - 1)] da.$$

So,

$$D_u L(u;\lambda)(\delta u) = \lambda_2 \delta u(0) + \int_0^\ell \left[\rho g\ \delta u_2(a) + \lambda_1(a)\frac{u'(a)}{|u'(a)|}\delta u'(a)\right] dx = 0\ , \forall \delta u.$$

Using an integration by parts:

$$\lambda_1(\ell)\frac{u'(\ell)}{|u'(\ell)|}\delta u(\ell) + \left(\lambda_2 - \lambda_1(0)\frac{u'(0)}{|u'(0)|}\right)\delta u(0) + \int_0^\ell \left[\rho g\ \delta u_2(a) - \left(\lambda_2(a)\frac{u'(a)}{|u'(a)|}\right)'\delta u'(a)\right] dx = 0, \forall \delta u.$$

Thus,

$$\lambda_1(\ell)\frac{u'(\ell)}{|u'(\ell)|} = 0;\ \lambda_2 = \lambda_1(0)\frac{u'(0)}{|u'(0)|};\ \left(\lambda_2(a)\frac{u'(a)}{|u'(a)|}\right)' = \rho g e_2\ on\ (0,\ell).$$

λ_1 is the tension, usually denoted by T, while $t = u'(a) = u'(a)/|u'(a)|$ is the unitary tangent. The internal efforts are $T = Tt$. Thus, by denoting $T_O = \lambda_2$:

$$T = Tt, t = u', |u'| = 1, \frac{dT}{da} - \rho g e_2 = 0\ on\ (0,\ell),$$

$$T(0) = T_O;\ T(\ell) = 0; u(0) = 0.$$ ■

7.8. Primal and dual problems

Equation [7.1] shows that:

$$J(u) = L(u;\lambda) = \inf_{v\in V} \sup_{\eta\in\Lambda} L(v;\eta) = \sup_{\eta\in\Lambda} \inf_{v\in V} L(v;\eta)\ .$$

Thus, we may adopt two different approaches:

1) The first approach consists of determining $u = \text{argmin}\ \{F(v) : v \in V\}$, $F(v) = \sup\{L(v;\eta): \eta \in \Lambda\}$ and, then, find the Lagrange's multipliers λ using the known u.

2) The second approach consists of determining $\lambda = \text{argmax}\ \{G(\eta) : \eta \in \Lambda\}$, $G(\lambda) = \inf\{L(v;\eta): v \in V\}$ and, then, finding u using the knowns λ.

The first approach consists of solving the problem $u = \arg\left(\inf_{v\in V}\left[\sup_{\eta\in\Lambda} L(v;\eta)\right]\right)$, which is referred to as the *primal problem*. The second one leads to the problem $\lambda = \arg\left(\sup_{\eta\in\Lambda}\left[\inf_{v\in V} L(v;\eta)\right]\right)$, which is referred to as the *dual problem*.

The solution of the primal problem is usually performed by using standard optimization methods, such as penalty methods, projected gradient or reduced gradient methods etc. When considering the dual problem, the main approach is furnished by *Uzawa's method*, which is a gradient's method applied to the solution of the dual problem: at each iteration number n, $\lambda^{(n-1)}$ is given and we determine $u^{(n-1)}$ such that $G\left(\lambda^{(n-1)}\right) = L\left(u^{(n-1)};\lambda^{(n-1)}\right)$. Then, we determine an ascent direction $d^{(n-1)}$ (see Table 7.2).

Constraint	$d^{(n-1)}$
$\psi(x,u(x)) \leq 0, \forall x \in \Omega$	$max\left\{0,\psi\left(x,u^{(n-1)}(x)\right)\right\}$
$\psi(x,u(x)) = 0, \forall x \in \Omega$	$\psi\left(x,u^{(n-1)}(x)\right)$
$I(u) \leq 0, I: V \to \mathbb{R}$	$max\left\{0,I\left(u^{(n-1)}\right)\right\}$
$I(u) = 0, I: V \to \mathbb{R}$	$I\left(u^{(n-1)}\right)$

Table 7.2. *Ascent directions for Uzawa's method*

Given a fixed step $\mu > 0$, we determine $\lambda^{(n)} = \lambda^{(n-1)} + \mu d^{(n-1)}$, which completes the step. Then, we increase n and we pass to new iteration. The global algorithm for Uzawa's method reads as:

1) An initial guess $\lambda^{(0)}$, a fixed step $\mu > 0$, a set of stopping conditions are given.

2) Set iteration number $n \leftarrow 0$.

3) Increase iteration number $n \leftarrow n + 1$.

4) Determine $u^{(n-1)} = \inf_{v \in V} L(v; \lambda^{(n-1)})$.

5) Determine $d^{(n-1)}$.

6) Determine $\lambda^{(n)} = \lambda^{(n-1)} + \mu d^{(n-1)}$.

7) Test for stopping conditions. If the decision is to continue the iterations, go to 3) or simply stop.

7.9. Matlab® determination of minimum energy solutions

The approximation of minimum energy solutions is performed according to the general principles presented in the preceding sections: we may consider a family $F = \{\varphi_p : 1 \leq p \leq k\}$, $u \approx \sum_{i=1}^{k} u_i \varphi_i$ and look for the unknowns $U = (u_1, \dots, u_k)$. As previously, the coefficients are arranged in a matrix $U = (U_{pj})_{\substack{1 \leq p \leq k \\ 1 \leq j \leq n}}$ such that $u_{pj} = (u_p)_j$. In this case, we look for a minimum of:

$$J_k(U) = J\big(u(U)\big), u(U) = \sum_{i=1}^{k} u_i \varphi_i = U^t \Phi, \Phi = (\varphi_1, \dots, \varphi_n)^t.$$

Notice that $U \to u(U)$ is linear. Let us set $\delta u = u(\delta U) = \sum_{i=1}^{k} \delta u_i \varphi_i$. We have:

$$g(\alpha) = J_k(U + \alpha \delta U) = J\big(\sum_{i=1}^{k} (u_i + \alpha \delta u_i) \varphi_i\big).$$

Then,

$$g'(0) = DJ(U^t \Phi)(U^t \delta\Phi).$$

Let $\mathcal{M}(k,n)$ be the set of the $k \times n$ real matrices. $T: \mathcal{M}(k,n) \to \mathbb{R}$ be given by:

$$T(\delta U) = DJ(U^t\Phi)(\delta U^t\Phi).$$

Since $\delta u \to DJ(u)(\delta u)$ is linear, we have:

$$T(\delta U) = \delta U^t DJ(U^t\Phi)(\Phi).$$

Moreover, the linearity of T implies that:

$$T(\delta U) = \sum_{p,j} \mathbb{T}_{pj}\delta U_{pj}.$$

We may evaluate $\mathbb{T}_{pj}$ by taking,

$$\delta U = \widehat{\delta U_{pj}},$$

$\left(\widehat{\delta U_{pj}}\right)_{rl} = 0$, if $p \neq r$ or $\neq l$; $\left(\widehat{\delta U_{pj}}\right)_{rl} = 1$, if $p = r$ and $j = l$.

Moreover,

$$g'(0) = \sum_{p,j} \frac{\partial J_n}{\partial U_{pj}}(U)\delta U_{pj},$$

so that:

$$\frac{\partial J_n}{\partial U_{pj}} = \mathbb{T}_{pj} = DJ(U^t\Phi)\left(\widehat{\delta U_{pj}}^t\Phi\right).$$

Let us illustrate such an evaluation: let us assume that:

$$J(u) = \int_a^b f\big(t, u(t)\big)dt.$$

where f_ℓ is evaluated by the subprogram `f_ell`, while the gradient $\left(\frac{\partial f_\ell}{\partial u_1}, \dots, \frac{\partial f_\ell}{\partial u_n}\right)$ is evaluated by the subprogram `gradf_ell`. Then, J is evaluated by the subprogram

```
function v = functional(f_ell, u,a,b)
fff = @(t) f_ell(t, u(t));
xlim.lower.x = a;
xlim.upper.x = b;
xlim.dim = 1;
v = subprogram_integration(fff,xlim).result;
end
```

DJ may be evaluated as follows:

```
function v = der_functional(gradf_ell,u,du,a,b)
fff = @(t) dot(gradf_ell(t, u(t)),du(t));
xlim.lower.x = a;
xlim.upper.x = b;
xlim.dim = 1;
v = subprogram_integration(fff,xlim).result;
end
```

J_n and its gradient are evaluated as in section 7.5.3.

```
function v = jn(U,basis,J)
u = @(t) basis.projection(U,t);
v = J(u);
end

function v = gradjn(U,basis,dJ)
u = @(t) basis.projection(U,t);
v = zeros(size(U));
for j = 1: size(U,2)
    for p = 1: size(U,1)
        c = zeros(size(U));
        c(p,j) = 1;
        du = @(t) basis.projection(c,t);
        v(p,j) = dJ(u,du);
    end;
end;
```

For instance, let us consider:

```
a = 0;
b = 1;
f_ell = @(t,u) sum(u.^2);
gradf_ell = @(t,u) 2*u;
```

```
k = 3;
phi = polynomial_basis(0,a,b,k);
bp = basis(phi);
U = [0 0 ;
     0 1 ;
     1 0;
     0 0
     ];
J = @(u) functional(f_ell, u,a,b);
v = jn(U,bp,J)
dJ = @(u,du) der_functional(gradf_ell,u,du,a,b);
gv = gradjn(U,bp,dJ)
```

This code produces: `v =0.4000` and

```
gv =
    0.6667    1.0000
    0.5000    0.6667
    0.4000    0.5000
    0.3333    0.4000
```

The exact value is `gv(i,1)` = $\frac{2}{i+2}$, `gv(i,2)` = $\frac{2}{i+1}$, which is in accordance with the results obtained.

Let us consider the unconstrained situation. In order to determine an approximate solution, we must find:

$$U = \arg\min\{J_n(V) : V \in \mathcal{M}(k,n)\}.$$

The evaluation of the gradient $\frac{\partial J_n}{\partial U_{pj}}$ allows the use of gradient-based optimization methods for the minimization of J_n. As an example, let us consider the situation where:

$$J(u) = \int_0^{2\pi} [(u_1(t) - \sin(t))^2 + (u_2(t) - \cos(t))^2] dt\,.$$

The exact solution is $u_1(t) = \sin(t)$, $u_2(t) = \cos(t)$. Let us consider the approximation by a polynomial family with = 5. We use f min search in order to minimize J_n. The results are exhibited in Figure 7.8.

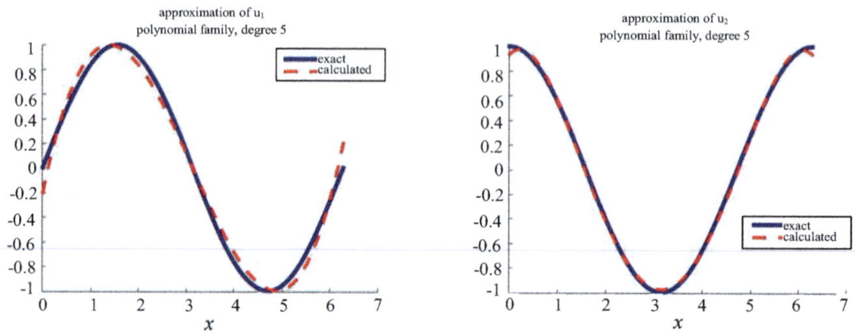

Figure 7.8. *Determination of the minimum of a functional*

Let us consider the situation where constraints must be taken into account. As previously observed, we must consider a penalty functional $P(u; \lambda)$ and the Lagrangian $L(u; \lambda) = J(u) + P(u; \lambda)$. The primal problem may be solved by considering the *penalty methods*: a penalty function for the primal problem is a function $p: V \rightarrow \mathbb{R}$ such that $p(u) = 0$, if $u \in C$; $p(u) > 0$, if $u \notin C$. A Typical penalty function for the constraint $\psi(t, u(t)) \leq 0 \; on \; \Omega$ is:

$$p(u) = \frac{1}{2} \int_{\Omega} [\psi^+]^2 dt \;, \psi^+ = max\{\psi(t, u(t)), 0\}.$$

A typical penalty function for the constraint $\psi(t, u(t)) = 0 \; on \; \Omega$ is:

$$p(u) = \frac{1}{2} \int_{\Omega} [\psi(t, u(t))]^2 dx \,.$$

For the primal problem, $\sup\{L(u; \lambda) : \lambda\}$ is attained for $\lambda \rightarrow +\infty$, so that we may consider a fixed value $\lambda > 0$ and set $J_\lambda(u) = L(u; \lambda) = J(u) + \lambda p(u)$. Then, we determine:

$$u_\lambda = \arg\min\{J_\lambda(u) : u \in V\}.$$

This problem is unconstrained and may be solved by the preceding method. As an example, let us consider restrictions $\psi_{\leq}(x, u(x)) \leq 0$ and $\psi_{=}(x, u(x)) = 0$. We assume that $\psi_{\leq}, \psi_{=}, \partial\psi_{\leq}/\partial u_i$ and

$\partial\psi_= / \partial u_i$ are evaluated by subprograms psineg, psieg, gradpsineg and gradpsieg, respectively. Then, we may evaluate the penalty by using the following subprogram:

```
function v = p_pen(u,lambda,psineg,psieg,a,b)
fff = @(t) penalty(t,u,psineg,psieg);
xlim.lower.x = a;
xlim.upper.x = b;
xlim.dim = 1;
v = subprogram_integration(fff,xlim).result;
v = lambda*v;
end
```

where:

```
function v = penalty(t,u,psineg,psieg)
aux = 0;
auxineg = psineg(t,u(t));
if ~isempty(auxineg)
   aux2 = (abs(auxineg)+auxineg)/2;
   aux = aux + sum(aux2.^2);
end;
auxeg = psieg(t,u(t));
if ~isempty(auxeg)
   aux = aux + sum(auxeg.^2);
end;
v = 0.5*aux;
return;
end
```

The Gâteaux differential of the penalty is evaluated as follows:

```
function v =
derp_pen(u,lambda,du,psineg,psieg,gradpsineg,gradps
ieg,a,b)
fff = @(t)
der_penalty(t,u,du,psineg,psieg,gradpsineg,gradpsie
g);
xlim.lower.x = a;
xlim.upper.x = b;
xlim.dim = 1;
v = subprogram_integration(fff,xlim).result;
v = lambda*v;
end
```

with

```
function v =
der_penalty(t,u,du,psineg,psieg,gradpsineg,gradpsieg)
aux = 0;
auxineg = psineg(t,u(t));
dauxineg = gradpsineg(t,u(t));
if ~isempty(auxineg)
   aux2 = (abs(auxineg)+auxineg)/2;
   aux =aux2'*dauxineg*du(t);
end;
auxeg = psieg(t,u(t));
dauxeg = gradpsieg(t,u(t));
if ~isempty(auxeg)
   aux = aux + auxeg'*dauxeg*du(t);
end;
v = aux;
return;
end
```

Let us consider:

```
function v = Jpen(u,lambda,J,P)
v = J(u) + P(u,lambda);
end

function v = der_Jpen(u,lambda,du,dJ,dP)
v = dJ(u,du) + dP(u,lambda,du);
end
```

We define the values of J_λ and DJ_λ as follows:

```
J = @(u) functional(f_ell, u,a,b);
P = @(u,lambda) p_pen(u,lambda,psineg,psieg,a,b);
Jlambda = @(u) Jpen(u,lambda,J,P);
dJ = @(u,du) der_functional(gradf_ell, u,du,a,b);
dP = @(u,lambda,du)
derp_pen(u,lambda,du,psineg,psieg,gradpsineg,gradps
ieg,a,b);
dJlambda = @(u,du) der_Jpen(u,lambda,du,dJ,dP);
```

Then, we may apply the method presented above for unconstrained situations. As an example, let us consider $\psi_{\leq}\big(t,u(t)\big) =$

$(\sin(t) - u_1(t), \cos(t) - u_2(t))$, $\psi_= (x, u(x)) = []$ (no equality constraint), $f_\ell(t, u) = u_1(t) - \sin(t) + u_2(t) - \cos(t)$. The exact solution is $u_1(t) = \sin(t)$, $u_2(t) = \cos(t)$. Let us consider the approximation by a polynomial family with $= 5$. We use `fminsearch` in order to minimize J_n. The results are exhibited in Figure 7.9.

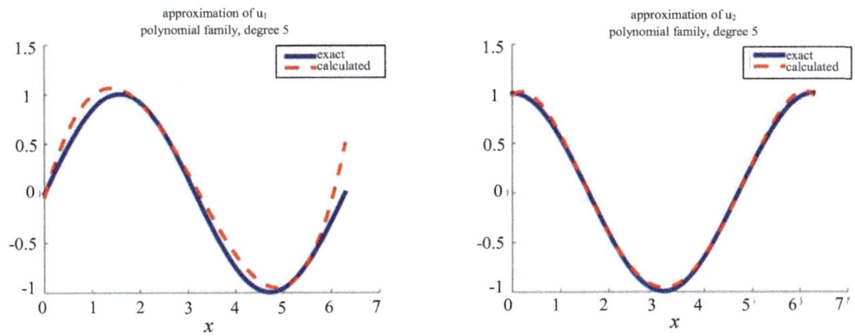

Figure 7.9. *Determination of the unconstrained minimum of a functional*

An alternative approach to primal problems is furnished by *barrier methods*: a barrier function is a function $b: V \to \mathbb{R}$ such that $b(u) \in \mathbb{R}$, if $u \in C$; $b(u) = +\infty$, if $u \notin C$ and the penalty is $P(u, \lambda) = \lambda b(u)$. We do not develop this approach in this book, but the practical implementation is analogous to those of penalty methods.

For the dual problem, $L(u; \lambda) = J(u) + P(u, \lambda)$ and we may use Uzawa's algorithm. Let us split the vector λ in $\lambda_\leq, \lambda_=$, corresponding to inequality and equality restrictions and stocked by variables `lambineg,lambeg`, respectively. Then, the penalty is evaluated as follows:

```
function v =
lag_pen(u,lambineg,lambeg,psineg,psieg,a,b)
fff = @(t)
lagpnlty(t,u,psineg,psieg,lambineg,lambeg);
xlim.lower.x = a;
xlim.upper.x = b;
```

```
xlim.dim = 1;
v = subprogram_integration(fff,xlim).result;
end

function v =
lagpnlty(t,u,psineg,psieg,lambineg,lambeg)
aux1 = 0;
aux2 = 0;
auxineg = psineg(t,u(t));
if ~isempty(auxineg)
   aux1 = lambineg(t)'*auxineg;
end;
auxeg = psieg(t,u(t));
if ~isempty(auxeg)
   aux2 =  lambeg(t)'*auxeg;
end;
v = aux1 + aux2;
return;
end
```

The Gâteaux differential of the penalty is evaluated as follows:

```
function v =
derlag_pen(u,lambda,du,psineg,psieg,gradpsineg,grad
psieg,a,b)
fff = @(t)
der_lagpnlty(t,u,du,psineg,psieg,gradpsineg,gradpsi
eg);
xlim.lower.x = a;
xlim.upper.x = b;
xlim.dim = 1;
v = subprogram_integration(fff,xlim).result;
end
```

with

```
function v =
der_lagpntly(t,u,du,psineg,psieg,gradpsineg,gradpsi
eg)
auxineg = psineg(t,u(t));
dauxineg = gradpsineg(t,u(t));
if ~isempty(auxineg)
   aux1 = lambineg(t)'*dauxineg*du(t);
else
```

```
    aux1 = [];
end;
auxeg = psieg(t,u(t));
dauxeg = gradpsieg(t,u(t));
if ~isempty(auxeg)
   aux2 = lambeg(t)'*dauxeg*du(t);
else
    aux2 = [];
end;
if isempty(aux1)
v = aux2;
elseif isempty(aux2)
    v = aux1;
else
    v = aux1 + aux2;
end;
return;
end
```

Then, we evaluate the Lagrangian $L(u;\lambda)$ as follows:

```
function v = Jpenlag(u,lambineg,lambeg,J,P)
v = J(u) + P(u,lambineg,lambeg);
end

function v =
der_Jpenlag(u,lambineg,lambeg,du,dJ,dP)
v = dJ(u,du) + dP(u,lambineg,lambeg,du);
end
```

In order to implement Uzawa's method, we may use

```
J = @(u) functional(f_ell, u,a,b);
P = @(u,lambineg,lambeg)
lag_pen(u,lambineg,lambeg,psineg,psieg,a,b);
Jlambda = @(u) Jpenlag(u,lambineg,lambeg,J,P);
dJ = @(u,du) der_functional(gradf_ell, u,du,a,b);
dP = @(u,lambda,du)
derlag_pen(u,lambineg,lambeg,du,psineg,psieg,gradps
ineg,gradpsieg,a,b);
dJlambda = @(u,du)
der_Jplag(u,lambineg,lambeg,du,dJ,dP);
```

At each step of Uzawa's method, we need to determine an unconstrained minimum of `Jlambda` and, then, use it to update the values of `lambineg,lambeg`. Uzawa's iterations may be implemented as follows: the Lagrange's multipliers are discretized on $\Omega = (a, b)$ by using a partition tp, such that $a = tp_1 < tp_2 < \cdots < tp_{n+1} = b$. Table `tp` contains the values of these points `tp(i)` $= tp_i$. The discrete values of the multipliers are stocked in the tables `Lineg` and `Leg`, with initial values `Lineg0` and `Leg0`: `Lineg(i)`$=\lambda_{\leq}(tp_i)$, `Leg(i)`$=\lambda_{=}(tp_i)$. An initial guess `U0` is given. Then, we use the code below:

```
Uopt = U0;
Lineg = Lineg0;
Leg = Leg0;
erruz = precuz + 1.;
nituz = 0;
while nituz < nituzmax && erruz > precuz
nituz=nituz+1;
if isempty(Lineg)
        lambineg = @(t) [];
else
        lambineg = @(t) interp1(tp,Lineg',t)';
end;
if isempty(Leg)
        lambeg = @(t) [];
else
        lambeg = @(t) interp1(tp,Leg',t)';
end;
    P = @(u, lambineg,lambeg)
lag_pen(u,lambineg,lambeg,psineg,psieg,a,b);
    Jlambda = @(u) Jpenlag(u, lambineg,lambeg,J,P);
    fobj = @(U) jn(U,bp,Jlambda);
    U_ini = Uopt;
    Uopt = fminsearch(fobj,U_ini);
    errineg = 0;
    if ~isempty(Lineg)
       auxineg = Lineg;
       for j=1:size(Lineg,2)
           aux = psineg(tp(j),u(tp(j)));
```

```
            for i = 1: size(Lineg,1)
                Lineg(i,j) = max([Lineg(i,j) + mu*aux(i),0]);
            end;
        end;
        errineg = norm(auxineg-Lineg);
    end;
    erreg = 0;
    if ~isempty(Leg)
        auxeg = Leg;
        for j=1:size(Leg,2)
            aux = psieg(tp(j),u(tp(j)));
            for i = 1: size(Leg,1)
                Leg(i,j) = Leg(i,j) + mu*aux(i);
            end;
        end;
        erreg = norm(auxeg-Leg);
    end;
    erruz = errineg + erreg;
end;
```

As an example, let us consider $\psi_{\leq}\big(t,u(t)\big) = \big(\sin(t) - u_1(t), \cos(t) - u_2(t)\big)$, $\psi_{=}\big(x,u(x)\big) = []$ (no equality constraint), $f_\ell(t,u) = |u(t)|^2$. The exact solution is $u_1(t) = \max\{\sin(t), 0\}$, $u_2(t) = \max\{\cos(t), 0\}$. Let us consider the approximation by a polynomial family with $= 10$. We use `fminsearch` in order to minimize J_n. The results are exhibited in Figure 7.10.

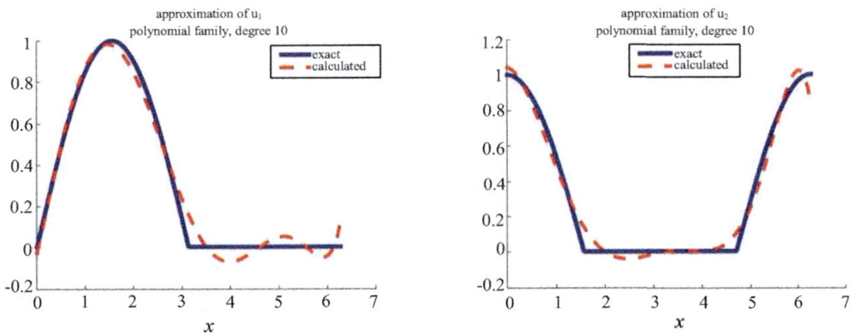

Figure 7.10. *Determination of the constrained minimum of a functional by Uzawa's method*

7.10. First-order control problems

Differential equations may be used as constraints. For instance, let us consider the situation where:

$$J(x,u) = \int_a^b f(t,x(t),u(t))dt;$$

$$C = \left\{(x,u) : x' = \frac{dx}{dt} = f_s(t,x(t),u(t)) \text{ on } (a,b), x(a) = x_a\right\}.$$

This situation is referred to as *first-order control problem*: u is the control x_a is the initial data and x is the state – it is determined by the control and the initial data. The differential equation is referred as *state equation* and J is usually called the *cost functional*. In this case, with $\lambda = (\lambda_1, \lambda_2)$,

$$P(x,u;\lambda) = \lambda_2(x(a) - x_a) + \int_a^b \lambda_1(t)\left(x' - f_s(t,x(t),u(t))\right)dt$$

and

$$L(x,u;\lambda) = \lambda_2(x(a) - x_a) + \int_a^b \left[f(t,x(t),u(t)) + \lambda_1(t)\left(x' - f_s(t,x(t),u(t))\right)\right]dt.$$

Thus,

$$D_x L(x,u;\lambda)(\delta x) = \lambda_2 \delta x(a) + \int_a^b \left[\left[\frac{\partial f}{\partial x}(t,x(t),u(t)) - \lambda_1(t)\frac{\partial f_s}{\partial x}(t,x(t),u(t))\right]\delta x + \lambda_1(t)\delta x'\right]dt = 0, \forall \delta x,$$

which shows that (using integration by parts)

$$\lambda_1'(t) = \frac{\partial f}{\partial x}(t,x(t),u(t)) - \lambda_1(t)\frac{\partial f_s}{\partial x}(t,x(t),u(t)) \text{ on } (a,b) \qquad [7.2]$$

$$\lambda_1(b) = 0; \ \lambda_2 = -\ \lambda_1(a).$$

Moreover,

$$D_u L(x, u; \lambda)(\delta u) = \int_a^b \left[\frac{\partial f}{\partial u}(t, x(t), u(t))\delta u - \lambda_1(t)\frac{\partial f_s}{\partial u}(t, x(t), u(t))\delta u\right] dt = 0, \forall \delta u,$$

which shows that:

$$\frac{\partial f}{\partial u}(t, x(t), u(t)) = \lambda_1(t)\frac{\partial f_s}{\partial u}(t, x(t), u(t)) \text{ on } (a, b). \qquad [7.3]$$

Finally, recall that:

$$x' = f_s(t, x(t), u(t)) \text{ on } (a, b), x(a) = x_a. \qquad [7.4]$$

Equations [7.2], [7.3] and [7.4] may be solved in order to determine the unknowns $(x, u; \lambda)$. Notice that the approach presented in section 7.4.4 shows that:

$$\delta x' = \frac{\partial f_s}{\partial x}(t, x, u)\delta x + \frac{\partial f_s}{\partial u}(t, x, u)\delta u \text{ on } (a, b), \delta x(a) = 0.$$

This last equation is adequate for numerical solution: analogously to the situations examined in 7.9, we may consider a family $F = \{\varphi_i : 1 \le i \le k\}$, $u \approx \sum_{i=1}^k u_i \varphi_i$ and look for the unknowns $U = (u_1, \dots, u_k)$. In this case, we look for a minimum of:

$$J_n(U) = J(x(U), u(U)), u(U) = \sum_{i=1}^k u_i \varphi_i,$$

where $x(U)$ is the solution of:

$$\frac{dx(U)}{dt} = f_s(t, x(U)(t), u(U)(t)) \text{ on } (a, b), x(U)(a) = x_a.$$

Let us use the same notation as in section 7.9:

$$u(U) = U^t \Phi, u(\delta U) = \delta u = \delta U^t \Phi.$$

Let δx be the solution of:

$$\frac{d\delta x}{dt} = \frac{\partial f_s}{\partial x}(t, x(U), u(U))\delta x + \frac{\partial f_s}{\partial u}(t, x(U), u(U))\delta u \text{ on } (a, b), \delta x(a) = 0.$$

The gradient of J_n is evaluated analogously to section 7.9:

$$\frac{\partial J_n}{\partial U_{pj}} = DJ(U^t\Phi)\left(\widehat{\delta U_{pj}}^{\,t}\Phi\right).$$

with

$$\left(\widehat{\delta U_{pj}}\right)_{rl} = 0, \text{ if } p \neq r \text{ or } \neq l\ ; \left(\widehat{\delta U_{pj}}\right)_{rl} = 1, \text{ if } p = r \text{ and } j = l\ .$$

We have:

$$\frac{\partial J_n}{\partial U_{ij}} = g'(0) = \int_a^b \left[\frac{\partial f}{\partial x}\big(t, x(U), u(U)\big)\widehat{\delta x}_{ij} + \frac{\partial f}{\partial u}\big(t, x(U), u(U)\big)\widehat{\delta u_{ij}}\right] dt\,,$$

where $\widehat{\delta u_{ij}} = \Phi^t \widehat{\delta U_{ij}}$ and

$$\frac{d\widehat{\delta x}}{dt} = \frac{\partial f_s}{\partial x}(t, x, u)\widehat{\delta x} + \frac{\partial f_s}{\partial u}(t, x, u)\widehat{\delta u} \text{ on } (a, b), \widehat{\delta x}(a) = 0.$$

In Matlab®, the state x may be determined as follows:

```
function x = state(fs,u,a,b,x0)
f = @(t,x) fs(t,x,u(t));
sol = ode45(f, [a,b],x0);
x = @(t) deval(sol,t);
end
```

δx may be evaluated as follows:

```
function dx = der_state(dfsdx,dfsdu,x,u,du,a,b,dx0)
df = @(t,dx) dfsdx(t,x(t),u(t))*dx +
dfsdu(t,x(t),u(t))*du(t);
sol = ode45(df, [a,b],dx0);
dx = @(t) deval(sol,t);
end
```

J may be evaluated as follows:

```
function v = functional(f,fs,u,a,b,x0)
x = state(fs,u,a,b,x0);
```

```
fff = @(t) f(t,x(t),u(t));
xlim.lower.x = a;
xlim.upper.x = b;
xlim.dim = 1;
v = subprogram_integration(fff,xlim).result;
end
```

DJ may be evaluated as follows:

```
function v =
der_functional(dfdx,dfdu,dfsdx,dfsdu,x,u,du,a,b,dx0)
dx = der_state(dfsdx,dfsdu,x,u,du,a,b,dx0);
fff = @(t)
dfdx(t,x(t),u(t))*dx(t)+dfdu(t,x(t),u(t))*du(t);
xlim.lower.x = a;
xlim.upper.x = b;
xlim.dim = 1;
v = subprogram_integration(fff,xlim).result;
end
```

J_n and its gradient are evaluated as in section 7.5.3.

```
function v = jn(U,basis,f,fs,a,b,x0)
u = @(t) basis.projection(U,t);
v = functional(f,fs,u,a,b,x0);
end

function v =
gradjn(U,basis,dfdx,dfdu,fs,dfsdx,dfsdu,a,b,x0,dx0)
u = @(t) basis.projection(U,t);
x = state(fs,u,a,b,x0);
v = zeros(size(U));
for i = 1: basis.nel
    du = @(t)  basis.phi.value(i,t);
    v(i) =
der_functional(dfdx,dfdu,dfsdx,dfsdu,x,u,du,a,b,dx0);
end;
```

Let us illustrate the use of these equations by considering the situation where:

$$f_s(t,x,u) = u \ , f(t,x,u) = \frac{u^2}{2m} + \alpha|x| + hx \ , x(0) = 0.$$

In this case,

$$\frac{\partial f_s}{\partial x} = 0; \frac{\partial f_s}{\partial u} = 1 ; \frac{\partial f}{\partial x} = h + \alpha sign(x); \frac{\partial f}{\partial u} = \frac{u}{m}$$

In Matlab®, we define this situation as follows:

```
f = @(t,x,u) u.^2/(2*m) + alfa*abs(x) + h*x;
dfdu = @(t,x,u) u/m ;
dfdx = @(t,x,u) alfa*sign(x) + h;
fs   = @(t,x,u) u;
dfsdu = @(t,x,u) 1 ;
dfsdx = @(t,x,u) 0;
```

The objective function is defined as:

```
fobj = @(U) jn(U,bp,f,fs,a,b,x0);
```

Its gradient is defined as:

```
gradfobj = @(U)
gradjn(U,bp,dfdx,dfdu,fs,dfsdx,dfsdu,a,b,x0,dx0);
```

We consider $m = 1$, $g = 9.81$, $\mu = 0.2$, $\alpha = \mu mg$, $h = mg/2$. The results furnished by a polynomial basis and by a trigonometrical basis (see section 2.2.6), both with $k = 5$, are shown in Figure 7.11.

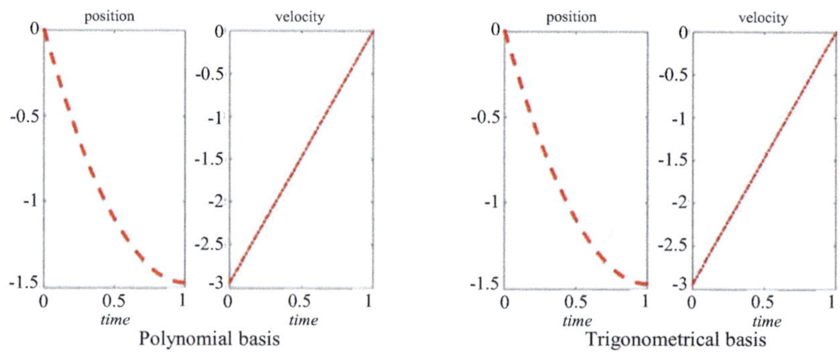

Figure 7.11. *Determination of a first-order control*

7.11. Second-order control problems

The situation where

$$J(x,u) = \int_a^b f\big(t, x(t), x'(t), u(t)\big)dt:$$

$$C = \left\{(x,u) : x'' = \frac{d^2x}{dt^2} = f_s(t,x,x',u) \text{ on } (a,b), x(a) = x_a, x'(a) = x'_a\right\}.$$

is referred to as *second-order control problem*. The approach presented in section 7.10 extends to this situation: here, $\lambda = (\lambda_1, \lambda_2, \lambda_3)$:

$$P(x,u;\lambda) = \lambda_2(x(a) - x_a) + \lambda_3(x'(a) - x'_a) + \int_a^b \lambda_1(t)\big(x'' - f_s(t,x,x',u)\big)dt$$

and

$$L(x,u;\lambda) = \lambda_2(x(a) - x_a) + \lambda_3(x'(a) - x'_a) + \int_a^b \big[f\big(t,x(t),x'(t),u(t)\big) + \lambda_1\big(x'' - f_s(t,x,x',u)\big)\big]dt\,.$$

Thus,

$$D_xL(x,u;\lambda)(\delta x) = \lambda_2\delta x(a) + \lambda_3\delta x'(a) + \int_a^b \left[\left[\frac{\partial f}{\partial x}(t,x,x',u) - \lambda_1\frac{\partial f_s}{\partial x}(t,x,x',u)\right]\delta x + \left[\frac{\partial f}{\partial x'}(t,x,x',u) - \lambda_1\frac{\partial f_s}{\partial x'}(t,x,x',u)\right]\delta x' + \lambda_1\delta x''\right]dt = 0, \forall\delta x\,,$$

In this case, an integration by parts furnishes:

$$\frac{\partial f}{\partial x}(t,x,u) - \lambda_1 \frac{\partial f_s}{\partial x}(t,x,x',u) +$$
$$\left(\lambda_1 \frac{\partial f_s}{\partial x'}(t,x,x',u) - \frac{\partial f}{\partial x'}(t,x,x',u)\right)' - \lambda_1'' = 0 \ on\ (a,b)$$

$$\lambda_1(b) = 0\,, \lambda_1'(b) = 0,$$

$$\lambda_3 = \lambda_1(a),\ \ \lambda_2 = -\lambda_1'(a) - \lambda_1(a)\frac{\partial f_s}{\partial x}(a, x_a, x_a', u(a)),$$

$$x'' = f_s(t,x,x',u)\ on\ (a,b), x(a) = x_a, x'(a) = x_a'.$$

These equations determine x and λ. Notice that the method presented in section 7.4.4 furnishes:

$$\delta x'' = \frac{\partial f_s}{\partial x}(t,x,x',u)\delta x + \frac{\partial f_s}{\partial u}(t,x,x',u)\delta u,\ \ \delta x(0) = \delta x'(0) = 0.$$

This situation may be reduced to the first-order control problem studied in section 7.10, by setting $X = \left(x, \frac{dx}{dt}\right)^t$. Then:

$$\frac{dX}{dt} = \begin{pmatrix} \frac{dx}{dt} \\ \frac{d^2x}{dt^2} \end{pmatrix} = \begin{pmatrix} \frac{dx}{dt} \\ f_s\left(t, x, \frac{dx}{dt}, u\right) \end{pmatrix} = F_s(t, X, u).$$

Thus, this problem may be solved by the approach introduced in section 7.10. Let us illustrate this approach by considering a linear oscillator:

$$\frac{d^2x}{dt^2} + \frac{c}{m}\frac{dx}{dt} + \frac{k}{m}x = u\ on\ \ (0,T), x(0) = 0,\ \frac{dx}{dt}(0) = 1.$$

Assume that we aim to force this oscillator to have a target phase: we look for the minimization of the objective

$$J(u) = \int_0^1 \left(x(t) - \frac{\cos(\omega_0 t)}{\omega_0}\right)^2 dt$$

In this case,

$$F_s(t,X,u) = \begin{pmatrix} X_2 \\ u - \frac{c}{m}X_2 - \frac{k}{m}X_1 \end{pmatrix}, \quad X_0 = \begin{pmatrix} 0 \\ 1 \end{pmatrix}.$$

which may be evaluated as follows:

```
fs = @(t,x,u) [x(2) ; u - c*x(2)/m - k*x(1)/m];
dfsdu = @(t,x,u) 1 ;
dfsdx = @(t,x,u) [-k/m, -c/m];
```

The objective corresponds to:

$$f(t,X,u) = \left(X_1 - \frac{\cos(\omega_0 t)}{\omega_0}\right)^2$$

and may be defined as follows (`om0` contains the value of ω_0):

```
f  = @(t,x,u) (x(1) - cos(om0*t)/om0)^2;
dfdu = @(t,x,u) 0 ;
dfdx = @(t,x,u) 2*[x(1) - cos(om0*t)/om0, 0];
```

The results obtained with $m = 1$, $k = 1$, $c = 0$, $\omega_0 = \sqrt{k/m}$ are shown in Figure 7.12.

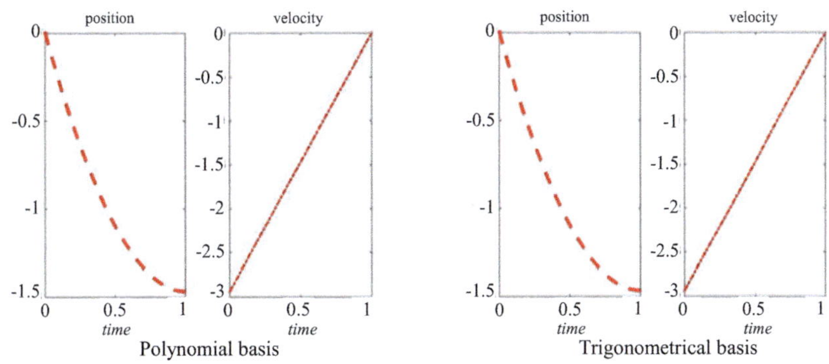

Figure 7.12. *Determination of a first-order control*

The results obtained with a polynomial family $\{\varphi_i(t) = t^{i-1}: 1 \leq i \leq n\}$, $n = 6$, $m = 1$, $k = 1$, $\omega_0 = \sqrt{k/m}$, $c = 0$ *and* 0.2, are shown in Figure 7.13.

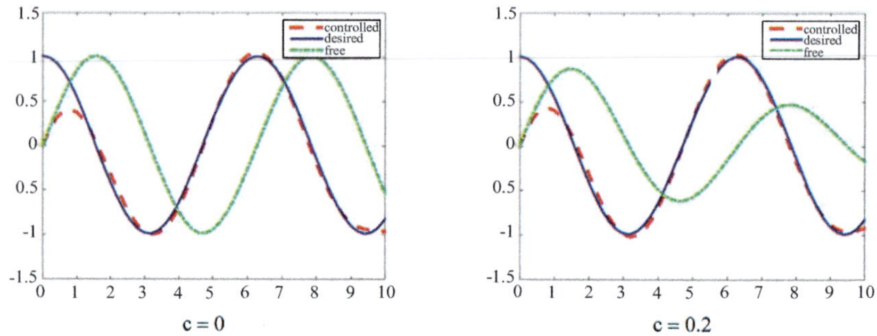

Figure 7.13. *Second-order control: a linear oscillator*

7.12. A variational approach for multiobjective optimization

Multiobjective optimization is often dated of the works of Francis Ysidro Edgeworth [EDG 81], which considers consumer decisions based on two utility functions – i.e. two objectives. Edgeworth pointed that the notion of optimum has no meaning in this situation, since the objectives may be contradictory. The reader may find in the literature the classical quote from Edgeworth about two objectives P and π ([EDG 81], p. 21):

> *It is required to find a point (x,y,) such that in whatever direction we take an infinitely small step, P and π do not increase together but that, while one increases, the other decreases.*

With this phrase, Edgeworth introduces the idea of a trade-off between the objectives and defines the set nowadays known as *Pareto Front*, associated with Vilfredo Federico Damaso Pareto, born in Paris from a mixed French/Italian couple. Pareto introduced the notion of ophelimity – concurrent of that of utility – and defined the maximum ophelimity as the situation where it is impossible to make a small displacement (Pareto was civil engineer) in order that the ophelimity of each individual increases or decreases [PAR 06]. The book of Pareto was the sequence of his course held at the University of Lausanne [PAR 96, PAR 97]. His work was translated in French [PAR 09] and popularized in English [HIC 34a, HIC 34b], then translated into English [PAR 71].

The name of Pareto was associated with multiobjective optimization by means of Pareto Fronts and Pareto Efficiency. A recent work [ZID 14] proposed a variational method for the determination of Pareto fronts: assuming that both the objective functions are non-negative, we look for an admissible curve that minimizes the area between the curve and the vertical or horizontal axis (see Figure 7.7). We present below this approach for the situation where we are interested in minimizing two objective functions $(f_1(x), f_2(x))$, but the approach generalizes to vectors of objectives having larger dimension.

As previously observed, in the framework of multiobjective optimization, we are mainly interested in the determination of the *Pareto front*, i.e. the set of all the possible compromises between the different objectives. A Pareto front gives the possible trade-offs between the objectives and is formed by the *non-dominated* or *efficient* points. Let us recall that we may define a partial order on $\mathbb{R}^n$ as follows:

$$X \preccurlyeq Y \Leftrightarrow X_i \leq Y_i \, for \, 1 \leq i \leq n \, .$$

$X \prec Y \Leftrightarrow X \preccurlyeq Y$ and there exists $i \in \{1, \dots, n\}$ such that $X_i < Y_i$.

DEFINITION 7.4.– We say that X is more efficient than Y or that X dominates Y if and only if $X \prec Y$.

Let $S \subset \mathbb{R}^n$. We say that X is Pareto-efficient on S if and only if $X \in S$ and *there is no other* element of S which is more efficient than X. The *Pareto front* or *Pareto Frontier* of S, denoted as $\Pi(S)$, is the set of the Pareto-efficients points of S:

$$\Pi(S) = \{X \in S : \nexists\, Y \in S \text{ such that } Y \prec X\}$$ ■

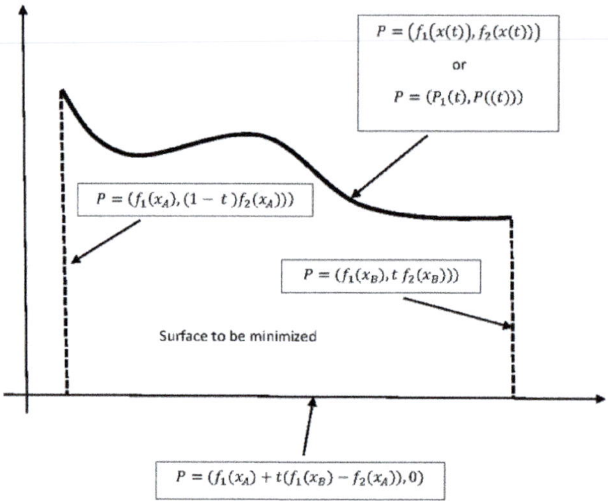

Figure 7.14. *Variational approach for multiobjective optimization (2 objectives)*

Let $f_i: \mathbb{R}^k \longrightarrow \mathbb{R}$, for $1 \le i \le n$. The Pareto front associated with the vector of objectives $f = (f_1, \dots, f_n)$ is $\Pi(S)$, where;

$$S = \{(f_1(y), \dots, f_n(y)) : y \in \mathbb{R}^k\}$$

Thus, the Pareto frontier is formed by the points $X = (f_1(x), \dots, f_n(x))$ such that *there is not any* $y \in \mathbb{R}^k$ such that $Y = (f_1(y), \dots, f_n(y))$ verifies $Y \prec X$.

Let us apply these ideas to the situation where two objectives are under consideration ($n = 2$) and the Pareto front is defined by a function $\varphi: \mathbb{R}^2 \rightarrow \mathbb{R}$ that splits the plane in two disjoint regions, such that:

$$S = \{P \in \mathbb{R}^2 : \varphi(P) \geq 0\}; R = \{x \in \mathbb{R}^2 : \varphi(P) < 0\}.$$

The common boundary separating the two regions is:

$$\Sigma = \{P \in \mathbb{R}^2 : \varphi(P) = 0\}.$$

Let us consider two points $A \neq B$ such that $\varphi(A) = \varphi(B) = 0$ (thus$\{A, B\} \subset \Sigma$). We consider the set:

$$\wp = \{P: (0,1) \rightarrow \mathbb{R}^2 : P \text{ } continuous;\ P(0) = A; P(1) = B; P(t) \in S, \forall t \in (0,1)\}$$

Let us introduce a functional $J: \wp \rightarrow \mathbb{R}$ such that:

$$P, Q \in \wp; P \leq Q \text{ } on \text{ } (0,1) \text{ } and \text{ } P < Q \text{ } on \text{ } I \subset (0,1), meas(I) > 0 \Rightarrow J(P) < J(Q).$$

Then

THEOREM 7.8.– Let $P^* = \arg\min\{J(P): P \in \wp\}$. Then $P^*(t) \in \Sigma$, for $t \in (0,1)$. ■

PROOF.– Since $P^* \in \wp$, we have $\varphi\big(P^*(t)\big) \geq 0$ on $(0,1)$. Assume that there exists a point $\bar{t} \in (0,1)$ such that $\varphi\big(P^*(\bar{t})\big) > 0$. We have $0 < t < 1$, since $P^*(0) = A \in \Sigma$ and $P^*(1) = B \in \Sigma$. Continuity of φ shows that there exists $\delta > 0$ such that $\varphi\big(P^*(t)\big) > 0$ for $t \in (\bar{t} - \delta, \bar{t} + \delta)$. Let us consider the function:

$$\theta(\alpha, t) = \varphi\left(P^*(t) - \alpha\binom{0}{1}\right) = \varphi(P_1^*(t), P_2^*(t) - \alpha\ ).$$

We have:

$$\theta(0, t) = \varphi\big(P^*(t)\big) > 0.$$

Let

$$\alpha(t) = \min\{1, \max\{\alpha > 0 : \theta(\alpha, t) > 0\}\}.$$

We have $\alpha(t) > 0, \forall t \in (\bar{t} - \delta, \bar{t} + \delta)$: indeed, if $\alpha(t) = 0$, then there exists a sequence $\{\alpha_n : n \in \mathbb{N}\}$ such that $\alpha_n > 0, \forall n$ and $\alpha_n \to 0$ for $n \to +\infty$, with $\theta(\alpha_n, t) \leq 0$, i.e.

$$\varphi\left(P^*(t) - \alpha_n \begin{pmatrix} 0 \\ 1 \end{pmatrix}\right) \leq 0.$$

Taking the limit for $n \to +\infty$, the continuity of φ shows that $\varphi(P^*(t)) \leq 0$. Thus, $0 < \varphi(P^*(t)) = 0$, which is a contradiction. Let us consider a regular function $\psi: (t - \delta, t + \delta) \to (0,1)$ such that $\psi \geq 0$, $\psi(t - \delta) = \psi(t + \delta) = 0$ and $\psi(s) = 1$ for $s \in (a, b), \bar{t} - \delta < a < b < \bar{t} + \delta$. ψ is extended by zero outside $(\bar{t} - \delta, \bar{t} + \delta)$ and we define:

$$P(t) = P^*(t) - \frac{1}{2}\psi(t)\alpha(t) \begin{pmatrix} 0 \\ 1 \end{pmatrix}$$

On the one hand,

$$P(t) \in S, \forall t \in (0,1) \ (since\ \psi(t)\alpha(t) \leq \frac{1}{2}\alpha(t) < \alpha(t)\ on\ (\bar{t} - \delta, \bar{t} + \delta))$$

So that $P \in \wp$. On the other hand,

$$P(t) \leq P^*(t), \forall t \in (0,1); P(t) < P^*(t), \forall t \in (a, b).$$

As a consequence, $J(P) < J(P^*)$. Or, $P^* = \arg\min\{J(P) : P \in \wp\}$, so that $J(P^*) \leq J(P) < J(P^*)$, which is a contradiction. ■

COROLLARY 7.2.– Let $S \subset \mathbb{R}^2$ be a non-empty closed set such that: $\exists m \in \mathbb{R}: X_1 \geq m \text{ and } X_2 \geq m, \forall X = (X_1, X_2) \in S$. Assume that S is regular in the sense that any continuous function defined on S has a continuous extension to $\mathbb{R}^2$. Let,

$$\Sigma = \{X \in S : \nexists Y \in S \text{ such that } Y < X\}.$$

Assume that Σ is a curve $P:(0,1) \to \mathbb{R}^2$ verifying $P(0) = A, P(1) = B$. Then,

$$P = \arg\min\{J(Q): Q \in \wp\}.$$ ■

PROOF.– Let

$$\varphi(P) = \max\{P_1 - Y_1, P_2 - Y_2 : Y \in S\}.$$

We have $\varphi(P) \leq \max\{P_1 - m, P_2 - m\}$. In addition,

$$P \in S, P \notin \Sigma \Rightarrow \exists \mathrm{Y} \in \mathrm{S} \text{ such that } \mathrm{Y} < P \Rightarrow \varphi(P) > 0,$$

$$P \in \Sigma \Rightarrow \nexists \mathrm{Y} \in \mathrm{S} \text{ such that } \mathrm{Y} < P \Rightarrow \varphi(P) = 0.$$

φ is continuous: indeed, if $\|P - Q\| \leq \delta$, then $-\delta \leq P_1 - Q_1 \leq \delta$ and $-\delta \leq P_2 - Q_2 \leq \delta$, so that:

$$Q_1 - Y_1 - \delta \leq P_1 - Y_1 \leq Q_1 - Y_1 + \delta;\ Q_2 - Y_2 - \delta \leq P_2 - Y_2 \leq Q_2 - Y_2 + \delta.$$

Thus,

$$\varphi(Q) - \delta \leq \varphi(P) \leq \varphi(Q) + \delta.$$

So, we have $\|P - Q\| \leq \delta \Longrightarrow |\varphi(P) - \varphi(Q)| \leq \delta$ and φ is continuous. Let us extend φ by a regular function outside S: the result yields from the theorem. ■

These results apply to the functional J defining the area between the horizontal axis and the curve.

For instance, we may use Stoke's theorem: the circulation of a field of vectors $\vec{v}$ along the closed curve C coincides with the flux of its rotational $\overrightarrow{rot}(\vec{v})$ on the surface $\mathcal{A}$ limited by the curve, i.e.,

$$\iint_{\mathcal{A}} \overrightarrow{rot}(\vec{v}).\vec{n}dS = \oint_C \vec{v}.d\vec{P}$$

Let us consider a vector field $\vec{v}$ such that $\overrightarrow{rot}(\vec{v}) = \vec{n} = \vec{e}_3$ for a curb on the plane $(0, \vec{e}_1, \vec{e}_2)$. Then:

$$\iint_{\mathcal{A}} 1 dS = \oint_C \vec{v}.d\vec{P}.$$

For instance:

$$\forall \gamma \in \mathbb{R} : \vec{v} = (-\gamma x_2, (1-\gamma)x_1, 0) \Rightarrow \overrightarrow{rot}(\vec{v}) = (0,0,1).$$

So, we may choose

$$\vec{v} = \frac{1}{2}(-x_2, x_1, 0); \vec{v} = (-x_2, 1, 0); \vec{v} = (1, x_1, 0).$$

In this case, the theorem and its corollary show that the solution is the curb $P^*(t): t \in (0,1)$ such that:

$$P^* = Arg\ Min \left\{ \oint_C v.dP : P \in K \right\}$$

$$K = \{P: [0,1] \to \mathbb{R}^2: P(0) = B, P(1) = A, P(t) \in S = Im(f) = f(\mathbb{R}^n)\}$$

$$x_A = Arg\ Min\ (f_1); x_B = Arg\ Min\ (f_2)$$

The second choice $\vec{v} = (x_2, 1, 0)$ corresponds to a curve generated by a function $P_2 = \eta(P_1)$. Then, $P_1(t) = t$ and $P_2 = \eta(t) = \eta(P_1)$. The third choice $\vec{v} = (1, -x_1, 0)$ corresponds a curve generated by a function $P_1 = \eta(P_2)$. In this case, $P_2(t) = t$ and $P_1 = \eta(t) = \eta(P_2)$.

EXAMPLE 7.8.– Let us consider $\vec{v} = \frac{1}{2}(x_2, -x_1, 0)$. On $(A, B): v.dP = (P_1 P_2' - P_2 P_1')/2$, so that:

$$\oint_C v.dP = cte + \frac{1}{2}\int_0^1 (P_1 P_2' - P_2 P_1') dt$$

We may simply look for $P^*(t): t \in (0,1)$ such that:

$$P^* = Arg\ Min \left\{ \int_0^1 (P_2 P_1' - P_1 P_2') dt : P \in K \right\}$$

An alternative consists of setting

$$P = (f_1(x(t)), f_2(x(t))).$$

Then,

$$P_1' = \nabla f_1(x)^t x', P_2' = \nabla f_2(x)^t x'$$

and

$$P_1 P_2' - P_2 P_1' = f_1(x)\nabla f_2(x)^t x' - f_2(x)\nabla f_1(x)^t x'$$

Under this formulation, we look for a curve $x^*(t): t \in (0,1)$ such that:

$$x^* = Arg\ Min \left\{\int_0^1 [x']^t T(x) : x \in C\right\}$$

$$T(x) = f_1(x)\nabla f_2(x) - f_2(x)\nabla f_1(x)$$

$$C = \{x: [0,1] \to \mathbb{R}^n : x(0) = x_B, x(1) = x_A\}$$

Euler–Lagrange's equations associated with this functional is:

$$\int_0^1 ([\delta x']^t T(x) + [x']^t [[\delta T(x)]^t \delta x]) = 0, \forall \delta x \textit{ such that } \delta x(0) = \delta x(1) = 0,$$

where

$$[\delta T(x)]^t \delta x = [(\delta x)^t \nabla f_1(x)]\nabla f_2(x) + f_1(x)[\nabla^2 f_2(x)\delta x] - [(\delta x)^t \nabla f_2(x)]\nabla f_1(x) + f_2(x)[\nabla^2 f_1(x)\delta x],$$

We notice that $\delta T(x)$is a matrix. By rewriting:

$$[x']^t [[\delta T(x)]^t \delta x] = [\delta x]^t \delta T(x) x',$$

We have:

$$\big(T(x)\big)' = \delta T(x) x'; \;\; x(0) = x_B, x(1) = x_A,$$

i.e.

$$\nabla T(x) x' = \delta T(x) x'; \;\; x(0) = x_B, x(1) = x_A.$$

■

EXAMPLE 7.9.– Let us consider $\vec{v} = (-x_2, 0, 0)$. On $(A, B): v.dP = -P_2 P_1'$, so that:

$$\oint_C v.dP = cte - \int_0^1 P_2 P_1' dt$$

Thus, we may determine $P^*(t): t \in (0,1)$ such that:

$$P^* = Arg\ Min \left\{ \int_0^1 P_2 P_1' dt : P \in K \right\}$$

We set, as in the preceding example,

$$P = (f_1(x(t)), f_2(x(t))).$$

Then,

$$P_1' = \nabla f_1(x)^t x', P_2' = \nabla f_2(x)^t x'$$

and

$$P_2 P_1' = f_2(x) \nabla f_1(x)^t x'$$

In this case, we look for a curve b$x^*(t): t \in (0,1)$ such that:

$$x^* = Arg\ Min \left\{ \int_0^1 [x']^t T(x) : x \in C \right\}$$

$$T(x) = -f_2(x) \nabla f_1(x)$$

$$C = \{x: [0,1] \to \mathbb{R}^n : x(0) = x_B, x(1) = x_A\}$$

The associated Euler–Lagrange's equation is:

$$\int_0^1 ([\delta x']^t T(x) + [x']^t [[\delta T(x)]^t \delta x]) = 0, \forall \delta x \ such\ that\ \delta x(0) = \delta x(1) = 0\,,$$

where,

$$[\delta T(x)]^t \delta x = -[(\delta x)^t \nabla f_2(x)] \nabla f_1(x) - f_2(x)[\nabla^2 f_1(x) \delta x],$$

Analogously, rewriting:

$$[x']^t[[\delta T(x)]^t\delta x] = [\delta x]^t\delta T(x)x',$$

we have:

$$\left(T(x)\right)' = \delta T(x)x';\ x(0) = x_B, x(1) = x_A,$$

i.e.

$$\nabla T(x)x' = \delta T(x)x';\ x(0) = x_B, x(1) = x_A. \qquad \blacksquare$$

EXAMPLE 7.10.– Let us consider $\vec{v} = (0, x_1, 0)$. On $(A, B)\colon v.dP = P_1P_2'$, so that:

$$\oint_C v.dP = cte - \int_0^1 P_1P_2'dt$$

Here, we look for $P^*(t)\colon t \in (0,1)$ such that:

$$P^* = Arg\ Min\ \left\{-\int_0^1 P_1P_2'dt : P \in K\right\}$$

with

$$P = (f_1(x(t)), f_2(x(t))).$$

We have:

$$P_1' = \nabla f_1(x)^t x', P_2' = \nabla f_2(x)^t x'$$

and

$$P_1P_2' = f_1(x)\nabla f_2(x)^t x'$$

In this case, we must determine a curve $x^*(t)\colon t \in (0,1)$ such that:

$$x^* = Arg\ Min\ \left\{\int_0^1 [x']^t T(x) : x \in C\right\}$$

$$T(x) = f_1(x)\nabla f_2(x)$$

$$C = \{x\colon [0,1] \to \mathbb{R}^n\colon x(0) = x_B, x(1) = x_A\}$$

The Euler–Lagrange's equation becomes:

$$\int_0^1([\delta x']^t T(x) + [x']^t[[\delta T(x)]^t \delta x]) = 0, \forall \delta x \textit{ such that } \delta x(0) = \delta x(1) = 0,$$

where,

$$[\delta T(x)]^t \delta x = [(\delta x)^t \nabla f_1(x)]\nabla f_2(x) + f_1(x)[\nabla^2 f_2(x)\delta x]\,.$$

Notice that $\delta T(x)$ is a matrix. By rewriting:

$$[x']^t[[\delta T(x)]^t \delta x] = [\delta x]^t \delta T(x) x',$$

we have:

$$\big(T(x)\big)' = \delta T(x)\, x'; \;\; x(0) = x_B, x(1) = x_A,$$

i.e.

$$\nabla T(x) x' = \delta T(x) x'; \;\; x(0) = x_B, x(1) = x_A.$$

■

7.13. Matlab® implementation of the variational approach for biobjective optimization

Let us assume that the multiobjective optimization problem involves two functions $f_1, f_2: \mathbb{R}^n \to \mathbb{R}$. Let:

$$x_i^* = \arg\min\{f_i(x): x \in \mathbb{R}^n\}$$

In regular situations, the Pareto frontier is a curb connecting the points $A = (f_1(x_1^*), f_2(x_1^*))$, $B = (f_1(x_2^*), f_2(x_2^*))$. Then, the Pareto front is given by $(f_1(x(t)), f_2(x(t))): t \in (0,1)$.

Let us introduce $x^0(t) = x_1^* + t(x_2^* - x_1^*)$ and $\chi(t) = x(t) - x^0(t)$. χ verifies $\chi(0) = \chi(1) = 0$. Assuming differentiability, it becomes natural to look for $\chi \in H_0^1(\Omega)$, $\Omega = (0,1)$. Thus, we may consider a family $\mathcal{F} = \{\varphi_i: 1 \le i \le k\}$ and $\chi \approx \sum_{i=1}^k \chi_i \varphi_i$. Then, we look for the coefficients $(\chi_1, \dots, \chi_k)$ such that the area limited by the curve $P(t) = \left(f\_1(x^0\text{ (t) } + \chi^t\Phi),\ f_2(x^0\text{ (t) } + \chi^t\Phi)\right)$ is minimal.

We represent the problem in a Matlab® class `biobjproblem.`, having properties as `f1, f2, dim.`:dim is an integer giving the dimension of the arguments of the objectives. `f1, f2` are structures defining the objectives. Their properties are `function` and `solution`. The last property contains x_i^*. For instance, `biobjproblem` may be defined as follows:

```
classdef biobjproblem
    properties
        f1
        f2
        dim
    end
    methods (Access = Public)
        function obj = biobjprob(f1,f2,dim)
            obj.f1 = f1;
            obj.f2 = f2;
            obj.dim = dim;
        end
    end
end
```

Let us consider a simple situation where $n = 1, f_1(x) = x^2\ , f_2(x) = (x-1)^2$. We define:

```
f1.function = @(x) x^2;
f2.solution = 0;
f1.function = @(x) (x-1)^2;
f2.solution = 1;
dim = 1;
pb = biobjproblem(f1,f2,dim)
```

These commands create pb, containing the definitiion of the multiobjective problem. We define x^0 as follows:

```
sol1 = pb.f1.solution;
sol2 = pb.f2.solution;
xzero = @(t) sol1 + t*(sol2 - sol1);
```

Let be given a vector χ and the family f as a `basis`. We generate a curb `c` as follows (`tspan` contains a partition of (0,1): for instance, `tspan(i)` $=\frac{i-1}{N}, 1 \leq i \leq N$)

```
functionp =
pareto_curb(chi,basis,xzero,problem,tspan)
c = @(t) basis.projection(chi,t) + xzero(t);
f1 = problem.f1.function;
f2 = problem.f2.function;
xa = zeros(problem.dim,length(tspan));
p1 = zeros(size(tspan));
p2 = zeros(size(tspan));
for i = 1: length(tspan);
xa(:,i) = c(tspan(i));
    p1(i) = f1(xa(:,i));
    p2(i) = f1(xa(:,i));
end;
p.x = p1;
p.y = p2;
p.curb = @(t) interp1(tspan,p.points,t);
return;
end
```

Assume that the area limited by the curve `p.curb` is evaluated by a subprogram `surface(p)`. Then, our objective becomes to determine the coefficients chi that minimize

```
fobj = @(chi) pareto_surface(chi,basis,xzero,problem,tspan)
```

where,

```
function v = pareto_surface(chi,basis,xzero,problem,tspan)
p = pareto_curb(chi,basis,xzero,problem,tspan);
v = surface(p);
return;
end
```

In the examples below, `surface(p)` evaluates the surface by using a trapezoidal rule involving `p.x`. as horizontal data and `p.y` as vertical data. The Pareto front obtained in the simple situation is shown in Figure 7.15.

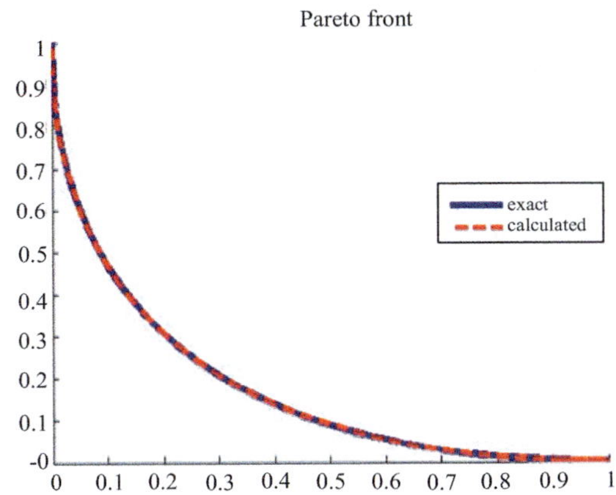

Figure 7.15. *Variational approach for a simple biobjective optimization problem*

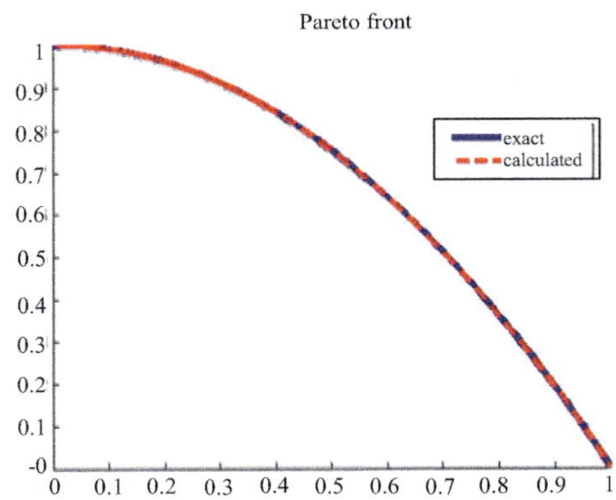

Figure 7.16. *Variational approach for a biobjective optimization problem*

A more complex situation is shown in Figure 7.16, where are exhibited the results obtained for $f_1(x) = |x_1|$, $f_2(x) = (1 + 9|x_2|)\left|1 - \left(\frac{f_1(x)}{1+9|x_2|}\right)\right|^2$.

7.14. Exercises

EXERCISE 7.1.– Let $\Omega \subset R^2$ be a regular bounded domain,

$$V = \{v = (v_1, v_2) \in [H^1(\Omega)]^2 \mid v = 0 \; on \; \partial\Omega\};$$

with the scalar product

$$(u, v)_V = \lambda(u, v)_{1,0} + \mu\big(div(u), div(v)\big)_0 .$$

We assume that $\lambda > 0$, $\lambda + 2\mu \geq 0$. Let $f = (f_1, f_2) \in [L^2(\Omega)]^2$, $F(v) = (f, v)_0$ and $J: V \to R$ be given by

$$J(v) = \frac{\lambda}{2}(v, v)_{1,0} + \frac{\mu}{2}\big(div(v), div(\mathrm{v})\big)_0 - F(v).$$

1) Show that

$$J(u + \alpha v) = J(u) + \alpha((u, v)_V - F(v)) + \alpha^2 J(v).$$

2) Determine $DJ(u)$.

3) Let $u^* = \underset{V}{ArgMin}\, J$. Show that $u^* = u$, where u is the solution of:

$$\lambda\Delta u + \mu\nabla\big(\mathrm{div}(v)\big) + f = 0 \quad \text{on } \Omega; \;\; u = 0 \text{ on } \partial\Omega$$

4) Show that J is strictly convex.

5) Conclude the uniqueness of u^*. ■

EXERCISE 7.2.– Let $S = \{v: (0,1) \to \mathbb{R}: v(0) = 1, v(1) = 0\}$. Let $J: S \to \mathbb{R}$ be a functional defined by $J(v) = \int_0^1 \big(t + v(t)\big)^2 dt$. Let $u = \underset{S}{\operatorname{argmin}}\, J$.

1) Verify that S is an affine subspace.

2) Determine $DJ(u)(v)$.

3) Characterize u as the solution of a variational equation.

4) Determine the differential equation and the boundary conditions satisfied by u.

5) Show that u does not exist. ■

EXERCISE 7.3.– Let $S = \{v: (0,1) \to \mathbb{R}: v(0) = 1, v(1) = 0\}$. Let $J: S \to \mathbb{R}$ be a functional defined by $J(v) = \int_0^1 \left[\frac{\varepsilon}{2}(v'(t))^2 + \left(t + v(t)\right)^2\right] dt$. Let $u_\varepsilon = \underset{S}{\operatorname{argmin}} J$.

1) Determine $DJ(u_\varepsilon)(v)$.

2) Characterize u_ε as the solution of a variational equation.

3) Determine the differential equation and the boundary conditions satisfied by u_ε.

4) Determine u_ε.

5) Analyze the limit of u_ε for $\varepsilon \to 0+$. ■

EXERCISE 7.4.– Let $S = \{v: (0,1) \to \mathbb{R}: v(0) = 1, v(1) = \mathrm{a}\}$. Let $J: S \to \mathbb{R}$ be a functional defined by $J(v) = \int_0^1 \left[\frac{1}{2}(v'(t))^2 + \frac{1}{4}\left(v(t)\right)^4 - f(t)v(t)\right] dt$. Let $u = \underset{S}{\operatorname{argmin}} J$.

1) Verify that S is an affine subspace.

2) Determine $DJ(u)(v)$.

3) Characterize u as the solution of a variational equation.

4) Determine the differential equation and the boundary conditions satisfied by u. ■

EXERCISE 7.5.– Let $S = \{v: (0,1) \to \mathbb{R}: v(0) = 1, v(1) = \mathrm{a}\}$. Let $J: S \to \mathbb{R}$ be a functional defined by $J(v) = \int_0^1 \left[t^2(v'(t))^2 + \frac{1}{5}\left(v(t)\right)^5 - f(t)v(t)\right] dt$. Let $u = \underset{S}{\operatorname{argmin}} J$.

1) Verify that S is an affine subspace.

2) Determine $DJ(u)(v)$.

3) Characterize u as the solution of a variational equation.

4) Determine the differential equation and the boundary conditions satisfied by u. ■

EXERCISE 7.6.– Find the curve defined by a function $u: (0,1) \to \mathbb{R}$ of fixed length $\ell > 0$, such that $u(0) = u(1) = 0$, which maximizes the area limited by the curb and the horizontal axis. For instance, set $\Omega = (0,1)$, $V = \{v: \Omega \to \mathbb{R}: v(0) = v(1) = 0\}$ and

$$I(v) =_0^1 \sqrt{1 + [v'(x)]^2} dx - \ell, J(v) = -_0^1 v(x) dx.$$

Then,

$$u = \operatorname*{argmin}_C J \ , \ \ C = \{v \in V: I(v) = 0\}.$$

1) Introduce the Lagrange multiplier η associated with $C$and construct the Lagrangian $L(u, \eta)$associated with this problem;

2) Determine the Gâteaux derivative $D_u L$;

3) Determine the differential equation and the boundary conditions satisfied by u;

4) Show that u defines an arc of circle. ■

EXERCISE 7.7.– Find the curve defined by a function $u: (0,1) \to \mathbb{R}$ connecting $(0,0)$ and $(1,0)$, such that it has the minimal length and respects an obstacle defined by a curb $\psi: (0,1) \to \mathbb{R}$ such that $\psi(0) \le$ and $\psi(1) \le 0$. (Set $V = \{v: (0,1) \to \mathbb{R}: v(0) = v(1) = 0\}, C = \{v \in V: v \ge \psi \text{ on } (0,1)\}$. Then, minimize $\int_0^1 \sqrt{1 + [u'(x)]^2} dx$ on C. ■

EXERCISE 7.8.– The motion of a point follows the equations:

$$\dot{x} = -u \ , \ \ x(0) = x_0$$

Find the control u satisfying $|u| \le 1$ on $(0, T)$ and $x(T) = x_1$ for which T is minimal. ■

EXERCISE 7.9.– Let us consider the second-order system:

$$\dot{x} = u \ , \ \dot{y} = u^2 \ ; \quad x(0) = y(0) = 0.$$

Find u such that $0 \leq u \leq 1$ on $(0, T)$, $x(T) = x_1$, $y(T) = y_1$ and $\gamma T^2 + \int_0^T f(t)u(t)dt$ is minimum ($\gamma > 0$, f known). ■

Bibliography

[ANG 07] ANG W.T., *A Beginner's Course in Boundary Element Methods*, Universal Publishers, Boca Raton, 2007.

[ARB 00] ARBOGAST L.F.A., *Du calcul des derivations*, Levrault Frères, Strasbourg, 1800.

[ARG 54] ARGYRIS J.H., "Energy theorems and structural analysis: a generalized discourse with applications on energy principles of structural analysis including the effects of temperature and non-linear stress-strain relations", *Aircraft Engineering and Aerospace Technology*, vol. 26, no. 10, pp. 347–356, 1954.

[ARG 55a] ARGYRIS J.H., "Energy theorems and structural analysis: a generalized discourse with applications on energy principles of structural analysis including the effects of temperature and non-linear stress-strain relations part I. General theory", *Aircraft Engineering and Aerospace Technology*, vol. 27, no. 10, pp. 42–48, 1955.

[ARG 55b] ARGYRIS J.H., "Energy theorems and structural analysis: a generalized discourse with applications on energy principles of structural analysis including the effects of temperature and non-linear stress-strain relations Part II. Applications to thermal stress problems and St. Venant torsion", *Aircraft Engineering and Aerospace Technology*, vol. 26, no. 12, pp. 410–422, 1955.

[ARI 38] ARISTOTLE, *Minor Works*, Loeb Classical Library 307, vol. 14, 1938.

[BAL 03] BALAKRISHNAN V., "All about the Dirac delta function", *Resonance*, vol. 8, no. 8, pp 48–58, 2003.

[BAN 22a] BANACH S., "Sur les opérations dans les ensembles abstraits et leur application aux équations intégrales", *Fundamenta Mathematicae*, vol. 3, no. 1, pp. 133–181, 1922.

[BAN 22b] BANACH S., *Théorie des opérations linéaires*, Monografie Matematyczne 1, Warszawa, 1922.

[BAN 24] BANACH S., TARSKI A., "Sur la décomposition des ensembles de points en parties respectivement congruentes", *Fundamenta Mathematicae*, vol. 6, no. 1, pp. 244–277, 1924.

[BAU 09] BAULDRY W.C., *Introduction to Real Analysis: An Educational Approach*, John Wiley & Sons, New York, 2009.

[BER 96] BERNOULLI J., "Problema novum mathematicis propositum", *Acta Eruditorum Lipsiae*, pp. 264–269, June 1696.

[BER 53a] BERNOULLI D., "Sur le mélange de plusieurs espèces de vibrations simples isochrones, qui peuvent coexister dans un même système de corps", *Histoire de l'Académie Royale des Sciences et des Belles-Lettres de Berlin*, no. 8, pp. 173–195, 1753.

[BER 53b] BERNOULLI D., "Réflexions et éclaircissemens sur les nouvelles vibrations des cordes exposées dans les Mémoires de l'Académie de 1747 & 1748", *Histoire de l'Académie Royale des Sciences et des Belles-Lettres de Berlin*, no. 9, pp. 147–172, 1753.

[BLO 11] BLOCH E.D., *The Real Numbers and Real Analysis*, Springer, 2011.

[BOO 59] BOOLE G., *A Treatise on Differential Equations*, MacMillan and Co., Cambridge, 1859.

[BRO 11] BROWER L.E.J., "Über Abbildung von Mannigfaltigkeiten", *Mathematische Annalen*, vol. 71, no. 1, pp. 97–115, 1911.

[BRO 16] BROWER L.E.J., "Normal coordinates in dynamical systems", *Proceedings of the London Mathematical Society*, Series 2, vol. 15, pp. 401–448, 1916.

[BUB 13] BUBNOV I.G., "Report on the works of Professor Timoshenko which were awarded the Zhuranskyi Prize", *Symposium of the Institute of Communication Engineers*, vol. 81, 1913.

[BUB 14] BUBNOV, I. G., *Structural Machanics of Shipbuilding*, Tipografiya Mor. Min., St Petersburg. Pt I, Sections 1-14, 1912; Pt 15-25, 1914.

[BUE 05] BUENO O., "Dirac and the dispensability of mathematics", *Studies in History and Philosophy of Modern Physics*, vol. 36, pp. 465–490, 2005.

[CAR 55] CARMICHAEL R., *A Treatise on the Calculus of Operation: Designed to Facilitate the Processes of the Differential and Integral Calculus and the Calculus of Finite Differences*, Longman, Brown, Green and Longmans, London, 1855.

[CAR 26] CARSON J.R., "The Heaviside operational calculus", *Bulletin of the American Mathematical Society*, vol. 32, no. 1, pp. 43–68, 1926.

[CAU 23] CAUCHY A.L., *Résumé des leçons sur le calcul infinitésimal*, Debure Frères, Libraires du Roi et de la bibliothèque du roi, 1823.

[CAU 27a] CAUCHY A.L., "Théorie de la propagation des ondes à la surface d'un fluide pesant d'une profondeur indéfinie", *Mémoires présentés par divers savants à l'Académie royale des sciences de l'Institut de France et imprimés par son ordre. Sciences mathématiques et physiques*, vol. 1, 1827.

[CAU 27b] CAUCHY A.L., "Mémoire sur les développements des fonctions en séries périodiques", *Mémoires de l'Académie des sciences*, vol. 6, p. 603, 1827.

[CAU 29] CAUCHY A.L., *Leçons sur le calcul différentiel*, Debure Frères, Libraires du Roi et de la bibliothèque du roi, 1829.

[CHL 87] CHLADNI E.F.F., *Entdeckungen über die Theorie des Klange*, Bey Weidmanns erben und Reich, 1787.

[CLA 42] CLAIRAUT A.C., "Sur l'intégration ou la construction des équations différentielles du premier ordre", *Mémoires de l'Académie Royale des Sciences*, pp. 293–323, 1742.

[COL 84] COLOMBEAU J.F., *New Generalized Functions and Multiplication of the Distributions*, North Holland, Amsterdam, 1984.

[COU 43] COURANT R., "Variational methods for the solution of problems of equilibrium and vibrations", *Bulletin of the American Mathematical Society*, vol. 49, pp. 1–23, 1943.

[DAL 43] D'ALEMBERT J.L.R, *Traité de dynamique*, Editions David, Paris, 1743.

[DAL 47] D'ALEMBERT J.L.R.., "Recherches sur la courbe que forme une corde tendue mise en vibration", *Histoire de l'Académie Royale des Sciences et des Belles-Lettres de Berlin*, no. 1, pp. 214–249, 1747.

[DAL 50] D'ALEMBERT J.L.R., "Addition au mémoire sur la courbe que forme une corde tenduë mise en vibration", *Histoire de l'Académie Royale des Sciences et des Belles-Lettres de Berlin*, no. 1, pp. 355–360, 1750.

[DEA 81] DEAKIN M.A.B., "The development of the Laplace transform, 1737–1937: I. Euler to Spitzer, 1737–1880", *Archive for History of Exact Sciences*, vol. 25, no. 4, pp. 343–390, 1981.

[DEA 82] DEAKIN M.A.B., "The development of the Laplace transform, 1737–1937: II. Poincaré to Doetsch, 1880–1937", *Archive for History of Exact Sciences*, vol. 26, no. 4, pp. 351–381, 1982.

[DEA 92] DEAKIN M.A.B., "The ascendancy of the Laplace transform and how it came about", *Archive for History of Exact Sciences*, vol. 44, no. 34, pp. 265–286, 1992.

[DEF 38] DE FERMAT P., *Methodus ad disquirendam maximam et minima*, letter to P. Mersenne, 1638.

[DEF 79] DE FERMAT P., *De tangentibus linearum curvarum*, Œuvres, Tome premier, Œuvres mathématiques diverses, Gauthier-Villars, Paris pp. 134–135, 1679.

[DEM 42] DE MAUPERTUIS P.L.M., "Loi du repos des corps", *Histoire de l'Académie Royale des Sciences et des Belles-Lettres*, pp. 170–176, 1742.

[DEM 44] DE MAUPERTUIS P.L.M., "Accord de différentes Loix de la Nature qui avoient jusqu'ici paru incompatibles", *Histoire de l'Académie Royale des Sciences et des Belles-Lettres*, pp. 417–426, 1744.

[DEM 48] DE MAUPERTUIS P.L.M., "Sur le mouvement d'un liquide visqueux emplissant l'espace", *Histoire de l'Académie Royale des Sciences et des Belles-Lettres*, pp. 267–294, 1748.

[DER 34] DE ROBERVAL, G., *Traité des Indivisibles*, 1634, postume publication in Ouvrages de Mathematique et de Physique, par Messieurs de l'Academie Royale des Sciences, Paris, 1634.

[DER 93] DE ROBERVAL, G., "Observations sur la composition des mouvements et sur le moyen de trouver les touchantes aux lignes courbes", in *Ouvrages de Mathematique et de Physique, par Messieurs de l'Academie Royale des Sciences*, Paris, pp. 69–11, 1693.

[DER 30] DE ROBERVAL G., "Observations sur la composition des mouvements et sur le moyen de trouver les touchantes des lignes courbes", *Mémoire de l'Académie Royale des Sciences*, vol. 6, pp. 1666–1699, 1730.

[DIR 29a] DIRICHLET P.G.L., "Sur la convergence des séries trigonométriques qui servent à représenter une fonction arbitraire entre des limites données", Crelle, *Journal für die reine und angewandte Mathematik*, vol. 4, pp. 157–169, 1829.

[DIR 29b] DIRICHLET P.G.L., "Solution d'une question relative à la théorie mathématique de la chaleur", Crelle, *Journal für die reine und angewandte Mathematik*, vol. 5, pp. 287–295, 1829.

[DIR 37] DIRICHLET P.G.L., "Über die darstellung ganz willkürlicher functionen durch sinusund cosinusreihen", *Repertorium der Physik*, vol. 1, pp. 146–174, 1837.

[DIR 58] DIRAC P., *The Principles of Quantum Mechanics*, Oxford University Press, 1958.

[DUB 73] DU BOIS-REYMOND P.D., "Über die Fourier'schen Reihen", *Nachrichten von der Gesellschaft der Wissenschaften zu Göttingen*, vol. 1873, pp. 571–584, 1873.

[DUR 12] DUREN P.L., *Invitation to classical analysis*, Pure and Applied Undergraduate Texts, American Mathematical Society, 2012.

[EAR 70] EARLE C.J., HAMILTON R.S., "A fixed point theorem for holomorphic mappings", *Proceedings of Symposia in Pure Mathematics, American Mathematical Society*, vol. 16, pp. 61–65, 1970.

[EDG 81] EDGEWORTH F.Y., *Mathematical Psychics*, C. Keagan Paul & Co., London, England, 1881.

[ERL 88] ERLICHSON H., "Galileo's work on swiftest descent from a circle and how he almost proved the circle itself was the minimum time path", *American Mathematical Monthly*, vol. 105, no. 4, pp. 338–347, 1988.

[EUL 32] EULER L., "Nova methodus innumerabiles aequationes differentiales secundi gradus reducendi ad aequationes differentiales primi gradus", *Commentarii academiae scientiarum Petropolitanae*, vol. 3, pp. 124–137, 1732.

[EUL 44] EULER L., "De constructione aequationum", *Commentarii academiae scientiarum Petropolitanae*, vol. 9, pp. 85–97, 1744.

[EUL 50a] EULER L., "Methodus aequationes differentiales altiorum graduum integrandi ulterius promota", *Commentarii academiae scientiarum Petropolitanae*, vol. 3, pp. 3–35, 1750.

[EUL 50b] EULER L., "Découverte d'un nouveau principe de la Mécanique", *Histoire de l'Académie Royale des Sciences et des Belles-Lettres de Berlin*, no. 6, pp. 185–217, 1750.

[EUL 53] EULER L., "Remarques sur les mémoires précédents de M. Bernoulli", *Histoire de l'Académie Royale des Sciences et des Belles-Lettres de Berlin*, no. 9, pp. 196–222, 1753.

[EUL 68] EULER L., *Institutionum calculi integralis volumen primum in quo methodus integrandi a primis principiis usque ad integrationem aequationum differentialium primi gradus pertractatur*, Petropoli impensis academiae imperialis scientiarum, 1768.

[EVA 10] EVANS, L.C., *Partial Differential Equations*, Graduate Studies in Mathematics, American Mathematical Society, vol. 19, Providence, RI, USA, 2010.

[FAE 49] FAEDO S., Un nuovo metodo de l'analisi esistenziale e quantitative dei problemi di propagazione'', *Annali della Scuola Normale di Pisa – Classe di Scienze*, vol. 1, no. 3, pp. 1–40, 1949.

[FAI 98] FAIRBANKS A., *The First Philosophers of Greece*, Kegan Paul, Trench, Trubner & Co. Ltd., 1898.

[FAN 52] FAN K., "Fixed-point and minimax theorems in locally convex topological linear spaces", *Proceedings of the National Academy of Sciences of U.S.A*, vol. 38, no. 2, pp. 121–126, 1952.

[FEY 85] FEYNMAN R., *The Character of Physical Law*, 12th edition, MIT Press, 1985.

[FOU 22] FOURIER J., *Théorie Analytique de la Chaleur*, Firmin-Diderot, Paris, 1822.

[FRE 04] FRECHET R.M., "Fonctions de lignes fermées", *Annales Scientifiques de l'Ecole Normale Supérieure*, series 3, vol. 21 pp. 557–571, 1904.

[FRE 06] FRECHET R.M., "Sur quelques points du calcul fonctionnel (PhD Thesis, Faculté des Sciences de Paris, 1905)", *Rendiconti del Circolo Matematico di Palermo*, vol. 22, no. 1, pp 1–72, 1906.

Printed and bound by CPI Group (UK) Ltd, Croydon, CR0 4YY

13/09/2026

14961022-0001

[FRE 07a] FRECHET R.M., "Sur les ensembles de fonctions et les opérations linéaires", *Comptes Rendus de l'Academie des Sciences de Paris*, vol. 144, pp. 1414–1416, 1907.

[FRE 07b] FRECHET R.M., "Généralisation d'un théorème de Weierstrass", *Comptes rendus de l'Académie des Sciences de Paris*, vol. 139, pp. 848–849, 1907.

[FRE 11] FRECHET R.M., "Sur la notion de différentielle", *Comptes rendus de l'Académie des Sciences de Paris*, vol. 152, pp. 845–1051, 1911.

[FRE 12] FRECHET R.M., "Sur la notion de différentielle dans le calcul fonctionnel", *Congrès Soc. Sav. Paris*, p. 45, 1912.

[FRE 14] FRECHET R.M., "Sur la notion de différentielle d'une fonction de ligne", *Transactions of the American Mathématical Society*, vol. 15, pp. 135–161, 1914.

[FRI 53] FRIEDRICHS K.O., "On the differentiability of the solutions of linear elliptic differential equations", *Communications on Pure and Applied Mathematics*, vol. 6, no. 3, pp. 299–326, 1953.

[FRI 44] FRIEDRICHS K.O., "The identity of weak and strong extensions of differential operators", *Transactions of the American Mathematical Society*, vol. 55, no. 1, pp. 132–151, 1944.

[GAL 38] GALILEO G., *Discorsi e Dimostrazioni Matematiche Intorno a Due Nuove Scienze*, Lodewijk Elzevir, Leiden, Netherlands, 1638.

[GAL15] GALERKIN B.G., "Rods and plates. Series occurring in various questions concerning the elastic equilibrium of rods and plates", *VestnikInzhenerovi Tekhnikov*, no. 19, pp. 897–908, 1915.

[GAN 13] GANDER M.J., WANNER G., "From Euler, Ritz, and Galerkin to modern computing", *Siam Review*, vol. 54, no. 4, pp. 627–666, 2013.

[GAT 13a] GÂTEAUX R.E., "Sur les fonctionnelles continues et les fonctionnelles analytiques", *Comptes Rendus de l'Academie des Sciences de Paris*, vol. 157, pp. 325–327, 1913.

[GAT 13b] GÂTEAUX R. E., "Sur la représentation des fonctionnelles continues", *Atti della Reale Accademia dei Lincei. Rendiconti della Classe di scienze fisiche, matematiche e naturali*, vol. 22, no. 2, pp. 646–648, 1913.

[GAT 14a] GÂTEAUX R. E., "Sur la représentation des fonctionnelles continues", *Atti della Reale Accademia dei Lincei. Rendiconti della Classe di scienze fisiche, matematiche e naturali*, vol. 23, no. 1, pp. 310–408, 1914.

[GAT 14b] GÂTEAUX R.E., "Représentation d'une fonctionnelle continue, satisfaisant à la condition du cycle fermé", *Atti della Reale Accademia dei Lincei. Rendiconti della Classe di scienze fisiche, matematiche e naturali*, vol. 23, no. 1, pp. 481–486, 1914.

[GAT 19a] GÂTEAUX R.E., "Sur la notion d'intégrale dans le domaine fonctionnel et sur la théorie du potentiel", *Bulletin de la Société Mathématique de France*, vol. 47, pp. 47–70, 1919.

[GAT 19b] GÂTEAUX R.E., "Fonctions d'une infinité de variables indépendantes", *Bulletin de la Société Mathématique de France*, vol. 47, pp. 70–96, 1919.

[GAT 22] GÂTEAUX R.E., "Sur diverses questions de calcul fonctionnel", *Bulletin de la Société Mathématique de France*, vol. 50, pp. 1–37, 1922.

[GEL 63] GELFAND I., SHILOV, G., *Generalized Functions*, *Theory of Differential Equations*, Academic Press, Boston, MA, vol. 3, 1963.

[GEL 64a] GELFAND I., SHILOV G., *Generalized functions*, *Properties and Operations*, Academic Press, Boston, MA, vol. 1, 1964.

[GEL 64b] GELFAND I., VILENKIN, N., *Generalized Functions*, *Applications of harmonic Analysis*, Academic Press, Boston, MA, vol. 4, 1964.

[GEL 66] GELFAND I., GRAEV, I., VILENKIN N., *Generalized Functions, Integral Geometry and Representation Theory*, Academic Press, Boston, MA, vol. 5, 1966.

[GEL 68] GELFAND I., SHILOV G., *Generalized Functions*, *Spaces of Fundamental and Generalized Functions*, Academic Press, Boston, MA, vol. 2, 1968.

[GOD 31] GODEL K., "Uber formal unentscheidbare Sätze der Principia mathematica und verwandter Systeme I", *Monatshefte für Mathematik und Physik"*, vol. 38, no. 1, pp. 173–198, 1931.

[GRA 83] GRAM J.P., "Ueber die Entwickelung reeller Funtionen in Reihen mittelst der Methode der kleinsten Quadrate", *Journal für die reine und angewandte Mathematik*, vol. 94 pp. 41–73, 1883.

[GRE 28] GREEN G., An essay on the application of mathematical analysis to the theories of electricity and magnetism, printed for the author by T. Wheelhouse, Nottingham, 1828.

[GRE 89] GREDIAC M., "Principe des Travaux Virtuels et Identification", *Comptes Rendus de l'Académie des Sciences*, vol. 302, pp. 1–5, 1989.

[HAD 02] HADAMARD J., "Sur les dérivées des fonctions de lignes", *Bulletin de la Société Mathématique de France*, vol. 30, pp. 40–43, 1902.

[HAD 03a] HADAMARD J., "Sur les opérations fonctionnelles", *Comptes Rendus de l'Académie des Sciences*, vol. 136, pp. 351–354, 1903.

[HAD 03b] HADAMARD J., *Leçons sur la propagation d'ondes et les équations de l'hydrodynamique*, Cours au Collège de France, Hermann, Paris, 1903.

[HAD 06] HADAMARD J., "Sur le principe de Dirichlet", *Bulletin de la Société Mathématique de France*, vol. 34, pp. 135–138, 1906.

[HAD 10] HADAMARD J., *Leçons sur le calcul des variations. Recueillies par Maurice Fréchet. Tome premier : Lavariation première et les conditions du premier ordre; les conditions de l'extremum libre*, Hermann, Paris, 1910.

[HAU 14] HAUSDORFF F., "Bemerkung über den Inhalt von Punktmengen", *Mathematische Annalen*, vol. 75, no. 3, pp. 428–433, 1914.

[HEA 73a] HEAVISIDE O., "On the best arrangement of Wheatstone's bridge for measuring a given resistance with a given galvanometer and battery", *Philosophical Magazine*, vol. 45, no. 298, pp. 3–8, 1873.

[HEA 73b] HEAVISIDE O., "On an advantageous method of using the differential galvanometer for measuring small resistances", *Philosophical Magazine*, vol. 45, no. 300, pp. 245–248, 1873.

[HEA 73c] HEAVISIDE O., "On duplex telegraphy", *Philosophical Magazine*, vol. 45, no. 302, pp. 426–432 1873.

[HEA 73d] HEAVISIDE O., "On the differential galvanometer", *Philosophical Magazine*, vol. 46, no. 308, pp. 469–472, 1873.

[HEA 74a] HEAVISIDE O., "On Wheatstone's bridge", *Philosophical Magazine*, vol. 47, no. 310, pp. 93–94, 1874.

[HEA 74b] HEAVISIDE O., "On telegraphic signalling with condensers", *Philosophical Magazine*, vol. 47, no. 314, pp. 426–434, 1874.

[HEA 93] HEAVISIDE O., "On operators in mathematical physics I", *Proceedings of the Royal Society London*, vol. 52, pp. 504–529, 1893.

[HEA 94] HEAVISIDE O., "On operators in mathematical physics II", *Proceedings of the Royal Society London*, vol. 54, pp. 105–143, 1894.

[HEL 97] HELMHOLTZ H., *Vorlesungen Über die Elektromagnetische Theorie des Lichts*, L. Voss, Hamburg, 1897.

[HER 16] HERMANN J., *Phoronomia, sive De viribus et motibus corporum solidorum et fluidorum*, R. & G. Wetstenios, Amsterdam, 1716.

[HIC 34a] HICKS J.R., ALLEN R.G.D., "A reconsideration of the theory of value. Part I", *Economica*, vol. 1, no. 1, pp. 52–76, 1934.

[HIC 34b] HICKS J.R., ALLEN R.G.D., "A reconsideration of the theory of value. Part II. A mathematical theory of individual demand functions", *Economica*, vol. 1, no. 2, pp. 196–219, 1934.

[HIL 04a] HILBERT D., *Grundzuge einer allgemeinen Theorie der linearen Integralgleichungen*, Part 1, Göttinger Nachrichten, pp. 49–91, 1904.

[HIL 04b] HILBERT D., *Grundzuge einer allgemeinen Theorie der linearen Integralgleichungen*, Part 2, Göttinger Nachrichten, pp. 213–259, 1904.

[HIL 05] HILBERT D., *Grundzuge einer allgemeinen Theorie der linearen Integralgleichungen*, Part 3, Göttinger Nachrichten, pp. 397–338, 1905.

[HIL 06a] HILBERT D., *Grundzuge einer allgemeinen Theorie der linearen Integralgleichungen*, Part 4, Göttinger Nachrichten, pp. 157–227, 1906.

[HIL 06b] HILBERT D., *Grundzuge einer allgemeinen Theorie der linearen Integralgleichungen*, Part 5, Göttinger Nachrichten, pp. 439–480, 1906.

[HIL 10] HILBERT D., *Grundzuge einer allgemeinen Theorie der linearen Integralgleichungen*, Part 6, Göttinger Nachrichten, pp. 355–419, 1910.

[HIL 12] HILBERT D., *Grundzuge einer allgemeinen Theorie der linearen Integralgleichungen*, B.G. Teubner, 1912.

[HOF 07] HOFFMAN K., *Analysis in Euclidean Space*, Dover, Mineola, NY, USA, 2007.

[KAK 41] KAKUTANI S., "A generalization of Brouwer's fixed point theorem", *Duke Mathematical Journal*, vol. 8, no. 3, pp. 457–459, 1941.

[KAT 02] KATSKADELIS J.T., *Boundary Elements: Theory and Applications*, Elsevier, 2002.

[KAT 13] KATZ M.G., TALL D., "A cauchy-Dirac delta function", *Foundations of Science*, vol. 18, no. 1, pp. 107–123, 2013.

[KIR 83] KIRCHHOFF K., "Zur Theorie der Lichtstrahlen", *Annalen der Physik*, vol. 254, no. 4, pp. 663–695, 1883.

[LAG 61a] LAGRANGE J.L., "Essai d'une nouvelle méthode pour déterminer les maxima et les minima des formules intégrales indéfinies", *Mélanges de Philosophie et de Mathématiques de la Soc. Roy. de Turin*, vol. 2, pp. 173–195, 1761.

[LAG 61b] LAGRANGE J.L., "Applications de la méthode exposée dans le mémoire précédent à la solution de différents problèmes de dynamique", *Mélanges de Philosophie et de Mathématiques de la Soc. Roy. de Turin* (Miscellania Taurinensia), vol. 2, pp. 196–298, 1761.

[LAG 64] LAGRANGE J.L., "Recherches sur la libration de la lune, dans lesquelles on tâche de résoudre la Question proposée par l'Académie Royale des Sciences pour le Prix de l'année 1764", *Prix de l'Académie Royale des Sciences Paris*, vol. 9, no. 1764, pp. 1–50, 1772.

[LAG 70] LAGRANGE J. L., "Mémoire sur l'utilité de la méthode de prendre le milieu entre les résultats de plusieurs observations, dans lequel on examine les avantages de cette méthode par le calcul des probabilités, et où l'on résout différents problèmes relatifs à cette matière", *Mélanges de Philosophie et de Mathématiques de la Soc. Roy. de Turin (Miscellania Taurinensia)*, vol. 5, pp. 173–223, 1770–1773.

[LAG 82] LAGRANGE J.L., "Théorie de la libration de la lune, et des autres Phénomènes qui dépendent de la figure non sphérique de cette Planète", *Histoire de l'Académie Royale des Sciences et des Belles-Lettres de Berlin*, 1780, classe de Mathématiques, pp. 203–309, 1782.

[LAG 88] LAGRANGE J.L., *Méchanique Analytique*, La veuve Desaint, Paris, 1788.

[LAG 09] LAGRANGE J.L., "Sur la théorie générale de la variation des constantes arbitraires dans tous les problèmes de la mécanique", *Mémoires de la première Classe de l'Institut de France*, vol. 6, pp. 771–805, 1809.

[LAG 10] LAGRANGE J.L., "Second Mémoire Sur la théorie générale de la variation des constantes arbitraires dans tous les problèmes de la mécanique", *Mémoires de la première Classe de l'Institut de France*, vol. 6, pp. 809–816, 1810.

[LAP 10] LAPLACE P.S., "Mémoire sur les approximations des formules qui sont fonctions de très grands nombres et sur leur application aux probabilités", *Mémoires de l'Académie des sciences*, vol. 10, pp. 1809–1810, 1810.

[LAP 12] LAPLACE P.S., *Théorie Analytique des probabilités*, Éditions Courcier, Paris, 1812.

[LEB 96] LEBEDEV L.P., VOROVICH I.I., GLADWELL G.M.L., *Functional Analysis: Applications in Mechanics and Inverse Problems*, Kluwer Academic Press, 1996.

[LEO 13] LEON S.J., BJORCK A., GANDER W., "Gram–Schmidt orthogonalization: 100 years and more", *Numerical Linear Algebra with Applications*, vol. 20, pp. 492–532, 2013.

[LER 34] LERAY J., "Les Loix du mouvement et du repos déduites d'un principe metaphysique", *Acta Mathematica*, vol. 63 , pp. 193–248, 1934.

[LAX 54] LAX P.D., MILGRAM A.N. , "Parabolic equations", in Contributions to the theory of partial differential equations, *Annals of Mathematics Studies*, Princeton University Press, Princeton, NJ, no. 33, pp. 167–190, 1954.

[MAX 73] MAXWELL J.C., *A Treatise on Electricity and Magnetism*, Clarendon Press, Oxford, 1873.

[NEW 87] NEWTON I., *Philosophiae Naturalis Principia Mathematica*, Jussu Societatis Regiæ ac Typis Joseph Streater, London, 1687.

[NEW 36] NEWTON I., *The Method of Fluxions and Infinite Series; with its Applications to the Geometry of Curve-lines*, Henry Woodfall, 1736.

[NIK 33] NIKODYM O., "Sur une classe de fonctions considérée dans l'étude du problèmede Dirichlet", *Fundamenta Mathematicae*, vol. 21, pp. 129–150, 1933.

[NIK 35] NIKODYM O., "Sur le principe du minimum", *Mathematica*, vol. 9, pp. 110–128, 1935.

[PAR 96] PARETO V., *Cours d'Economie Politique*, Lausanne, 1896.

[PAR 97] PARETO,V., *Cours d'Economie Politique, Part II*, Lausanne, vol. 2, 1897.

[PAR 06] PARETO V., *Manuale di economia politica con una introduzione alla scienza sociale*, Societa Editrice Libraria, Milano, 1906.

[PAR 09] PARETO, V., *Manuel d'Economie Politique*, translated by Alired Bonnet, V. Girard and E. Brière, Paris, 1909.

[PAR 71] PARETO V., *Manual of Political Economy*, translated by Ann S. Schwier, Augustus M. Kelley Publishers, Clifton, NJ, 1971.

[PAT 98] PATRICIA R.-L., "La moindre action comme lien entre la Philosophie naturelle et la mécanique analytique: continuités d'un questionnement", *Revista de la Sociedad Española de Historia de las Ciencias y de las Técnicas*, vol. 21, no. 40, pp. 439–484, 1998.

[PEN 02] PENDRICK G.J., *Antiphon the sophist, the fragments*, Cambridge University Press, 2002.

[PET 40] PETROV G.I., "Application of Galerkin's method to the problem of stability of flow of a viscous fluid", *Journal of Applied Mathematics and Mechanics*, vol. 4, no. 3, pp. 3–12, 1940.

[POI 08] POISSON S.D., "Mémoire sur la propagation de la chaleur dans les corps solides", *Nouveau Bulletin des sciences par la Société Philomatique de Paris*, vol. 1, pp. 112–116, no. 6, March 1808.

[POI 16] POISSON S.D., "Mémoire sur la manière d'exprimer les fonctions par des séries de quantités périodiques et sur l'usage de cette transformation dans la resolution de différens problemes", *Journal de l'École Royale Polytechnique, Cahier 18*, vol. 11, pp. 417–489, 1816.

[POI 23a] POISSON S.D., "Mémoire sur la distribution de la chaleur dans les corps solides", *Journal de l'École Royale Polytechnique*, vol. 12, pp. 1–144, 1823.

[POI 23b] POISSON S.D., "Addition au Mémoire précédent, et au Mémoire sur la manière d'exprimer les fonctions par des séries de quantités périodiques", *Journal de l'École Royale Polytechnique*, vol. 12, pp. 145–162, 1823.

[RAY 94] RAYLEIGH J.W.S., *Theory of Sound*, Macmillan, London, vol. 1, 1894.

[RAY 96] RAYLEIGH J.W.S., *Theory of Sound*, Macmillan, London, vol. 2, 1896.

[RIE 67] RIEMANN G., "Ueber die Darstellbarkeit einer Function durch eine trigonometrische Reihe", *Abhandlungen der Königlichen Gesellschaft der Wissenschaften zu Göttingen*, vol. 13, 1867.

[RIE 07] RIESZ F., "Sur une espèce de géométrie analytique des systèmes de fonctions sommables", *C. R. Acad. Sci. Paris*, vol. 144, pp. 1409–1411, 1907.

[RIE 09] RIESZ F., "Sur les opérations fonctionnelles linéaires", *C. R. Acad. Sci. Paris*, vol. 149, pp. 974–977, 1909.

[RIT 08] RITZ W., "Über eine neue Methode zur Lösing gewisser Variationsprobleme der mathematischen Physik", *J. Reine Angew. Math.*, vol. 135, pp. 1–61, 1908.

[RIT 09] RITZ W., "Theorie der Transversalschwingungen einer quadratischen Platte mit freien Raendern", *Annalen der Physik.*, vol. 333, no. 4, pp. 737–786, 1909.

[RYL 66] RYLL-NARDZEWSKI C., "On fixed points of semi-groups of endomorphisms of linear spaces", *Proc. 5th Berkeley Symp. Probab. Math. Stat.*, University of California Press, vol. 2, no. 1, pp. 55–61, 1966.

[SAT 59] SATO M., "Theory of hyperfunctions, I", *J. Fac. Sci. Tokyo*, vol. 1, no. 8, pp. 139–193, 1959.

[SAT 60] SATO M., "Theory of hyperfunctions, II", *J. Fac. Sci. Tokyo*, vol. 1, no. 8, pp. 387–437, 1960.

[SCH 30] SCHAUDER J.P., "Der Fixpunktsatz in Funktionalräumen", *Studia Math.*, vol. 2, pp. 171–180, 1930.

[SCH 45] SCHWARTZ L., "Généralisation de la notion de fonction, de dérivation, de transformation de Fourier et applications mathématiques et physiques", *Annales Univ. Grenoble*, vol. 21, pp. 57–74, 1945.

[SCH 50] SCHWARTZ L., *Théorie des distributions*, Hermann, Paris, vol. 1, 1950.

[SCH 51] SCHWARTZ L., *Théorie des distributions*, Hermann, Paris, vol. 2, 1951.

[SCH 54] SCHWARTZ L., "Sur l'impossibilité de la multiplication des distributions", *C. R. Acad. Sci. Paris*, vol. 239, pp. 847–848, 1954.

[SCM 07a] SCHMIDT E., "Zur Theorie der linearen und nichtlinearen Integralgleichungen. I. Teil: Entwicklung willk ürlicher Funktionen nach Systemen vorgeschriebener", *Math. Ann.*, vol. 63, no. 4, pp. 433–476, 1907.

[SCM 07b] SCHMIDT E., "Zur Theorie der linearen und nichtlinearen Integralgleichungen. Zweite Abhandlung: Auflösung der allgemeinen linearen Integralgleichung", *Math. Ann.*, vol. 64, no. 2, pp. 161–174, 1907.

[SCM 08] SCHMIDT E., "Zur Theorie der linearen und nichtlinearen Integralgleichungen. III. Teil: Uber die Auflösung der nichtlinearen Integralgleiohung und die Verzweigung ihrer Lösungen", *Math. Ann.*, vol. 65, no. 3, pp. 370–399, 1908.

[SER 14] SERVOIS J.F., "Analyse Transcendante. Essai sur un nouveau mode d'exposition des principes du calcul différential", *Annales de Mathématiques pures et appliquées*, vol. 5, pp. 93–140, 1814.

[SIM 12] SIMONYI K., *A Cultural History of Physics*, CRC Press, 2012.

[SMA 74] SMART D.R., *Fixed Point Theorems*, Cambridge University Press, 1974.

[SOB 30] SOBOLEV S.L., "Sur l'equation d'onde pour le cas d'un milieu heterogene isotrope", *Dokl. Akad. Nauk SSSR*, no. 7, pp. 163–167, 1930.

[SOB 31] SOBOLEV, S.L., "Wave equation for inhomogeneous medium", *First Inter. Session of Sci. Advance of Seismol. Institute of Sci. Acad. USSR*, no. 1, pp. 16–18, 1931.

[SOB 34a] SOBOLEV S.L., "On the integration of wave equation in inhomogeneous medium", *Leningrad: Izdat. Akad. Nauk USSR*, 1934.

[SOB 34b] SOBOLEV S.L., "Nouvelle mèthode de rèsolution du problème de Cauchy pour les èquations aux dèrivèes partielles de second ordre", *C. R. Acad. Sci. URSS*, vol. 1, no. 8. pp. 435–438, 1934.

[SOB 35a] SOBOLEV S.L., "Le problème de Cauchy dans l'espace des fonctionnelles", *C. R. Acad. Sci. URSS*, vol. 3, no. 7, pp. 291–294, 1935.

[SOB 35b] SOBOLEV S.L., "La théorie générale de la diffraction des ondes sur les surfaces de Riemann", *Trav. Inst. Math. Steklov*, vol. 9, pp. 39–105, 1935.

[SOB 38] SOBOLEV S.L., "Sur un théorème d'analyse fonctionnelle", *C. R. Acad. Sci. URSS*, vol. 20, no. 1, pp. 5–9, 1938.

[SOB 64] SOBOLEV S.L., *Applications de l'analyse fonctionnelle à la physique mathématique*, Akademie-Verlag, Berlin, 1964.

[SOU 10] SOUZA DE CURSI J.E., SAMPAIO R., *Modeling and Convexity*, ISTE Ltd, London and Elsevier, Cambridge, 2010.

[SOU 15] SOUZA DE CURSI J.E., SAMPAIO R., *Uncertainty Quantification and Stochastic Modeling with Matlab*, ISTE Press, London and Elsevier, Oxford, 2015.

[TAR 36] TARSKI A., "Der Wahrheitsbegriff in den formalisierten Sprachen", *Studia Philosophica*, vol. 1, pp. 261–405, 1936.

[TIK 35] TIKHONOV A., "Ein Fixpunktsatz", *Mathematische Annalen*, vol. 111, pp. 767–776, 1935.

[TIM 13] TIMOSHENKO S.P., "Sur la stabilité des systèmes élastiques", *Annales des Ponts et Chaussées*, vol. 9, pp. 496–566, 1913.

[TUR 56] TURNER M.J., CLOUGH R.W., MARTIN H.C. *et al.*, "Stiffness and deflection analysis of complex structures", *Journal of the Aerospace Sciences*, vol. 25, no. 9, pp. 805–823, 1956.

[VAN 50] VAN DER POL B., BREMMER H., *Operational Calculus Based on the Two-Sided Laplace Integral*, Cambridge University Press, 1950.

[VAR 25] VARIGNON P., *Nouvelle mécanique, ou Statique, dont le projet fut donné en M.DC.LXXXVII*, Claude Jombert, Paris, 1725.

[VIR 19] VIRGILIUS S., *Aeneid*, between 29 and 19 B.C.

[VIT 05] VITALI G., *Sul problema della misura dei gruppi di punti di una retta*, Tipogr. Gamberini e Parmeggiani, Bologna, 1905.

[VOL 87] VOLTERRA V., "Sopra le funzioni che dipendono da altre funzioni. I", *Rend. LinceiSer.*, vol. 3, no. 4, 1887.

[VON 32] VON NEUMANN, J., *Mathematische Grundlagen der Quantenmechanik*, Grundlehren der mathematischen Wissenschaften, Springer, Berlin, 1932.

[WEI 44] WEIL H., "David Hilbert and his mathematical work", *Bull. Amer. Math. Soc.*, vol. 50, no. 9, pp. 612–654, 1944.

[WIE 26] WIENER N., "The operational calculus", *Mathematische Annalen*, vol. 95, pp. 557–584, 1926.

[WIN 02] WINSTON NEWSON M.F., "Hilbert: mathematical problems", *Bull. Am. Math. Soc.*, vol. 8, no. 10, pp. 437–479, 1902.

[ZIE 64a] ZIENKIEWICZ, O.C., "Finite element method of analysis for arch dam shells and comparison with finite difference procedures", *Proceedings of Symposium on Theory of Arch Dams*, University of Southampton, 1964. (Theory of Arch Dams Pergamon Press: Oxford, pp. 123–139, 1965).

[ZIE 64b] ZIENKIEWICZ O.C., "The finite element method for analysis of elastic isotropic and orthotropic slabs", *Proceedings – Institution of Civil Engineers*, pp. 471–488, 1964.

Index

L, M, N

O, P, R, S, T

V, W

Other titles from

iSTE

in

Numerical Methods in Engineering

2015

JEBAHI Mohamed, ANDRÉ Damien, TERREROS Inigo, IORDANOFF Ivan
Discrete Element Method to Model 3D Continuous Materials

2014

BECKERS Benoit, BECKERS Pierre
Reconciliation of Geometry and Perception in Radiation Physics

BERGHEAU Jean-Michel
Thermo-Mechanical Industrial Processes: Modeling and Numerical Simulation

BONNEAU Dominique, FATU Aurelian, SOUCHET Dominique
4-volume series
Hydrodynamic Bearings – Volume 1
Mixed Lubrication in Hydrodynamic Bearings – Volume 2
Thermo-hydrodynamic Lubrication in Hydrodynamic Bearings – Volume 3
Internal Combustion Engine Bearings Lubrication in Hydrodynamic Bearings – Volume 4

DESCAMPS Benoît
Physical Form Finding of Lightweight Structures

2013

YASTREBOV Vladislav A.
Numerical Methods in Contact Mechanics

2012

DHATT Gouri, LEFRANÇOIS Emmanuel, TOUZOT Gilbert
Finite Element Method

2011

CHINESTA Francisco *et al.*
Natural Element Method for the Simulation of Structures and Processes

DAVIM Paulo J.
Finite Element Method in Manufacturing Processes

POMMIER Sylvie, GRAVOUIL Anthony, MOËS Nicolas, COMBESCURE Alain
Extended Finite Element Method for Crack Propagation

2010

SOUZA DE CURSI Eduardo, SAMPAIO Rubens
Modeling and Convexity

2008

BERGHEAU Jean-Michel, FORTUNIER Roland
Finite Element Simulation of Heat Transfer

EYMARD Robert
Finite Volumes for Complex Applications V

FREY Pascal, GEORGE Paul-Louis
Mesh Generation – 2nd edition

GAY Daniel, GAMBELIN Jacques
Modeling and Dimensioning of Structures

2005

BENKHALDOUN Fayssal
Finite Volumes for Complex Applications IV